Phytohormones in Plant Biotechnology and Agriculture

Phytohormones in Plant Biotechnology and Agriculture

Proceedings of the NATO-Russia Workshop held in Moscow, 12–16 May 2002

Edited by

Ivana Macháčková

Institute of Experimental Botany of the Academy of Sciences of the Czech Republic, Prague, Czech Republic

and

Georgy A. Romanov

Institute of Plant Physiology, Russian Academy of Sciences, Moscow, Russia

KLUWER ACADEMIC PUBLISHERS

DORDRECHT / BOSTON / LONDON

A C.I.P. Catalogue record for this book is available from the Library of Congress.

ISBN 1-4020-1723-5

Published by Kluwer Academic Publishers,
P.O. Box 17, 3300 AA Dordrecht, The Netherlands.

Sold and distributed in North, Central and South America
by Kluwer Academic Publishers,
101 Philip Drive, Norwell, MA 02061, U.S.A.

In all other countries, sold and distributed
by Kluwer Academic Publishers,
P.O. Box 322, 3300 AH Dordrecht, The Netherlands.

Printed on acid-free paper

Printed in the Netherlands.

TABLE OF CONTENTS

Cytokinins

Auxins

Complex hormonal regulation of developmental processes

PREFACE

Phytohormone research is a crucially important area of plant sciences. Phytohormones are one of the key systems integrating metabolic and developmental events in the whole plant and the response of plants to external factors. Thus, they influence the yield and quality of crops. During the last decade we have slowly begun to understand the molecular mechanisms underlying phytohormone action, largely as a result of the rapid developments that have been made internationally in the field of plant molecular genetics. Putative receptor proteins for ethylene (1993-95), brassinosteroids (1997) and cytokinins (2001) have been identified and the genes that encode them cloned. Primary response genes and elements of hormonal signal transduction have also been identified for most known phytohormones. There is now little doubt that phytohormones, like their animal counterparts, function as signal molecules and create a signalling network in the whole plant organism.

The *in vivo* activity of hormones depends, among other things, on their rate of biosynthesis and metabolism, and on their transport into and out of target cells. Consequently, genes and enzymes involved in these processes are of particular interest. In recent years a number of genes encoding enzymes for the synthesis, modification and degradation of different phytohormones have been cloned and identified, as have genes encoding proteins involved in phytohormone transport and its regulation. Some classes of phytohormone have been shown to participate in stress reactions and can increase the resistance of plants to unfavorable environmental factors. All the developments mentioned above provide potentially powerful tools for the creation of new plant forms with desirable properties for modern biotechnology, agriculture and horticulture. Indeed, new transgenic or mutant plants have recently been produced showing altered hormonal status or sensitivity. Such plants are now under intensive investigation in laboratories in many parts of the world.

Recent advances in phytohormone research and practical applications of this research were discussed at the NATO-Russia Advanced Research Workshop "Phytohormones in Plant Biotechnology and Agriculture", held in May 2002 at the Institute of Plant Physiology of the Russian Academy of Sciences, Moscow. The Workshop was dedicated to the centenary of the birth of the famous Russian plant physiologist, Academician Mikhail Chailakhyan (1902-2002). Chailakhyan pioneered investigations into the hormonal regulation of flowering, and of the flowering hormone "florigen". He quickly appreciated the potential for the application of the results of phytohormone research in agriculture and horticulture, and was one of the first to use gibberellin treatments in seedless grape production.

This book contains papers written by lecturers and selected participants in the Workshop. The main topics centre around: molecular aspects of phytohormone action and transport; phytohormones and flowering; phytohormones and stress; the use of phytohormones in plant biotechnology and agriculture; and the use of transgenic plants. In our opinion, the book reflects not only current progress and achievements in phytohormone research, but also the high degree of international co-operation and friendship that were a feature of the Workshop.

We would like to sincerely thank the North Atlantic Treaty Organization, the sponsors of the workshop, for their support. This enabled the workshop to achieve a high international standard. We are especially grateful to the NATO Scientific Affair Division, represented by Dr. Alain Jubier who personally attended the first day of the Workshop. We also acknowledge with thanks the Russian Ministry of Industry, Science and Technology and Professor Nikolaus Amrhein (ETH Zurich, Switzerland) for their financial support and Professor Vladimir Kuznetsov, Director of the Moscow' Institute of Plant Physiology for providing all necessary facilities for workshop organisation.

Ivana Macháčková and Georgy Romanov,
Co-Directors of the Workshop

ACKNOWLEDGEMENTS

Financial contribution of the following companies, which enabled to publish these Proceedings, is gratefully acknowledged:

Spectronex, spol. s r.o.
Nad Ostrovem 1119/7
Praha 10

BIO-RAD spol. s r.o.
Černická 772/19
Praha 4

BioTech a.s.
V Ochozu 20
Praha 10

P-LAB a.s.
Říčanská 10
Praha 10

Pragolab s.r.o.
Kovářská 4
Praha 9

AMEDIS, spol. s r.o.
Bobkova 786
Praha 9

Further, the invaluable help of Marie Kapounová, Dr. Jan Martinec, Dr. Eva Zažímalová and Mgr. Lenka Vítová of the Institute of Experimental Botany CAS, Prague in preparation of this book is gratefully acknowledged.

COLOUR PLATES

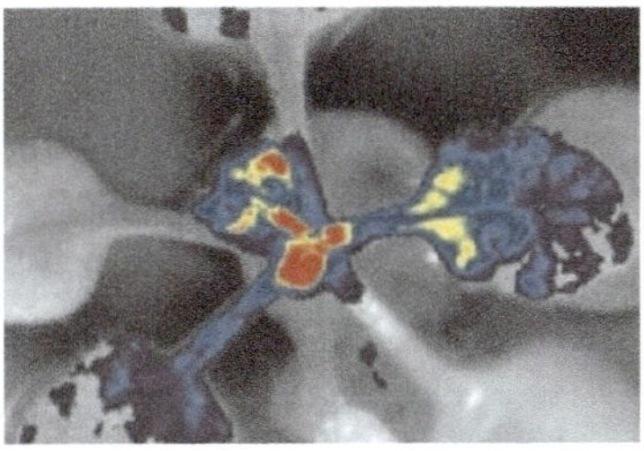

Davies et al.: Figure 13. The false-colour image of light emitted by transgenic Arabidopsis plants containing the firefly luciferase (LUC) reporter gene coupled to the AtGA20ox1 promoter demonstrates that the GA20ox1 gene is expressed most strongly in the shoot apex and in young developing leaves. (From P. Hedden, IAACR-BBSRC annual report, 2000.)

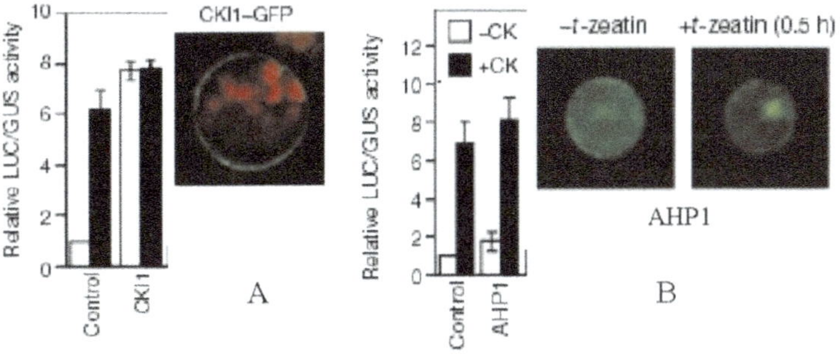

Davies et al.: Figure 18. (A) Left: Cytokinin signalling is initiated by CKI1, a histidine protein kinase receptor, in the absence of cytokinin. Protoplasts were cotransfected with CKI1 and the ARR6-LUC reporter gene and the luciferase activity recorded. Right: The CKI1 (as detected in protoplasts transfected with CKI1-GFP) is localized at the plasma membrane. (B) Cytokinin signalling intermediate AHP acts as a shuttle between the cytoplasm and nucleus in cytokinin signalling. Left: Overexpression of AHP (as detected by the ARR6-LUC reporter) does not affect cytokinin signalling; CK is still needed. However CK causes a relocation of AHP (as detected in protoplasts were transfected with AHP1-GFP) to the nucleus (right). (From Hwang and Sheen, 2001.)

Bou et al.: Figure 1. Slender phenotype and tuberization response of the antisense phyB lines. Only the phyB lines tuberize in LD. These plants have paler leaves and anelongated stems due to increase internode length.

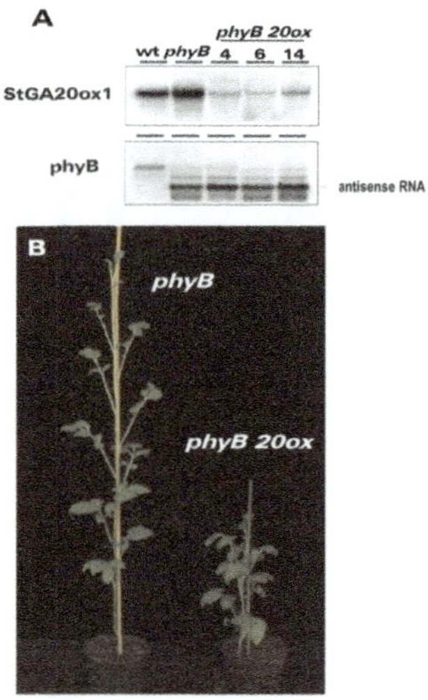

Bou et al.: Figure 4. Phenotype of the phyB 20ox antisense lines. (A) Down-regulated expression of the phyB and StGA20ox1 transcripts in the phyB 20ox lines. (B) Dwarf phenotype of the phyB 20ox mutants. The leaves of these plants show still a pale phenotype, which indicates that this trait is not dependant on Gas.

FLOWERING

HORMONAL REGULATION OF PLANT DEVELOPMENT IN THE STUDIES BY M. KH. CHAILAKHYAN

In Dedication of the 100th Anniversary of his Birth

N.P. AKSENOVA
Timiryazev Institute of Plant Physiology, Russian Academy of Sciences, Botanicheskaya 35, Moscow, 127276 Russia;
Fax: 7(095) 977-80 –18; e-mail: ifr@ippras.ru

March 21, 2002, marked the 100th birthday of Mikhail Khristoforovich Chailakhyan (1902-2002). M. Kh. Chailakhyan was a well-known scientist in the field of hormonal regulation of plant growth and development. His research activity extended more than sixty years and included a wide range of problems devoted to fundamental aspects of plant ontogeny. His works included investigations of the mechanisms of flowering, tuberization, sex expression, and the integrity of the plant organism. He published more than four hundred papers and monographs. In this short report it is impossible to cover all of his works, so here only some of the most characteristic trends of his investigations will be described.

M. Kh. Chailakhyan, after graduating from Erevan University in 1931, he became a postgraduate student at the laboratory of Biochemistry and Plant Physiology, then located in Leningrad. After moving to Moscow, his laboratory was transformed into Timiryazev Institute of Plant Physiology, USSR Academy of Sciences (IFR). In 1934, Chailakhyan defended his candidate thesis on a comparison of the development of winter and spring cereal varieties. In 1935, he became the head of the Laboratory of Growth and Development of the IPP (Institute of Plant Physiology). In 1939, he defended his doctoral thesis on the problem of hormonal regulation of plant development, and in 1968 he became the member of the USSR Academy of Sciences. He worked in the Laboratory of Growth and Development for 56 years, until his last days (Chailakhyan, 1980).

One of the main subjects of Chailakhyan's studies was the hormonal regulation of development in higher plants. He paid special attention to the problem of photoperiodic, temperature and hormonal control of a crucial stage in the development of higher plants: their transition from the vegetative state to flowering. These issues prevailed in his investigations throughout his life as a researcher. M.Kh. Chailakhyan started to develop this subject during the early 1930-ies. His first model plant was *Chrysanthemum*, a short-day plant, which remains vegetative in long days and flowers in short days. He exposed to short days either the whole plants, or only

I. Macháčková and G.A. Romanov (eds.), Phytohormones in Plant Biotechnology and Agriculture, 3-7

the shoot tips of leafless plants, or only the leaves on the intact plants. Plants flowered only if the leaves were exposed to short days. So he concluded, that a floral stimulus, which he named florigen, is formed in leaves and moves to buds, where it causes floral morphogenesis. Experiments were repeated with other short-day and long-day plants with similar results. In interspecies graftings it was also shown that leafless tops from vegetative plants of one photoperiodic group could flower under the influence of flowering plants of another photoperiodic group. Thus, florigen was similar in diverse species and in plants belonging to different photoperiodic groups. In 1937 M. Kh. Chailakhyan summed up his views on the regulation of flowering in a monograph entitled "Hormonal Theory of Plant Development" (Chailakhyan, 1937), which evoked active response from researchers of plant development in many countries. It is worth noting that young Anton Lang purposely translated this book into German in order that his teacher, Prof. G. Melchers, could read it (Lang, 1994). Further development of florigen concept resulted from experiments with gibberellin treatment of long- and short-day plants. His friendship with Prof. A.Lang, who carried out similar experiments in United States and gave first milligrams of gibberellic acid to Chailakhyan, helped him very much in performing these experiments. Different species were treated with gibberellic acid. Experiments showed that gibberellin caused flowering only of typical rosette long-day plants under unfavourable short days, but did not stimulate flowering of short-day plants under long days. These experiments and analysis of endogenous gibberellins and other hormones in long- and short-day plants brought Chailakhyan to the conclusion that florigen consists of two groups of hormones: gibberellins and anthesins (Chailakhyan, 1961). He proposed that short-day plants can synthesize gibberellins in short days as well as in long days in quantities sufficient for their flowering. But they needed short days for synthesis of anthesins. On the contrary, long-day plants had enough anthesins both in short and long days but they needed long days to produce enough gibberellins.

Gibberellins are familiar substances, but anthesins are still not identified chemically in spite of numerous efforts. However, an anthesin-like physiological effect of purified extract from the tobacco plants was found. Extracts were made not only from the leaves of flowering Maryland Mammoth tobacco but also from the leaves of vegetative Sylvestris plants growing in short days (Chailakhyan et al., 1977, Chailakhyan et al., 1984). Both extracts caused flower differentiation in seedlings of short-day plant *Chenopodium rubrum* kept in continuous light, and caused full flowering in older plants of this species.

Further development of florigen theory was related to the analysis of hormonal regulation of flowering in plants belonging to different ecological groups: in winter and spring varieties, in ephemers and in forms with long juvenile phases of both neutral and photoperiod - sensitive plants. As a result a general concept of autonomous and induced mechanisms of floral control was proposed (Chailakhyan, 1988). It was concluded that in different plant species the synthesis of each florigen component - gibberellins or anthesins – can proceed in two different ways: either by autonomous pathway or by induction due to external signals, such as day length, temperature, and others. Autonomous or inducible control of gibberellin and anthesin synthesis depended on evolution and adaptation of particular plant species to ecological habitats.

The florigen theory was based on physiological experiments. But it might induce some ideas for searching the genetic mechanisms controlling the transition of plants from vegetative phase to flowering. At present, the florigen theory elaborated by Chailakhyan attracts attention of molecular biologists in relation to physiological function of large groups of structural and regulatory genes that control the onset of flowering.

Chailakhyan worked not only on the problem of plant flowering regulation. He also studied hormonal and photoperiodic control of another reproductive process – tuberization. His general approach to this problem included studying the effects of long and short days and various hormones and their ratios on growth and development of stolons and tubers. The most original results were obtained with graftings (Chailakhyan, 1984). Shoots of long-day, short-day or neutral tobacco plants were taken as scions and potato species with a strict short-day tuberization response served as stocks. Tuberization of potato stock was observed only under photoperiod favourable for flowering of tobacco scion. For example, the scion of *Nicotiana sylvestris* flowered in long days and remained vegetative in short days. The stock of *Solanum demissum* also only produced tubers in long days. Experiments have shown that flowering and tuberization processes have some common steps of regulation.

Chailakhyan and his co–workers, Profs. Khryanin, Culafic and others, performed many interesting experiments on the hormonal regulation of sex expression in dioecious and monoecious plants with male and female flowers. It was shown with *Cannabis sativa* plants that cytokinins formed in roots promote the appearance of female flowers, whereas gibberellins formed in leaves promote the formation of male flowers. If the roots were partly removed from the cuttings of *Cannabis* plants, then male flowers were preferentially formed. But if leaves were partly removed then female flowers were mostly formed. The addition of benzyladenine to the cultural medium with young seedlings of *Cannabis* favoured the formation of female plants whereas the addition of gibberellic acid promoted the formation of male plants. Similar results were obtained with maize and cucumber. It was concluded that sex expression in plants of both types might vary depending on hormonal balance and cytokinins and gibberellins were the most prominent regulators (Chailakhyan and Khryanin, 1982). This conclusion is important not only theoretically but also practically since some cultivated plants, e.g. cucumber, watermelons, and maize have pistilate and staminate flowers.

Along with his basic research, Chailakhyan paid serious attention to the practical application of phytohormones. As early as the1940s, Chailakhyan together with Dr. Turetskaya worked out the method for application of auxins and their synthetic analogs for rooting of some economic fruit and forest trees. He was the first person to begin in USSR to develop ways for applying the gibberellins to agricultural crops. Together with Profs. Smirnov and Sarkisova, and Dr. Manankov, he elaborated the procedures for increasing the yield of seedless Khishmish varieties of grapes. Together with Drs Milyaeva and Azizbeckova he found the way to increase the yield of alkaloid crocine in flowers of *Crocus sativa* with the aid of gibberellic acid. Chailakhyan initiated the research activity aimed at application of retardants for prevention of cereal lodging, and together with Prof. Prusakova he was successful in this area. In Chailakhyan's concluding monograph published in 1988 (Chailakhyan, 1988) a special chapter concerns the theoretical basis of the use of phytohormones in agriculture.

M. Kh. Chailakhyan was always an active propagandist for basic knowledge and modern discoveries in plant physiology. In the period from the 1930ies to the 1950ies he was widely known in scientific circles as a firm and consistent fighter against the pseudoscientific views of T. D. Lysenko and his supporters. This was the reason why he was repressed and twice (in 1939 and in 1948) expelled from his position as a head of the Laboratory of Growth and Development. Chailakhyan was deeply respected by scientists of the USSR. This respect was engendered by his firm convictions, exceptional honesty, simplicity, wisdom and personal charm.

M. Kh. Chailakhyan always considered the direct contacts and discussions with the plant physiologists from leading laboratories as very important and helpful for the development of science. He was not only a member of the USSR Academy of Sciences and the Armenian Academy of Sciences, but also a member of the German Academy Leopoldina, a corresponding member of the American Society of Plant Physiologists and the American Botanical Society. He was a honorary member of the Bulgarian Botanical Society, the Indian Society of Plant Physiologists and the International Association of Plant Growth Regulators. He was also honorary Professor of Rostock University and a honorary member of Ulm University (Bazyan and Chailakhyan, 2002). M.Kh. Chailakhyan had working contacts with plant physiologists in many countries of the world. He formed the scientific school comprising numerous successors and followers in Russia, the States of former USSR, Bulgaria, Czech Republic, Yugoslavia, United States, Israel and other countries.

RFERENCES

Bazyan, L. and Chailakhyan, M. (eds) (2002) *Portrait of a Scientist in the Interiere*, SIP MRA, Moscow, in Russ., 194 pp.

Chailakhyan, M.Kh. (1937) *Hormonal Theory of Plant Development*, Akad. Nauk USSR, Moscow, in Russ., 198 pp.

Chailakhyan, M.Kh. (1961) Principles of ontogenesis and physiology of flowering in higher plants, *Canad. J. Bot.* **39**, 1817-1841.

Chailakhyan, M.Kh. (1980) *Materials to Bibliography of USSR Scientists*, Nauka, Moscow, in Russ., pp. 3-25.

Chailakhyan, M.Kh. (1984) *Photoperiodic and Hormonal Control of Tuber Formation in Plants*, Nauka. Moscow, in Russ., 68pp.

Chailakhyan, M.Kh. (1988) *Control of the Higher Plant Flowering*. Nauka. Moscow. in Russ., 559 pp.

Chailakhyan, M.Kh., Grigorieva, N.Y. and Lozhnikova, V.N. (1977) The effect of extracts from the leaves of flowering tobacco plants on the flowering of *Chenopodium rubrum* seedlings and plantlets, *Soviet Plant Physiol*. **24**,773-776.

Chailakhyan, M.Kh. and Khryanin, V.N. (1982) *Sexuality of Plants and Its Hormonal Regulation*, Nauka, Moscow, in Russ., 173 pp.Chailakhyan, M.Kh., Lozhnikova, V.N., Grigorieva, N.Y. and Dudko, N.D. (1984) Effect of extracts from the leaves of vegetating tobacco plants on the flowering of *Chenopodium rubrum* seedlings and plantlets, *Dokl. Acad. Nauk USSR* **274**, 1276-1280.

Lang, A. (1994) Hormonal regulation of plant flowering, in: *The First Chailakhyan's Reading*, ONTI Puschino, pp. 27-58.

THE PLANT HORMONES: FROM ORIGINAL CONCEPT TO A MOLECULAR FLOWERING

PETER J. DAVIES
Cornell University, Ithaca, NY 14853 USA

1. THE DEVELOPMENT OF THE PLANT HORMONE CONCEPT AND EARLY WORK.

The plant hormone concept probably derives from observations of morphogenic and developmental correlations by Sachs between 1880 and 1893. He suggested that "Morphological differences between plant organs are due to differences in their material composition" and postulated the existence of root-forming, flower forming and other substances that move in different directions through the plant (Went and Thimann, 1937).

At about the same time Darwin (Darwin, 1880) was making his original observations on the phototropism of grass coleoptiles that led him to postulate the existence of a signal that was transported from the tip of the coleoptile to the bending regions lower down. After further characterizations by several workers of the way in which the signal was moved Went in the Netherlands finally was able to isolate the chemical by diffusion from coleoptile tips into agar blocks, which, when replaced on the tips of decapitated coleoptiles, resulted in the stimulation of the growth of the decapitated coleoptiles, and their bending when places asymmetrically on these tips. This thus demonstrated the existence of a growth promoting chemical that was synthesized in the coleoptile tips, moved basipetally, and when distributed asymmetrically resulted in a bending of the coleoptile away from the side with the higher concentration. This substance was originally named *Wuchsstoff* by Went, and later this was changed to *auxin*. After some false identifications the material was finally identified as the simple compound indoleacetic acid, universally known as IAA. (Wildman, 1997) gives a wonderful account of this early work with some chemical identifications of the nature of the actual characterized artifactual compounds.

Soon after the discovery of auxin, Chailakhyan, working in Russia, carried out his classic work demonstrating flowering was also regulated by a transported signal (Fig. 1) (Chailakhyan, 1975). He demonstrated that exposure of a single chrysanthemum leaf to short days would bring about flowering (Fig. 2) and suggested that some sort of chemical communication occurred. Grafting experiments demonstrated transmission of floral stimulus which was named *florigen*. Its transmission characteristics indicated that it was transported in the phloem. By 1958 the florigen concept had been modified to be a multi-component

I. Macháčková and G.A. Romanov (eds.), Phytohormones in Plant Biotechnology and Agriculture, 9-30

signal, one part of which may be gibberellins (Fig. 3).

Figure 1. Chailakhyan in his greenhouse in the early 1930's. (From Chailakhyan, 1975.)

Figure 2. Early experiments demonstrating the transported nature of the floral stimulus. The chrysanthemum plants flowered whenever leaves were placed in light-tight bags so that they received short photoperiods when the rest of the plant was in long days. (From Chailakhyan, 1975)

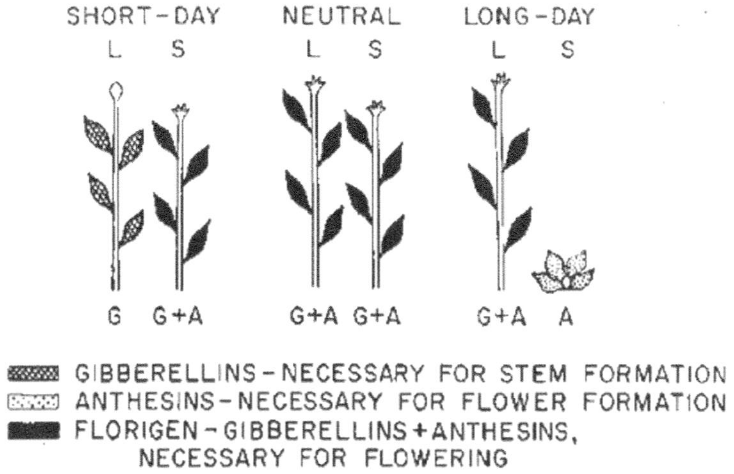

Figure 3.Chailakhyan's later scheme to explain the composition of florigen in long- and short-day plants. (From Chailakhyan, 1975.)

2. THE MEANING OF A PLANT HORMONE

This brings us to the meaning of the term Plant Hormone. In their book *Phytohormones* Went and Thimann (Went and Thimann, 1937) define a hormone as a substance which is transferred from one part of an organism to another. However this is before the full range of what we now consider plant hormones was known. For example, ethylene is not transported; is it therefore a plant hormone? The term hormone was first used in medicine about 100 years ago for a stimulatory factor, but it has come to mean a transported message. In fact the word *hormone,* however, comes from the Greek, where its meaning is to *stimulate.* Thus the origin of word itself does not require the notion of transport *per se.* Or, in the words of Carl Price, "It matters not whether or not we consider ethylene to be a plant hormone; bananas do!"

In view of the fact that not all plant hormones are transported I have proposed a new definition of plant hormones as follows: *Plant hormones are a group of naturally occurring, organic substances which influence physiological processes at low concentrations.* The processes influenced consist mainly of growth, differentiation and development, though other processes, such as stomatal movement, may also be affected. Plant hormones may be transported from one part of the plant to another but others, such as ethylene, may act in the cells or tissue where it is produced (Davies, 1995).

The way a plant hormone influences growth and development depends on:
1) The location of the hormone: this is affected by movement or transport.
2) The amount present: regulated by biosynthesis, degradation and conjugation.

3) The sensitivity (or responsiveness) of the tissue: this involves the presence of receptors and signal- transduction chain components.

All of the above are active areas of current research that will be considered at this volume. Examples of each will be introduced below and some will be considered in more detail in later chapters.

3. EFFECTS DETERMINED BY THE LOCATION OF THE HORMONE

The location of a hormone within a tissue is a most important regulator of the way in which that tissue responds. This is particularly true of auxin whose asymmetric presence in growing stems and roots determines the lateral differences in cellular extension rate leading to tropistic curvatures. We still have no reliable way of determining the location of native auxin within tissues. However the powerful technique of marker genes regulated by promoters of genes whose transcription is known to be turned on by auxin have given a method of visualizing the differences in auxin concentrations in tissues. For example the promoter of an IAA-induced gene fused to GUS gives a blue color in the presence of IAA in responsive tissues (Li *et al.*, 1991). This technique nicely demonstrates that auxin is redistributed in response to gravitropic stimuli. When a young shoot is placed on its side the blue color, indicating the presence of auxin, is found on the lower side of the stem (Fig. 4). This side elongates faster causing a bending.

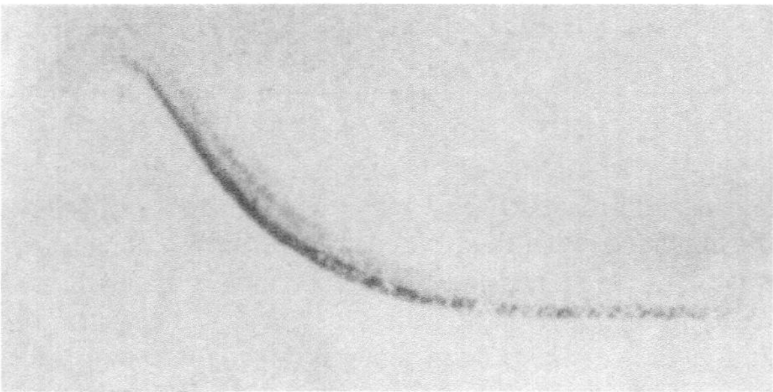

Figure 4. The location of auxin can be determined from the expression of GUS resulting in the presence of a blue color (here black) when a plant is transformed with the gene for GUS driven by the promoter for an early auxin activated gene. In a gravistimulated shoot auxin is located on the lower side of the stem. (From Hagen, 1995.)

3. 1. Hormone movement or transport

The polar auxin movement of auxin enables it to regulate growth in the subapical regions. IAA is synthesized in young apical tissues and then is transported basipetally to the growing zones of the stem, and more distantly, the root. The major stream is via the procambial strands and then the cells of the vascular cambium. In the root tip the downward transport in the central stele reverses in the cortical tissues (Davies and Mitchell, 1972) to give what has been termed an "inverted umbrella". The mechanism of IAA transport involves a pH difference across the cell membrane allowing the movement of un-ionized IAA into the cell, an uptake carrier and an efflux carrier for ionized IAA at the lower side of the cell (Fig. 5) (Rubery, 1987).

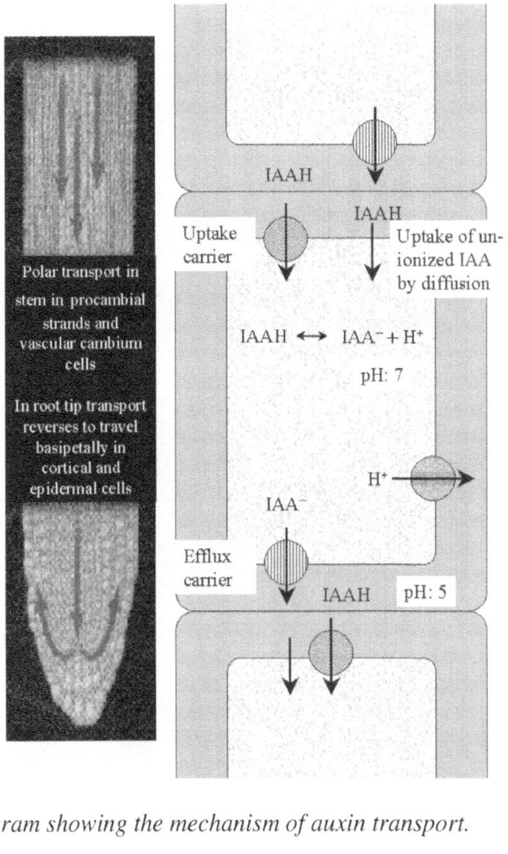

Figure 5. Diagram showing the mechanism of auxin transport.

Recent advances have now localized several of these carriers, most notably the efflux carrier(s). This has been achieved through the use of Arabidopsis *pin* mutants, so named because they often have an an inflorescence stem resembling a pinhead, rather than the normal inflorescence. *Atpin1* mutants have a contorted inflorescence stem and abnormal vascular tissue patterns in the stem (Fig. 6) This is correlated with a greatly reduced level of basipetal auxin transport in the inflorescence tissues (Gälweiler *et al.*, 1998). *AtPIN1* gene analysis revealed that the gene product is membrane localized. When the gene product was made fluorescently labeled it was shown to occur occurs in continuous vertical cell strands in vascular bundle and to be located at the basal end of elongated, parenchymatous xylem cells (Fig. 7). This is the same as predicted and previously found for the auxin efflux carrier. The *Atpin2* mutant influences root gravitropism so that the roots of these mutants fail to respond to gravity. The AtPIN2 protein functions as a transmembrane component of the auxin carrier complex and regulates direction of auxin efflux in cortical and epidermal root cells (Müller *et al.*, 1998).

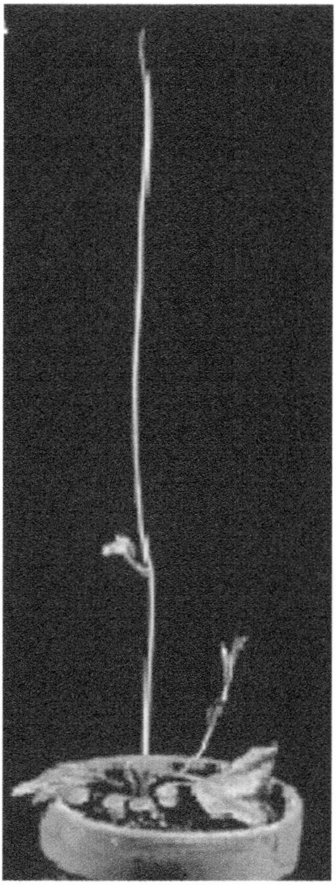

Figure 6. An Atpin1 mutant of Arabidopsis. (From Gälweiler et al., 1998.)

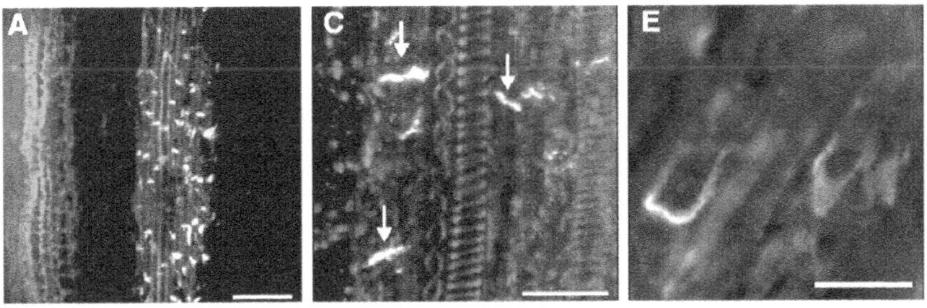

Figure 7. AtPIN1 immunolocalization in longitudinal Arabidopsis stem tissue sections using anti-AtPIN1 fluorescent antibodies (here bright white). The AtPIN1 signals are found in continuous vertical cell strands in vascular bundles (A), at the basal end of elongated, parenchymatous xylem cells in the neighborhood of vessel elements (C & E). Size bars represent 100 μm (A) or 25 μm (C & E). (From Gälweiler et al., 1998.)

Localization of the *ATPIN2* gene product with a fluorescent antibody showed it to be located in the membranes of root cortical and epidermal cell files just behind the root apex, with a basal and outer lateral distribution towards the elongation zones (Fig. 8), again matching the previously determined pathways of basipetal auxin transport. *Atpin3* mutant hypocotyls are defective in gravitropic as well as phototropic responses and root gravitropism. Expression of the *PIN3* gene in *PIN3:GUS* transgenic seedlings shows the PIN3 protein to also be located in the cell membranes of the auxin transporting tissues. What was unique about PIN3 is that it was found to be relocated in root columella cells in response to a change in the gravity vector; one hour after roots were turned sideways the PIN3 protein relocalized from the basipetal cell membrane to the upper cell membrane with respect to gravity (Fig. 9) (Friml *et al.*, 2002), so perhaps PIN3 is important in redirecting auxin laterally in response to gravity and thus leading to auxin regulation of the gravitropic response. The possible mechanism of PIN protein relocalization has been revealed by observation of the location of the fluorescently-labelled protein in the presence of metabolic inhibitors. Treatment with brefeldin (BFA), a vesicle trafficking inhibitor, leads to the detection of PIN1 in vesicles in 2h. When the BFA is washed out the PIN1 returns to its previous membrane localization (Geldner et al., 2001) (Fig. 10). This implies a constant recycling of the PIN proteins between the correct localization in the cell membrane and internal vesicles, with the gravity vector possibly determining the site of reintegration of the auxin transporting PIN auxin efflux carriers. The cycling of both PIN1 and PIN3 was demonstrated to be actin-dependent.

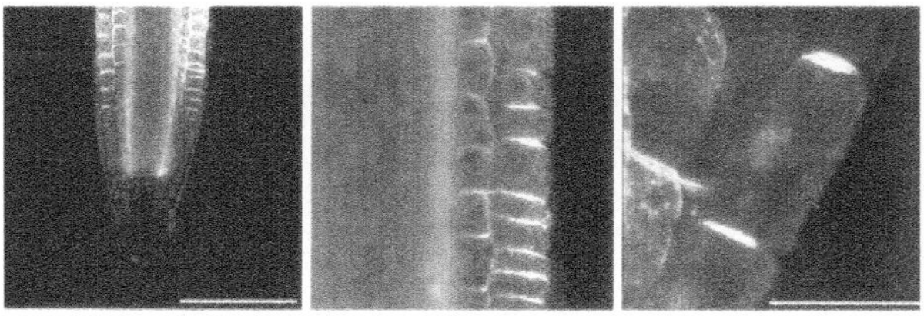

Figure 8. AtPIN2 immunolocalization in Arabidopsis root tip sections using anti-AtPIN1 fluorescent antibodies (here bright white). The AtPIN1 signals are found in membranes of root cortical and epidermal cell files, with a basal (and outer lateral) distribution towards the elongation zones. Size bars represent 100 μm (A) or 25 μm (C & E). (From Müller et al., 1998.)

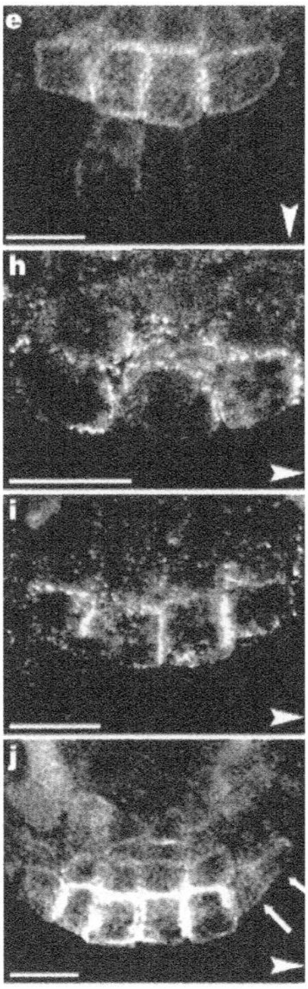

Figure 9. Gravity-dependent asymmetric relocation of the PIN3 protein in Arabidopsis root columella cells after a change in the gravity vector before (e) after 2 min (h) and 10 min (i). After 1 h the PIN3 localization domain expands in lateral root cap (indicated by arrows) and in columella initials (j). Scale bars, 10 μm. The apparent gravity vectors are indicated by arrowheads. (From Friml et al., 2002.)

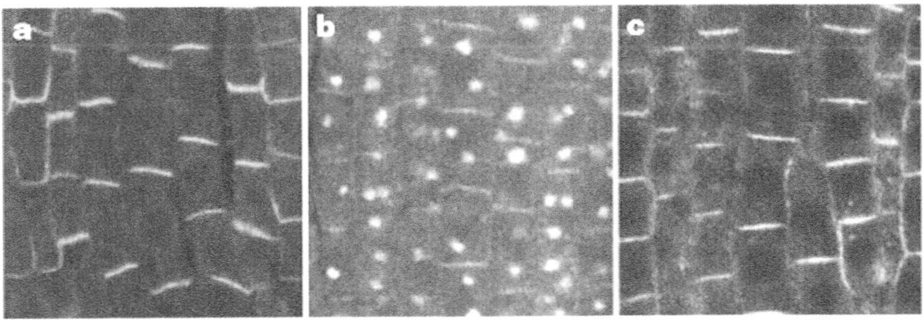

Figure 10. Reversible inhibition of PIN1 recycling Arabidopsis seedling roots by treatment with brefeldin (BFA), a vesicle trafficking inhibitor. a) Two-hour buffer control showing PIN1 in cell membranes. b) Treatment with 50 μM BFA for 2 h showing PIN1 in vesicles. c) Treatment with 50 μM BFA for 2 h followed by 2 h washing out; PIN1 Returns to its cell membrane localization. Scale bars, 1 μM. (From Geldner et al., 2001.)

4. EFFECTS DETERMINED BY THE AMOUNT OF HORMONE PRESENT: BIOSYNTHESIS, DEGRADATION AND CONJUGATION

There are numerous examples of the regulation of plant growth by hormonal concentration. One of the best characterized has been that of bolting in spinach (*Spinacea oleracea*) by Zeevaart and co-workers who have convincingly demonstrated that GA_1 naturally increases in response to long days to produce bolting (Zeevaart *et al.*, 1993).

Another clear example is that of Mendel's tall and dwarf pea, which can also be accounted for in terms of the level of GA_1 present. Mendel's tallness gene (wild-type tall = *Le*; dwarf mutant = *le-1*) has been shown to code for the enzyme GA 3β-hydroxylase, which catalyzes the conversion of $GA_{20}$to GA $_1$by the addition of a hydroxyl group to carbon 3 of GA_{20} (Fig. 11). This gene has been isolated and sequenced (Lester et al., 1997). The *le* mutant was shown to differ from the wild type *Le* by having a single base change from G to A at base 685, which led to a single amino acid change of an alanine to a threonine at amino acid 229. The activity of *Le* and *le-1* recombinant GA 3β-hydroxylases in E. coli were assayed by HPLC. The le mutant enzyme was shown to possess only 1/20 of the activity found in the Le wild-type enzyme in converting radiolabelled GA_{20} into GA_1 (Fig. 12), so would account for the decrease in GA_1 content and the shorter stature of the dwarf plants. The mutation is close to iron-binding motif of other 2-oxoglutarate-dependent dioxygenases so a change in iron binding may be the cause of the decreased enzyme activity.

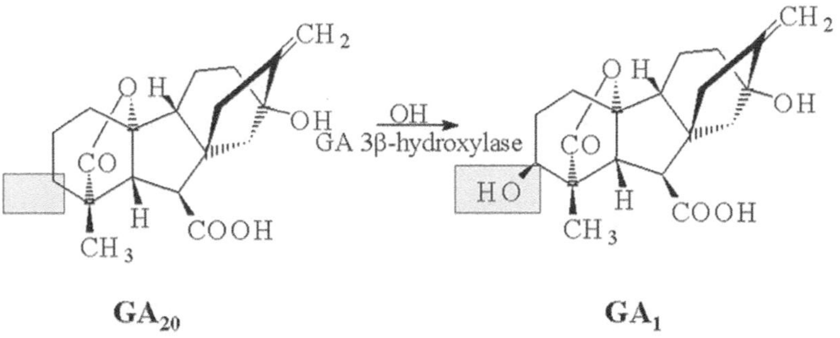

Figure 11.Conversion of GA_{20} to GA_1 by GA 3β-hydroxylase, which adds a hydroxyl group (OH) to carbon 3 of GA_{20}

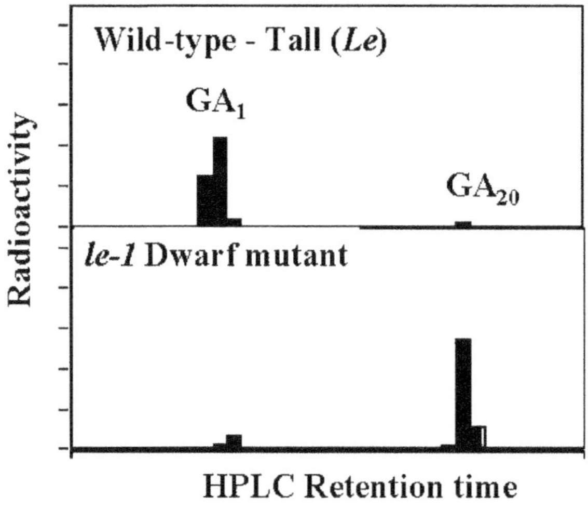

Figure 12. Activity of Le and le-1 GA 3β-hydroxylases in transformed E. coli transformed as determined by the conversion of radiolabelled GA_{20} to GA_1, assayed by HPLC. (From Lester et al., 1997.)

The site of biosynthesis also has a bearing on where the higher concentrations of hormone are present. As a neat demonstration Hedden and co-workers transformed Arabidopsis with the *AtGA20ox1* promoter coupled the luciferase reporter gene and then visualized the light output by the plants. Figure 13 (see color plates) shows the apical and young leaf location of GA 20-oxidase activity, which is involved in the conversion of GA_{53} to GA_{20}, and thus GA biosynthesis.

5. EFFECTS DETERMINED BY THE SENSITIVITY (OR RESPONSIVENESS) OF PLANT TISSUES TO HORMONES

Changes in hormone responsiveness (sensitivity) during development

The sensitivity of stem elongation in gibberellin-deficient nana (na) peas to gibberellin is different in dark-grown and light grown seedlings with the light grown plants being less sensitive, so that dark grown seedlings elongate more in response to applied GA_1 than do light grown seedlings (Fig. 14) (Reid, 1988). During development under constant conditions sensitivity can also change. Proton movement across the membrane of plasma-membrane vesicles is detected by a drop in the membrane electrical potential, and this is affected by auxin. The optimal concentration of auxin to produce this effect changes throughout development in Nicotiana (tobacco) dropping by two and a half orders of magnitude during floral induction (Fig. 15) (Santoni *et al.*, 1991).

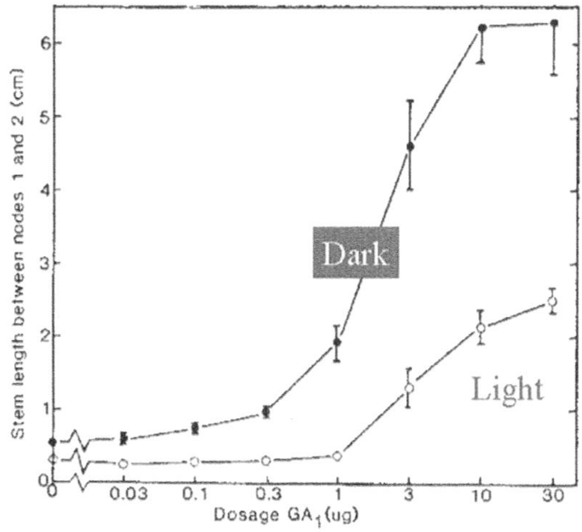

Figure 14. The stem elongation response of light- and dark-grown gibberellin-deficient nana (na) peas to applied gibberellin. (From Reid, 1988.)

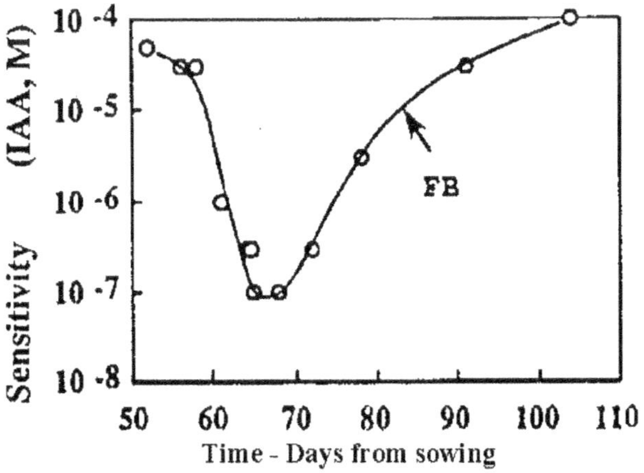

Figure 15. The optimal concentration of auxin to produce a drop in the membrane electrical potential throughout development in Nicotiana in relation to the time of the start of flowering. (From Rossignol et al., 1991.)

5. 1. Gibberellin Signal Transduction in Stem Elongation Shown by mutants inArabidopsis GA

We care now starting to elucidate the gibberellin signal transduction pathway through the use of Arabidopsis mutants (Fig. 16). In a GA-insensitive mutant (*gai-1*) growth is much reduced. In mutants lacking GA because of a mutation blocking GA biosynthesis a second mutation (*rga*) partially restores growth. Some mutants are also extra tall without gibberellin: the *Spy* mutant behaves as if treated with GA. *Spy* and *rga* show that stem elongation is normally repressed and what GA does is to negate the repression. GAI and RGA, which are very similar, act in the absence of GA to suppress growth. Whether mutations in these genes result in the promotion of growth or an insensitivity to GA depends on the locus of the mutation in the protein - whether it is in the repression domain or the regulatory domain respectively (Davies, 2002). SPY, which is also a negative regulator, enhances the effect of GAI and RGA. GA acts to block the actions of SPY, GAI and RGA. GAI and RGA turn out to be transcription factors, whereas SPY functions as a second messenger. The null response is a hormone type response, but it is normally repressed by a member of the signal transduction chain, so that you get the minus-hormone morphology. Binding of GA hormone to the receptors turns off this inhibition. The negative regulator RGA protein is found in the nucleus. This has been visualized by attaching the gene to produce the green fluorescent protein to that for RGA (Fig. 17) (Silverstone *et al.,* 2001). In the absence of GA, experimentally brought about by the GA biosynthesis inhibitor Paclobutrazole, the amount of RGA in the nucleus increases, whereas it is severely decreased in the presence of GA. This is brought about because the RGA is degraded in the presence of GA so that GA negates an inhibition of growth.

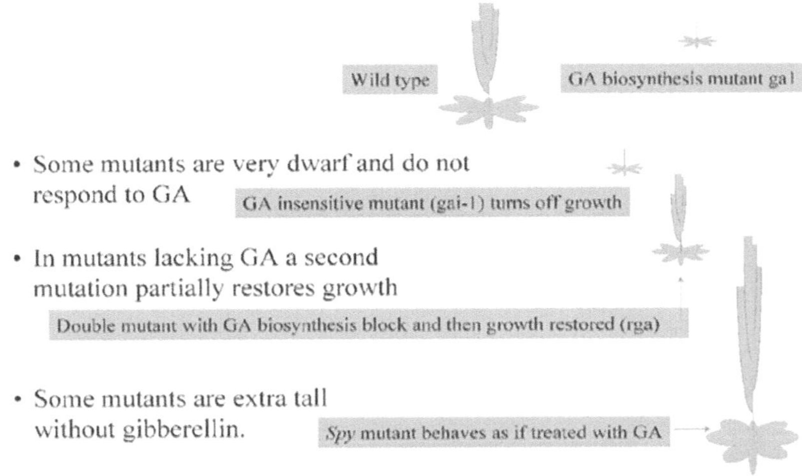

Figure 16. Gibberellin signal transduction in stem elongation as shown by mutants in Arabidopsis.

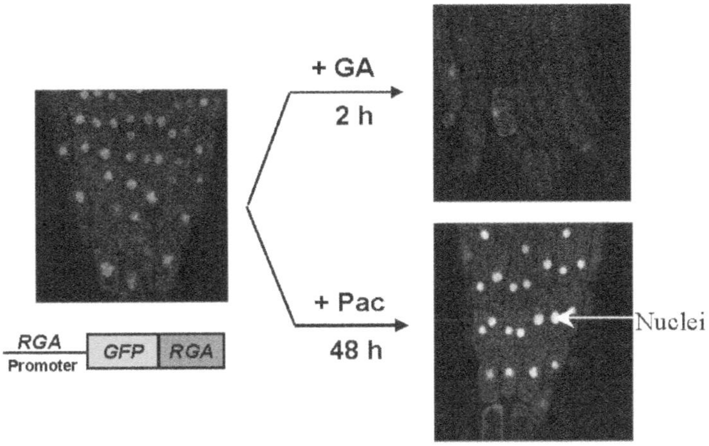

Figure 17. The effect of GA and paclobutrazole (a GA biosynthesis inhibitor) on GFP-RGA level in nuclei. RGA is a nuclear transcription factor that is degraded in the presence of GA. (From Silverstone et al. 2001. Diagram by T-P Sun)

5. 2. Cytokinin Signal Transduction

Cytokinins have long been a hormone for which plant biologists have searched for an action product or pathway; now this long search is producing results. First Arabidopsis genes switched on by cytokinin were identified. The promoter of one such Arabidopsis cytokinin primary response gene, encoding the response regulator 6 (*ARR6*), was fused to the firefly luciferase (*LUC*) gene. This was used to transform protoplasts, so that the luciferase, which is detected by its light emission, was produced in the presence of cytokinin. Sequences encoding suspected signalling intermediates transfected into protoplasts and the luciferase production measured. CKI1, a membrane localized histidine protein kinase, was found to induce luciferase even in the absence of cytokinin (Fig.18a, see color plates). Another cytokinin signalling intermediate AHP1 moves to the nucleus in the presence of cytokinin (Fig. 18b, see color plates) (Hwang and Sheen, 2001). Ectopic expression of CKI1 (and other signaling intermediates) is sufficient to promote cytokinin responses in transgenic tissues so that tissues expressing these genes develop as if they had been treated with cytokinins (Fig. 19).

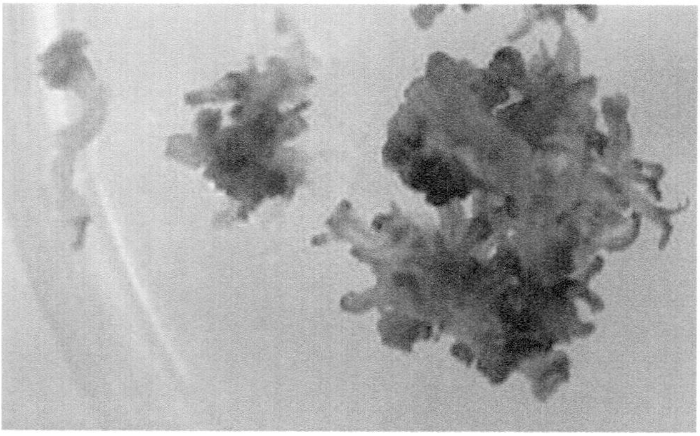

Figure 19. Ectopic expression of CKI1 (and other signaling intermediates) is sufficient to promote cytokinin responses in transgenic tissues. Tissue culture with: left: vector alone; center: IAA alone; right: tissues transformed with CKI1 growing in the presence of IAA. (From Hwang and Sheen, 2001.)

6. REGULATION BY BOTH CHANGING HORMONE LEVELS AND CHANGES IN SENSITIVITY

As a dark grown seedling emerges into the light its growth rate slows down substantially, despite the fact that light grown plants have higher levels of hormones. This turns out to be regulated by a change in both hormone level and sensitivity. When dark-grown pea seedlings transferred to light their GA_1 level drops rapidly due to metabolism of GA_1. This then increases to a higher level, similar to that in light grown plants, over the next 4 days (Fig. 20a) (O'Neill *et al.*, 2000). At the same time sensitivity to applied GA of pea seedlings transferred to the light rapidly falls, so that the elongation rate of plants in the light is lower than in the dark even though their GA_1 content is higher (Fig. 20b).

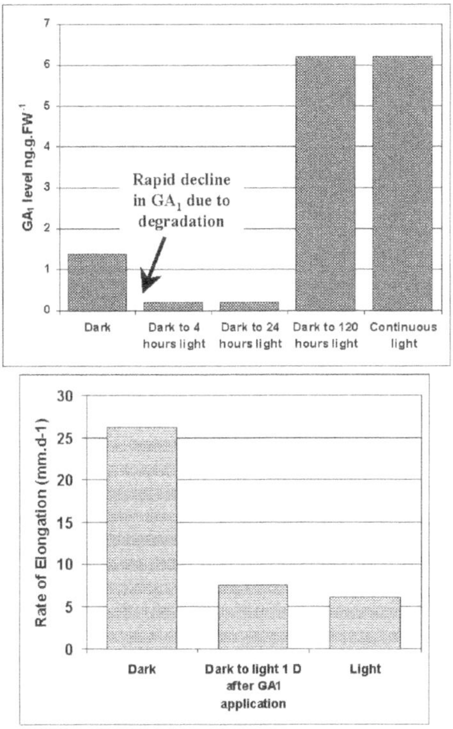

Figure 20. (A)The change in the level of GA_1in germinating pea seedlings as they are transferred from darkness into light and (B) the change in their elongation response to applied GA_1 in darkness and following transfer to light. (Drawn from data in O'Neill et al., 2000.)

7. INTEGRATED EFFECTS OF MULTIPLE HORMONES

We tend to examine the effects of plant hormones on an individual basis, but the growth of a plant is in response to all signals, internal and external, including the net effect or multiple hormones. These hormones may vary in level and responsiveness and also produce numerous often complicated interactions.

As an example pea height varies because of differences in GA_1 content and gibberellin sensitivity (signal transduction components): nana has very low GA_1 because an early step in GA biosynthesis is blocked; dwarf has a low level of GA_1;

wild-type tall has a high level of GA_1; whereas "slender" is ultra tall regardless of GA_1 content and probably represents a mutation in a signal transduction component (Fig 21a). However auxin level also tracks height, including in slender plants that are tall regardless of GA_1 content (Fig. 21b) (Law and Davies, 1990). Auxin levels also rise following treatment of dwarf plants with GA_1 (Barratt and Davies, 1997).

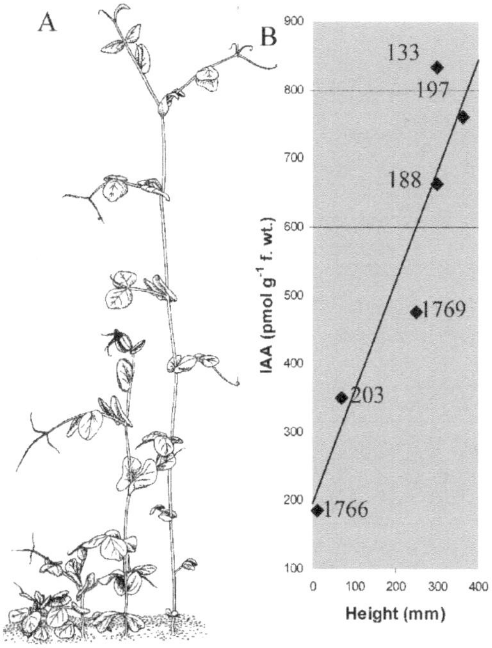

Figure 21. Pea height varies because of differences in GA_1 content and gibberellin sensitivity (signal transduction components) (A) From left to right: nana (na)- very low GA_1 (early step in GA biosynthesis blocked); dwarf (le-1) - low GA_1; wild-type tall (Le) - high GA_1; slender (la cry^s)- regardless of GA_1 content. (B) However auxin level also tracks height, including in slender plants that are tall regardless of GA_1 content because of the presence differing GA biosynthesis gene alleles. Line numbers are shown: 1766 - nana; 203 - dwarf; 1769 - wild-type tall; 188, 133, 197 - all slender with very low, low and high GA_1 respectively. (From data of Law and Davies, 1990.)

Not only does gibberellin promote IAA biosynthesis, but auxin promotes gibberellin biosynthesis. Decapitation of pea shoots inhibits, and IAA promotes, the step between GA_{20} to GA_1 in subapical internodes as detected by the metabolism of applied radiolabelled GA_{20}. Decapitation also reduces, and IAA restores, endogenous GA_1 content. This is because IAA up-regulates GA 3β-hydroxylase transcription, and down-regulates that of GA 2-oxidase, which degrades GA_1 (Fig.

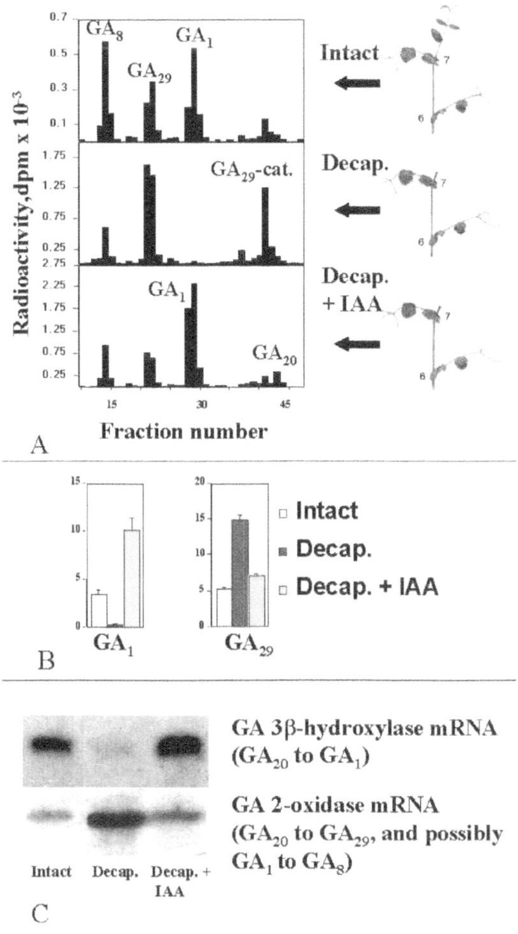

Figure 22. (A) The conversion of radiolabelled GA_{20} into other GAs, including growth-active GA_1 in the upper stem of an intact pea plant (top), a decapitated plant (middle), or a decapitated plant to which IAA in lanolin has been applied to the cut stem tip (bottom). (B) The levels of GA_1 and GA_{29} in the upper stems of intact, decapitated and decapitated pea plants to which auxin was applied. (C) Northern blots of the level of GA 3β-hydroxylase and GA 20-oxidase mRNA in pea stems under the above conditions. From Ross et al., 2000; drawings provided by J. Ross.)

22) (Ross et al., 2000). We can conclude that IAA coming from the apical bud promotes and is required for GA_1 biosynthesis in subtending internodes (Fig. 23) (Ross *et al.*, 2000; Ross an O'Neill, 2001). Thus the growth of young stems is the net result of both IAA and GA_1, with considerable cross-talk occurring between the two hormones.

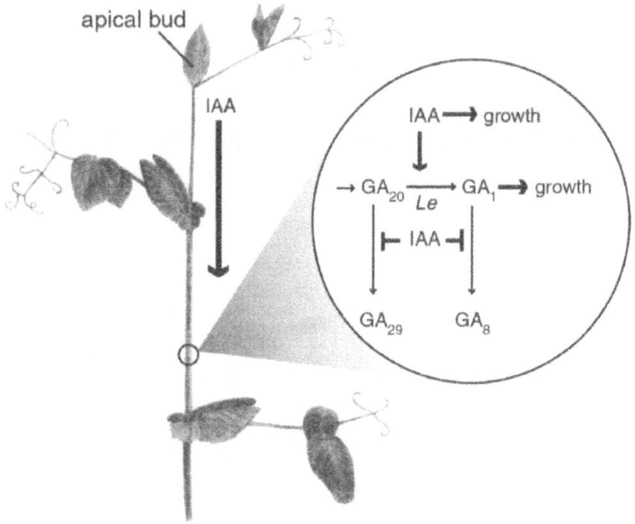

Figure 23. Auxin coming from the stem apex promotes the biosynthesis of GA_1 in subtending internodes and retards its degradation. (From Ross and O'Neill, 2001.)

We have not yet started to understand complex hormone interactions. For example the brassinosteroid-deficient pea genotype *lkb* appears dwarf, but does not respond to applications of GA. It does, however, respond strongly to IAA applications, and, surprisingly, in the presence of IAA it responds to GA applications (Fig. 24) (Yang *et al.*, 1996)!

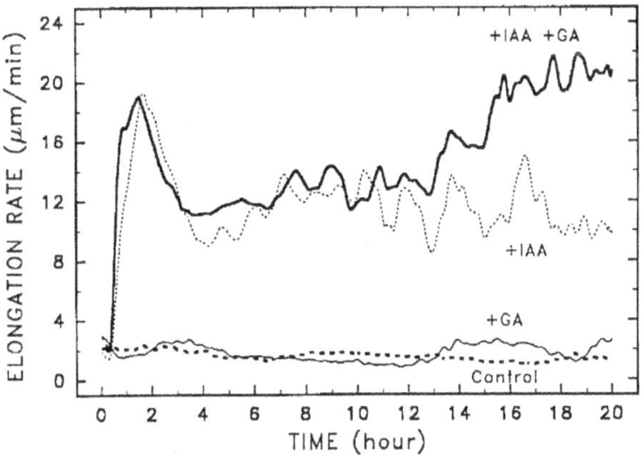

Figure 24. Dwarf lkb plants, which are brassinolide deficient and have a sub-normal auxin level, do not respond to GA, but will do so in the presence of added IAA. The figure shows lkb stem elongation responses to exogenous IAA (light dots), GA_3 (light solid line) or both (dark solid line), as compared to a control treatment (heavy dashed line). (From Yang et al., 1996)

8. BACK TO FLORIGEN

Florigen has not yet been found. We need therefore to ask whether florigen exists: as one opinion has put it: "Flowering is a religion based on the totally unfounded dogma of florigen" (author uncertain). There is no doubt that the flowering signal can be transmitted (Aukerman and Amasino, 1998). However a biochemically defined substance with florigenic activity has never been isolated. The reasons for this may be:

1) The difficulty of bioassay;

2) Florigen may be a mixture or balance of hormones and other substances, and these may vary between species; Gibberellin will promote flowering in many but not all cases.

3) It may be a substance not extracted by the normal methods. Consequently, it has been proposed that florigen might be not a single molecule but a combination of different known hormones and metabolites or assimilates that can vary among plant species. For example in Fuschia, sucrose is a flowering stimulus though it is not the long day florigen, and flowering, induced by LD or sucrose, is inhibited by gibberellin. The inhibitory action of GA on flowering involves diversion of assimilate away from the shoot apex and into the elongating stem (King and Ben, 2001) and this volume). In *Sinapis alba* and Arabidopsis the leaf exudate was enriched in glutamine during the inductive LD, so perhaps amino acids play a role in the transition to flowering (Corbesier *et al.*, 2001).

One possibility is that florigen could involve nucleic acid or protein signaling via the phloem. mRNA made in the companion cell has been shown to move into the sieve tube via plasmodesmata and travel in the phloem sieve tubes to be translated in the cells at the sink (Xoconostle *et al.*, 1999).

8. 1. Genetic studies are helping elucidate the floral stimulus pathway

Genes for a floral stimulus, floral inhibitor and for responsiveness to the floral stimulus have been identified in pea and Arabidopsis, but have not led to the identification of the signal. The indeterminate (*id1*) gene in maize may be involved in floral signal transmission, though maize is non- photoperiodic. An *id1* homozygous mutant plant remains in a state of vegetative growth for a prolonged period of time (Fig. 25) (Colasanti and Sundaresan, 2000). The expression pattern of *id1* is unlike the expression patterns of any of the flowering- time genes found in Arabidopsis: its mRNA is detected only in young leaves. The presence of *id1* in leaves, and its absence in the shoot apex, suggests that the *id1* gene does not mediate the transition to flowering. It could have a role in regulating the production of a floral stimulus or in repressing the activity of a floral inhibitor, though regulation of signal transmission, possibly by plasmodesmata, has also been suggested.

Further molecular studies promise to elucidate the florigen complex.

Figure 25. An id1 mutant maize plant (right) remains in a state of vegetative growth for a prolonged period of time and makes many more leaves than a normal plant (left). ID1 acts in the leaves so could represent a flowering signal. (From Colasanti and Sundaresan, 2000.)

REFERENCES

Aukerman, M.J. and Amasino, R.M. (1998) Floral induction and florigen, *Cell* **93,** 491-494.

Barratt, N.M. and Davies, P.J. (1997) Developmental changes in the gibberellin-induced growth response in stem segments of light-grown pea genotypes, *Plant Growth Regul.* **21**, 127-134.

Chailakhyan, M.Kh. (1975) Forty years of research on the hormonal basis of plant development-some personal reflection, *The Botanical Review* **41,** 7-29.

Colasanti, J. and Sundaresan, V. (2000) 'Florigen' enters the molecular age: long-distance signals that cause plants to flower, *Trends Biochem. Sci.* **25**, 236-240.

Corbesier, L., Havelange, A., Lejeune, P., Bernier, G. and Perilleux, C. (2001) N content of phloem and xylem exudates during the transition to flowering in Sinapis alba and Arabidopsis thaliana, *Plant Cell Environ.* **24**, 367-375.

Darwin, C.: The power of movement in plants. London, 1880.

Davies, P.J. (1995) The plant hormones: Their nature, occurrence, and functions, in: P.J. Davies (ed.), *Plant Hormones: Physiology, Biochemistry and Molecular Biology*, Kluwer Academic Publishers, Dordrecht, The Netherlands, , pp. 1-12.

Davies, P.J. (2002) Gibberellins: regulators of plant height, in: L.Taiz, and E. Zeiger (eds.), *Plant Physiology*, Sinauer Associates, Sunderland, MA, Pp. 461-492.

Davies, P.J. and Mitchell, E.K. (1972) Transport of indoleacetic acid in intact roots of *Phaseolus coccineus, Planta* **105,** 139-154.

Friml, J., Wisniewska, J., Benkova, E., Mendgen, K. and Palme, K.(2002) Lateral relocation of auxin efflux regulator PIN3 mediates tropism in Arabidopsis, *Nature* **415**, 806-809.

Gälweiler, L., Guan, C., Müller, A., Wisman, E., Mendgen, K., Yephremov, A. and Palme, K. (1998) Regulation of polar auxin transport by AtPIN1 in Arabidopsis vascular tissue, *Science* **282**, 2226-2230.

Geldner, N., Friml, J., Stierhof, Y.D., Juergens, G. and Palme, K. (2001) Auxin transport inhibitors block PIN1 cycling and vesicle trafficking, *Nature* **413,** 425-428.

Hagen, G. (1995) The control of gene expression by auxin, in P.J. Davies (ed.), *Plant Hormones: Physiology, Biochemistry and Molecular Biology*, Kluwer Academic Publishers, Dordrecht, The Netherlands, Pp. 228-245.

Hwang, I. and Sheen, J. (2001) Two-component circuitry in Arabidopsis cytokinin signal transduction, *Nature London* **413**, 383-389.

King, R.W. and Ben, T.Y. (2001) A florigenic effect of sucrose in Fuchsia hybrida is blocked by gibberellin-induced assimilate competition, *Plant Physiol.* **125**, 488-496.

Law, D.M. and Davies, P.J. (1990) Comparative indole-3-acetic acid levels in the slender pea and other pea phenotypes, *Plant Physiol.* **93**, 1539-1543.

Lester, D.R., Ross, J.J., Davies, P.J. and Reid, J.B. (1997) Mendel's stem length gene (*Le*) encodes a gibberellin 3 -hydroxylase, *Plant Cell* **9**, 1435-1443.

Li, Y., Hagen, G. and Guilfoyle, T.J. (1991) An auxin-responsive promoter is differentially induced by auxin gradients during tropisms, *Plant Cell* **3,** 1167-1176.

Müller, A., Guan, C., Gälweiler, L., Taenzler, P., Huijser, P., Marchant, A., Parry, G., Bennett, M., Wisman, E. and Palme, K. (1998) AtPIN2 defines a locus of Arabidopsis for root gravitropism control, *EMBO J.* **17**, 6903-6911.

O'Neill, D.P., Ross, J.J. and Reid, J.B. (2000) Changes in gibberellin A_1 levels and response during de-etiolation of pea seedlings, *Plant Physiol.* **124**, 805-812.

Reid, J.B. (1988) Internode length in *Pisum*. Comparison of genotypes in the light and dark, *Physiol. Plant.* **74,** 83-88.

Ross, J.J., O'Neill, D.P., Smith, J.J., Kerckhoffs, L.H.J. and Elliott, R.C. (2000) Evidence that auxin promotes gibberellin A_1 biosynthesis in pea, *Plant Jour.* **21**, 547-552.

Ross, J. and O'Neill, D. (2001) New interactions between classical plant hormones, *Trends in Plant Science* **6**, 2-4.

Rossignol, M., Santoni, V., Francois, J.M. & Vansuyt G.: Changing membrane sensitivity to auxin during development.- in Progress in Plant Growth Regulation, C.M. Karssen, L.C. van Loon, D Vreugdenhil, eds. 207-213, Kluwer Academic, Dordrecht, 1991

Rubery, P.H. (1987) Auxin transport, in: P.J. Davies (ed.), *Plant Hormones and Their Role in Plant Growth and Development*, Kluwer, Boston, pp. 341-362.

Silverstone, A.L., Jung, H.S., Dill, A., Kawaide, H., Kamiya, Y. and Sun, T.P. (2001) Repressing a repressor: Gibberellin-induced rapid reduction of the RGA protein in Arabidopsis, *Plant Cell* **13,** 1555-1565.

Went, F.W. and Thimann, K.V. (1937) *Phytohormones*, Macmillan, New York.

Wildman, S.G.(1997) The auxin-A, B enigma: Scientific fraud or scientific ineptitude?, *Plant Growth Regul.* **22**, 37-68.

Xoconostle, C.B., Xiang, Y., Ruiz, M.R., Wang, H.L., Monzer, J., Yoo, B.C., McFarland, K.C., Franceschi, V.R. and Lucas, W.J. (1999) Plant paralog to viral movement protein that potentiates transport of mRNA into the phloem, *Science* **283**, 94-98.

Yang, T., Davies, P.J. and Reid, J.B. (1996) Genetic dissection of the relative roles of auxin and gibberellin in the regulation of stem elongation in intact light-grown peas, *Plant Physiol.* **110**, 1029-1034.

Zeevaart, J.A.D., Gage, D.A. and Talon, M. (1993) Gibberellin A_1 is required for stem elongation in spinach, *Proc. Natl. Acad. Sci. USA* **90**, 7401-7405.

THREE ROLES FOR GIBBERELLIN IN FLOWERING

ROD KING
CSIRO Plant Industry,
PO Box 1600, Canberra, ACT 2601,Australia

1. INTRODUCTION

Seasonal flowering of plants involves responses to many environmental signals including changes in irradiance, temperature and daylength. The range and complexity of environmental inputs is summarized by Thomas and Vince-Prue (1997) but integrating this information into plant response has only begun recently. Genetic studies of flowering time in *Arabidopsis* (Simpson and Dean, 2002) highlight a network of interacting pathways involving plant response to daylength, vernalization, photosynthetic input and gibberellin (GA) and a fifth, autonomous pathway, whereby the plant progresses to flowering in the absence of external signals.

"Perception" of low temperature (vernalization) is likely to be in meristematic cells of the shoot apex (see ref in Metzger, 1995). By contrast, daylength and photosynthetic responses involve leaves and, therefore, signal(s) must be transported from the leaf to the shoot apex, the term florigen being proposed more than 65 years ago by Mikhail Chailakhyan for this transported factor (Chailakhyan, 1937).
Chailakhyan's florigen theory, also known as the hormonal theory of flowering in plants, met stiff ideological resistance in Russia when in 1938 he presented his findings to the Scientific Council of the USSR Academy of Sciences as his doctoral dissertation. As he recalled 50 years later (Chailakhyan, 1988), "the book met with favorable responses from reviewers" but one reviewer, T.D. Lysenko, " in broken, brief, and harsh phrases often unconnected with each other, attempted to demonstrate the complete unsoundness of the book, saying that the florigen theory is the same thing as the phlogiston theory of the Middle Ages and unconditionally concluding in a raised voice that the author of the book does not even deserve a candidates degree, much less a doctoral degree."

Fortunately, Chailakhyan was not dissuaded by this criticism and the wider scientific community responded positively to the theory with studies in the 1950s first considering auxin as a likely candidate and then that the gibberellins were florigen (see Lang, 1965). Later Chailakhyan expanded the hypothesis to include two components in time, first gibberellins and then "anthesins" (Chailakhyan, 1958) and his studies in the late 1970s confirmed earlier evidence of transmissible

I. Macháčková and G.A. Romanov (eds.), Phytohormones in Plant Biotechnology and Agriculture, 31-39

inhibitors as well (Lang *et al.*, 1977). The nearest he came to florigen involved a patent in 1983 on a method of preparing florigenically active plant extracts but this approach has yet to provide a positive chemical identification.

Controversy continues over claims of the existence of florigen(s) and, in particular, that they may be gibberellins (GAs). However, recent studies of GAs in *L. temulentum*, reviewed here, provide a continuous trail of evidence for gibberellins as one of a possible "complex" of floral stimuli/inhibitors in long day (LD) responses of grasses of temperate origin. What happens in short day (SD)-responsive plants is unclear. Furthermore, in rosette forms of dicots, LD may increase GAs but they may regulate stem bolting rather than flowering (Zeevaart, 1983).

Where GA promotes flowering it may activate expression of GA-regulated genes which control cell division e.g. *CDC2* and the *LEAFY* gene, the latter playing a central role in floral organ initiation and development. By contrast, with some dicot species, rather than promoting flowering, GA application can be inhibitory, but there are specific prerequisites for such response and, in one case, this is related to the diversion of sugars to the growing stem and away from the shoot apex.

2. GIBBERLLINS PROMOTE FLOWERING: A CONTINUOUS TRAIL FROM THE LEAF TO THE SHOOT APEX

GAs during floral evocation LD exposure of leaves of dicots (see Zeevaart, 1983; Pharis and King, 1985; Xu *et al.*, 1997) and monocots (Junttila *et al.*, 1997; Gocal *et al.*, 1999; King, Moritz and Evans, unpublished) may cause an increase in GA synthesis, GA_{19} being converted to GA_{20} and, thence, to bioactive GAs. The last steps of the GA biosynthesis and of subsequent catabolism are shown in Figure 1 and summarized in the many recent reviews (Hedden and Phillips, 2000). For example, in leaves of *L. temulentum*, when the critical LD photoperiod length for flowering is reached (16 h after the beginning of the light period) the contents of GA_{20} and of a bioactive product, GA_5, rise 4 to 5-fold in LD relative to leaves in SD and harvested at the same time. GA_1 increased 3-fold in LD, but some hours after the increase in GA_5. Especially in dicots it is known that such LD-induced increases in GA content lead to bolting and, possibly, flowering and this occurs with increased expression of biosynthetic enzymes shown diagrammatically in Figure 1 (see Zeevaart, 1983; Xu *et al.*, 1997). Similarly, transgenic lines overexpressing either of the two GA biosynthetic genes (the 20-oxidase or the 3-oxidase) show enhanced flowering along with greater stem growth (Zeevaart, unpublished MSU Plant Res. Lab. Ann. Report, 2000). Conversely, overexpressing the catabolic 2-oxidase will reduce the content of bioactive GAs so causing dwarfing as seen in rice (Sakamoto *et al.*, 2001) but effects on flowering need examination.

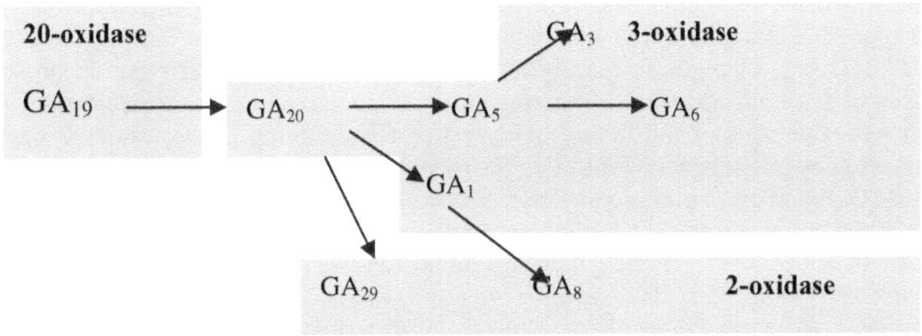

Figure 1. The last steps in synthesis of bioactive 13-hydroxylated GAs showing the two important synthetic enzymes (in bold) and the subsequent catabolic enzyme (2-oxidase) which inactivates GAs.

The increases in LD of leaf GAs provide a picture consistent with GAs acting as LD floral stimuli. However, to establish that GAs are long day floral stimuli it is also essential to show matching increases in them or in their bioactive products in the shoot apex and this has now been achieved with the grass *L. temulentum* using very sensitive GCMS techniques along with the inclusion of matching deuterated GAs as internal standards (King *et al.*, 2001; 2003).

On the day after the single LD the shoot apex content of the highly florigenic GA_5 doubles (King *et al.*, 2001), GA_6 rising in parallel and to a similar concentration (King *et al.*, 2002). The timing of the increase in shoot apex GA_5 content was the same in three separate experiments, the peak level being evident about 12 h after the maximum in leaf GA_5 content.

Both GA_5 and GA_6 are florigenic when only 1 or 5 µg is applied to the leaf of plants held in non-inductive SD (Evans *et al.*, 1990; King *et al.*, 2001; 2002) and their lack of effect on stem elongation, except at high doses, parallels the response to a LD. Also, when supplied via agar, GA_5 causes floral initiation of apices excised from vegetative plants (King *et al.*, 1993). Furthermore, the effective threshold concentration of GA_5 required in the agar matches very closely the maximum estimated endogenous GA_5 concentration in the apices. Most interestingly, when supplied via agar a 2-3-fold increase in GA_5 concentration over the threshold value was sufficient to obtain a maximum flowering response (King *et al.*, 1993). Thus, doubling the endogenous GA content would have been florally effective.

Endogenous GA_5 and GA_6 are presumably transported from the leaf blade through the leaf sheath and into the shoot apex. Although such detail is not finally established, a number of observations support this claim. Firstly, [$^2H_4$] GA_5 applied to the *L. temulentum* leaf blade is transported intact to the shoot apex (King *et al.*, 2001). Secondly, if more [$^2H_4$] GA_5 is applied to the leaf, the amount transported to the apex and the flowering response increase in parallel (King *et al.*, 2001). Thirdly, based on the speed of transport of floral stimulus in *L. temulentum* (1-2 cm h^{-1}; Evans and Wardlaw, 1966), the time of increase in endogenous GA_5 and GA_6 at the apex (12 h after the leaf increase) fits with the time required to transport the GA over a distance of 10-12 cm. Thus, these florally inductive GAs would begin arriving at the shoot apex of *L. temulentum* on the morning after the inductive LD. These calculations also fit with estimates of the timing of arrival of floral stimuli at the shoot apex when assessed by excising shoot apices for *in vitro* culture at various times before and after the LD. Floral evocation is complete when isolated shoot apices can proceed to initiate spikelets in the absence of further leaf input, and this begins on the morning after the LD as apex GAs approach their maximum and is complete within about 12 h (McDaniel *et al.*, 1991; King *et al.*, 1993).

Thus, taken as a whole, these results make a strong case for specific GAs including GA_5 and GA_6 as sufficient LD stimuli for the first event(s), floral evocation, in *L. temulentum*.

GAs during inflorescence development At inflorescence initiation, 4-6 d after a single LD, there are clear increases at the shoot apex in both GA_1 and GA_4 (King *et al.*, 2001 and unpublished). That applied GA_4 could now enhance flowering and that LAB19998, an cyclohexanedione-class GA biosynthetic inhibitor inhibits flowering when applied at this time (4 to 6 days), shows the importance of this late LD-induced GA increase in regulating endogenous levels of GA_1 and GA_4, (see King *et al.*, 2001). Earlier, as discussed below, $GA_{1/4}$ are apparently excluded from the vegetative shoot apex.

3. GIBBERELLINS ARE INACTIVE IF SENSITIVE TO CATABOLIC ENZYMES

It was unexpected that GA_1 and GA_4 were undetectable in the vegetative and early floral apex (King *et al.*, 2001) although detected in leaves (Gocal et al., 1999). However, their absence from the shoot apex was clear because added internal standards were detected in the same analysis. Thus, GA_1 and GA_4 must be actively excluded from the vegetative and early floral shoot apex whereas there is ready access for GA_5 and GA_6. This restriction may be enzymatic involving a GA 2-oxidase. In rice such a 2-oxidase expresses highly in vegetative and early floral plants in the cells of vascular and provascular tissue where they terminate just below the shoot apex (Sakamoto *et al.*, 2001). Applying this finding to *L. temulentum*, GAs would be transported in the vascular bundles to a point just below (>1 mm) the shoot apex, and if protected against inactivation by 2-oxidases they would pass into the 0.6 mm shoot apex. Later, however, at inflorescence formation, this localized ring of 2-oxidase disappears in rice, and if it does likewise in *L. temulentum* then GA_1 and GA_4 contents in the shoot apex could increase, as they do.

The implication that florigenic activity reflects the extent to which a GA is protected from 2β-hydroxylation is supported by the florigenicity for *L. temulentum* of applied GA_5, GA_6, GA_{32}, GA_3 and 2,2-dimethyl GA_4 (Pharis *et al.*, 1987; Evans *et al.*, 1990). Compared with GA_1 and GA_4 all these GAs should be structurally protected, but with disappearance of localized 2-oxidase activity as at inflorescence initiation in rice, then GA_4 should now reach the shoot apex and promote floral development as we have found (King *et al.*, 2001).

4. GIBBERELLIN-REGULATED GENES EXPRESS AT THE SHOOT APEX

One early change in metabolism at the induced shoot apex involves a rise in the incorporation of ^{32}P into nucleotides (Rijven and Evans, 1967) and of ^{35}S into proteins (Evans and Rijven, 1967). Based on autoradiography, incorporation was most prominent in both the dome of the shoot apex and at the lateral sites where the spikelet primordia are visibly initiated a day or two later (Knox and Evans, 1968), after which floral morphogenesis ensues (Knox and Evans, 1966). Later, by the afternoon of the first day after the LD, expression of the cell cycle gene *LtCDC2* increases dramatically (Gocal, 1997). Then, a day later (about 24h after the end of the LD), there is increased expression of a *L. temulentum AP1*-like gene, *LtMADS2*, a gene functionally related to the *Arabidopsis AP1* gene (Gocal *et al.*, 2001a). The increase in *LtCDC2* is especially important as it is considered an indicator of GA-regulation, particularly in studies with graminaceous species (Sauter *et al.*, 1995). Thus, these early changes at the shoot apex immediately after the LD are not only likely to be relevant and important for floral evocation, but they support the possibility that GA_5 and GA_6 really are florigens for *L. temulentum*.

Other genes that should define the timing of GA action include a *GAMYB* gene and the *LFY* gene. Studies with *Arabidopsis* have established that the *LFY* gene is required for inflorescence initiation, that applied GA_3 directly activates the *LFY* promoter, and that this response is blocked and flowering delayed in a mutant with much reduced GA biosynthesis (Blázquez *et al.*, 1997; 1998). Moreover, at least one of the *Arabidopsis* GAMYB transcriptional activators, AtMYB33, binds to an 8 bp motif in the *LFY* promoter both *in vivo* (Blázquez and Weigel, 2000) and *in vitro* (Gocal *et al.*, 2001b). Mutating 6 bp of this promoter motif removed most of its MYB binding capacity. *GAMYB* genes act directly in the GA signal transduction path (Gubler *et al.*, 1995) and with GA treatment *Arabidopsis* not only flowers but there is an increase in expression of *AtMYB33* at its shoot apex (Gocal *et al.*, 2001b). Thus, these findings provide evidence of a direct link between GA and *LFY* expression, this latter gene binding in turn to the promoter of a gene regulating inflorescence specification (e.g. *AP1*) and to the promoters of genes which regulate floral organ identity.

With *L. temulentum*, the *LtGAMYB* gene is expressed in the vegetative shoot apex but is up-regulated substantially when the inflorescence is forming five or more days after the LD. This expression is initially in the terminal and lateral spikelet sites, then decreases but increases again when the stamen primordia are initiating

(Gocal *et al.*, 1999). As discussed above, levels of GA_1 and GA_4 at the shoot apex rise dramatically at inflorescence formation and applied GA_4 also becomes florally active for *L. temulentum* (King *et al.*, 2001). Thus, GAs may regulate the increased expression of *LtGAMYB* at this time. On the other hand, *LtLFY*, another GA-regulated gene (Blázquez *et al.*, 1997; 1998), is not expressed until 4 to 5 days later than *LtGAMYB* (Gocal *et al.*, 2001a). However, this information for *LtLFY* depends on in situ expression analysis and there was no basal level of gene expression so it is difficult to estimate the first time of *LtLFY* expression.

5. GIBBERELLINS WHICH CAUSE EXCESSIVE STEM ELONGATION CAN INHIBIT FLOWERING OF SOME DICOT SPECIES

That GA may inhibit flowering is paradoxical given the emphasis above on promotion of flowering by GAs but such inhibition apparently involves a GA-mediated nutrient diversion and may be of particular relevance to the flowering of fruit and forestry trees (Moncur and Hasan, 1994; Monselise and Goldschmidt, 1982) as well as to flowering of some ornamentals including *Bougainvillea* (Hackett and Sachs, 1967) and *Fuchsia hybrida* (Sachs and Bretz 1961). Practical application has followed with growth retardants being used to promote flowering by inhibiting GA biosynthesis and dramatically so in *Eucalyptus nitens* where years are cut from the flowering time of breeding stock (Moncur and Hasan, 1994).

Our measurements of sucrose content of the shoot apex of *Fuchsia* indicate that GA inhibits its early flowering response in LD by diverting assimilates away from the apex and into the growing stem. Furthermore, it is certain that flowering of *Fuchsia hybrida* is regulated by sucrose supply because apex sucrose content and flowering increased in parallel with increase in light intensity in non-inductive short days (King and Ben-Tal, 2001).

6. CONCLUSIONS

As summarized in Figure 2, a number of Chailakhyan"s postulates about florigen are matched by changes detected in gibberellins first in long-day-exposed leaves (King, Moritz and Evans, unpublished) and then at the shoot apex (King *et al.*, 2001). Thus, for this and, likely, other temperate grasses, gibberellins meet the criteria of a chemical signal controlling flowering. However, a complete understanding of hormonally–regulated flowering must accommodate multiple actions of GAs. While GA_5 and GA_6 appear important for floral evocation, GA_1 and GA_4 become important for promoting inflorescence development. GA action at the molecular level apparently involves activation of specific GA-regulated genes but only after inflorescence formation. A distinct action whereby GAs inhibits flowering (Fig. 2) adds complexity to explaining flowering responses but such inhibition is restricted to a limited number of species. Furthermore, this inhibition by GA involves regulation of sucrose supply to the shoot apex, an action which is quite separate from promotion by GA of floral evocation and development.

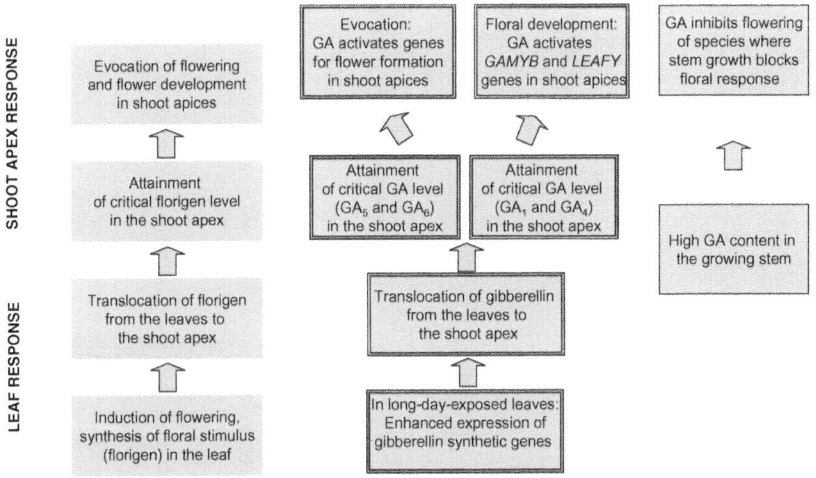

Figure 2. A model for the pathway from daylength perception in the leaf to flowering at the shoot apex. Features of Chailakhyan's hypothesis (on the left: from Milyaeva and Romanov, 2002) are compared with experimentally identified roles for gibberellins (centre). In a restricted number of species gibberellin may inhibit rather than promote flowering (on the right). Such inhibition involves a clearly different and indirect mechanism of action whereby sugars are diverted from the shoot apex and so flowering is inhibited.

REFERENCES

Blázquez, M.A. and Weigel, D. (2000) Integration of floral inductive signals in *Arabidopsis*, *Nature* **404**, 889-892.

Blázquez, M.A., Soowal, L.N., Lee, I. and Weigel D. (1997) *LEAFY* expression and flower initiation in *Arabidopsis*, *Development* **124**, 3835-3844.

Blázquez, M.A., Green, R., Nilsson, O., Sussman, M.R. and Weigel, D. (1998) Gibberellins promote flowering of *Arabidopsis* by activating the *LEAFY* promoter. *Plant Cell* **10**, 791-800.

Chailakhyan, M. Kh. (1937) Concerning the hormonal nature of plant development processes, *C.R.(Dokl.) Acad. Sci. URSS*. **16**, 227-230.

Chailakhyan, M. Kh.(1958) Hormonal factors in the flowering of plants, *Soviet Plant Physiol.* **5**, 547-563.

Chailakhyan, M. Kh. (1988) 50 years of the hormonal theory of plant development, *Soviet Plant Physiol.* **35**, 466-480.

Evans, L.T. and Rijven, A.H.G.C. (1967) Inflorescence initiation in *Lolium temulentum* L. XI Early increases in the incorporation of ^{32}P and ^{35}S by shoot apices during induction, *Aust. J. Biol. Sci.* **20**,1033-1042.

Evans, L.T. and Wardlaw, I.F. (1966) Independent translocation of ^{14}C-labelled assimilates and of the floral stimulus in *Lolium temulentum, Planta* **68**, 319-326.

Evans, L.T., King, R.W., Chu, A., Mander, L.N. and Pharis, R.P. (1990) Gibberellin structure and florigenic activity in *Lolium temulentum*, a long day plant, *Planta* **182**, 97-106.

Gocal, G.F.W. (1997) *Molecular biology of floral evocation in Lolium temulentum.* PhD Dissertation. Australian National University, Canberra, 181 pp.

Gocal, G.F.W., Poole, A.T., Gubler, F., Watts, R.J., Blundell, C. and King, R.W. (1999) Long-day up-regulation of a *GAMYB* gene during *Lolium temulentum* inflorescence formation, *Plant Physiol.* **119**, 1271-1278.

Gocal, G.F.W., King, R.W., Blundell, C.A., Schwartz, O.M., Andersen, C.H. and Weigel, D. (2001a) Evolution of floral meristem identity genes. Analysis of *Lolium temulentum* genes related to *APETALA1* and *LEAFY* of *Arabidopsis, Plant Physiol.* **125**, 1788-1801.

Gocal, G.F.W., Sheldon, C.C., Gubler, F., Moritz, T., Bagnall, D.J., MacMillan, C.P., Li, S.F., Parish, R.W., Dennis, E.S., Weigel, D. and King, R.W. (2001b) *GAMYB-like* genes, flowering, and gibberellin signaling in *Arabidopsis, Plant Physiol.* **127**, 1682-1693.

Gubler, F., Kalla, R., Roberts, J.K. and Jacobsen, J.V. (1995) Gibberellin-regulated expression of a *myb* gene in barley aleurone cells: evidence for *Myb* transactivation of a high-pI alpha -amylase gene promoter, *Plant Cell* 7, 1879-1891.

Hackett,W.P. and Sachs, R.M. (1967) Chemical control of flowering in *Bougainvillea* 'San Diego Red", *Proc. Amer. Soc. Hortic. Sci.* **90**, 361-364.

Hedden, P. and Phillips, A.L. (2000) Gibberellin metabolism: new insights revealed by the genes, *Trends in Plant Sci.* **5**, 523-530.

Junttila,O., Heide, O.M., Lindgard, B. and Ernsten, A. (1997) Gibberellins and the photoperiodic control of leaf growth in *Poa pratensis, Physiol. Plant.* **101**, 599-605.

King, R.W. and Ben-Tal, Y. (2001) A florigenic effect of sucrose in *Fuchsia* and its inhibition by gibberellin-induced assimilate competition, *Plant Physiol.* **125**, 1-9.

King, R.W., Blundell, C. and Evans, L.T. (1993) The behaviour of shoot apices of *Lolium temulentum in vitro* as the basis of an assay system for florigenic extracts, *Aust. J. Plant Physiol.* **20**, 337-348.

King, R.W., Evans, L.T., Mander, L.M., Moritz, T., Pharis, R.P. and Twitchin, B. (2003) Synthesis of gibberellin GA_6 and examination of its role in flowering of *Lolium temulentum, Phytochem.* **62**, 77-82.

King, R.W., Moritz, T., Evans, L.T., Junttila, O. and Herlt, A.J. (2001) Long-day induction of flowering in *Lolium temulentum* involves sequential increases in specific gibberellins at the shoot apex, *Plant Physiol.* **127**, 624-632.

Knox, R.B. and Evans, L.T. (1966) Inflorescence initiation in *Lolium temulentum* L. VIII. Histochemical changes at the shoot apex during induction, *Aust. J. Biol. Sci.* **19**, 233-245.

Knox, R.B. and Evans, L.T. (1968) Inflorescence initiation in *Lolium temulentum* L. XII. An autoradiographic study of evocation in the shoot apex, *Aust. J. Biol. Sci.* **21**, 1083-1094.

Lang, A. (1965) Physiology of flower initiation, in W. Ruhland (ed.), *Encyclopedia of Plant Physiology XV/1*, Springer Verlag, Berlin, pp. 1380-1536.

Lang. A., Chailakhyan, M.Kh. and Frolova, I.A. (1977) Promotion and inhibition of flower formation in a day-neutral plant in grafts with a short-day and a long-day plant, *Proc. Natl. Acad. Sci.* USA **74**, 2412-2416.

McDaniel, C.N., King, R.W. and Evans, L.T. (1991) Floral determination and *in vitro* floral differentiation in isolated shoot apices of *Lolium temulentum* L., *Planta* **185**, 9-16.

Metzger, J.D. (1995) Hormones and reproductive development, in P.J. Davies (ed.), *Plant Hormones.Physiology, Biochemistry and Molecular Biology*, Kluwer Academic, Dordrecht, pp 617-648.

Milyaeva, E.L. and Romanov, G.A. (2002) Molecular genetics returns to basic postulates of the florigen theory, *Russian Plant Physiol.* **49**, 438-444.

Moncur, M.W. and Hasan, O. (1994) Floral induction in *Eucalyptus nitens, Tree Physiol.* **14**, 1303-1312.

Monselise, S.P. and Goldschmidt, E.E. (1982) Alternate bearing in fruit trees, *Hortic. Rev.* **4**, 128-73.

Pharis, R.P. and King, R.W. (1985) Gibberellins and reproductive development in seed plants, *Annu. Rev. Plant Physiol.* **36**, 517-568.

Pharis, R.P., Evans, L.T., King, R.W. and Mander, L.N. (1987) Gibberellins, endogenous and applied, in relation to flower induction in the long-day plant *Lolium temulentum, Plant Physiol.* **84**, 1132-1138.

Rijven, A.H.G.C. and Evans, L.T. (1967) Inflorescence initiation in *Lolium temulentum* L. X. Changes in ^{32}P incorporation into nucleic acids of the shoot apex at induction, *Aust. J. Biol. Sci.* **20**, 13-24.

Sachs, R.M. and Bretz, C.F. (1961) The effect of daylength, temperature and gibberellic acid upon flowering of *Fuchsia hybrida, Amer. Soc. Hortic. Sci.* **80**, 581-588.

Sakamoto, T., Kobayashi, M., Itoh, H., Tagiri, A., Kayano, T., Tanaka, H., Iwahori, S. and Matsuoka, M. (2001) Expression of a gibberellin 2-oxidase gene around the shoot apex is related to phase transition in rice, *Plant Physiol.* **125**, 1508-1516.

Sauter, M., Mekhedov, S.L. and Kende, H. (1995) Gibberellin promotes histone H1 kinase activity and the expression of *cdc2* and *cyclin* genes during the induction of rapid growth in deepwater rice internodes, *Plant J.* **7**, 623-632.

Simpson, G.G. and Dean, C. (2002) *Arabidopsis*, the Rosetta stone of flowering time?, *Science* **296**, 285-289.

Thomas, B. and Vince-Prue, D. (1997) *Photoperiodism in plants*, Academic Press, London, 428 pp.

Xu, YunLing., Gage, D.A. and Zeevaart, J.A.D. (1997) Gibberellins and stem growth in *Arabidopsis thaliana*. Effects of photoperiod on the *GA4* and *GA5* loci, *Plant Physiol.* **114**, 1471-1461.

Zeevaart, J.A.D. (1983) Gibberellins and flowering, in A. Crozier (ed.), *The Biochemistry and Physiology of Gibberellins*, Vol 2., Praeger, New York, pp. 333-373.

ADDITIONAL NOTES

The content of this review is based on two other recent reviews by the author.

THE ROLE OF GIBBERELLIN-REGULATED GENE EXPRESSION DURING ANTHER AND EARLY FLOWER DEVELOPMENT IN TOMATO

G.W.M. BARENDSE AND K.J.P.T. VAN DEN HEUVEL
Department of Cell Biology, University of Nijmegen, Toernooiveld 1, 6525 ED Nijmegen, The Netherlands.

The role of hormones, especially auxins, during fruit and seed development is well established and the role of gibberellins has been studied previously (Barendse *et al.*,1991). The work of Jacobsen and Olszewski (1991) associated a special role for gibberellin (GA) in tomato by studying early flower and anther development in the GA deficient *gib-1* mutant. The GA effects obtained with this mutant could not be achieved with other plant hormones.

The role of the plant hormone gibberellin (GA) during early flower development has been studied by the use of the GA biosynthesis *gib-1* mutant of tomato (Van den Heuvel, 2000). This *gib-1* mutant is a monogenic recessive mutant in which GA-biosynthesis is blocked at the first step prior to *ent*-kaurene formation, an essential step in the GA-biosynthetic pathway. This mutation results in GA deficiency which leads to the inability to germinate, dwarf growth and arrest of flower development. Application of GA largely reverses the mutant phenotype to wild type. The developmental arrest occurs prior to meiosis when the anthers contain pollen mother cells that are in the G1 phase of the premeiotic interphase.

In this study the role of GA was investigated on the molecular level during the premeiotic flower development by the isolation of transcripts of GA regulated genes and the effect of GA on gene expression. The isolated cDNAs have been used as molecular markers to investigate the hormone response with respect to (1) the translocation on endogenous GAs to the flower and (2) the role during cellular processes like cell growth and DNA replication. The hypothesis that the cellular basis of the phenotype of the mutant may be caused by reduced cell elongation and cell division was tested.

The cDNAs corresponding to GA regulated mRNAs were isolated by differential screening. Four cDNAs were characterized on the basis of their genetic sequence and appeared to code for histon H1 (*leH1*) and three variants of histon H2B (*leH2B-1, -2, -3*). The corresponding protein structures showed strong similarities to earlier described histones (Van den Heuvel *et al.*, 1999b). The gene expression of the histones *leH1* and *leH2B* is developmentally dependent regulated in the leaves. GA regulated expression was found in the leaves of the mutant, but not in wild type leaves. The enhanced gene expression by GAs, especially those synthesized via the

I. Macháčková and G.A. Romanov (eds.), Phytohormones in Plant Biotechnology and Agriculture, 41-47

early 13-hydroxylation pathway, is possibly related to enhanced DNA replication in the leaf cells. Enhanced chromosome replication may indicate increased cell division and/or enhanced DNA level in cells, which both may depend on the cell elongation induced by the applied hormone.

The kinetics of GA regulated and development dependent expression was also studied with a cDNA (*tgas118*). Although the mRNA corresponding with the isolated cDNA clone has been localised in all investigated tissue types of the wild type plant, it appeared that the accumulation of this mRNA was especially high in flower tissues during flower development. In contrast, the transcription decreases in the developing flower buds of the *gib-1* mutant if no exogenous GA is applied. The expression of this mRNA is enhanced within hours after application in leaves as well as in the flower bud. *In situ* hybridization of the mRNA in the *gib-1* mutant flower buds has revealed that application of GA induces a comparable localisation of the transcription as occurs in the wild type flower buds. The possible corresponding protein shows strong similarities with gamma-thionins. These proteins have a function during the defense of plants against pathogens. Besides a genetic similarity it appeared also that gene expression is regulated by among others wounding, so that also the protein TGAS118 may have a comparable function.

Two mRNAs which show a specific expression in flowers and which show no transcription in other tissue types have been analysed (Van den Heuvel *et al.*, 2002). One mRNA (*tgas105*) codes for a protein which shows similarity with the cystein-rich extensin proteins, while the other mRNA (*tgas100*) has only significant homology with an anther specific gene from *Anthirhinum majus.* The kinetics of the GA regulated transcription differs strongly for both genes as shown by Northern blot analyses.

The *tgas100* mRNA shows first expression at 48 hours after GA application in the flower buds of the *gib-1* mutant, while transcription of the *tgas105* mRNA enhances already strongly at 8 hours after GA application. *In situ* analyses of the localisation of transcription in the miscellanous cell types has revealed that the *tgas100* mRNA specifically accumulates in the tapetum cells of the anthers. The mRNA *tgas105* is localised in several cell types of the anthers, including in cells of the sepals and the ovary as also established by the northern blot hybridizations (Fig. 1).

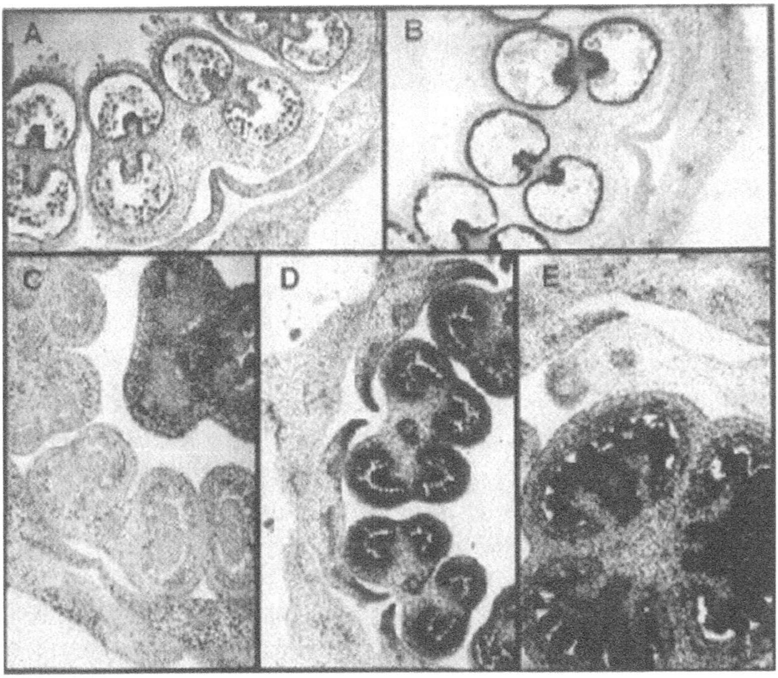

Figure 1. Localisation of tgas100 and tgas105 transcripts in wild type flower buds by in situ hybridizations using DIG-labelled RNA probes. Expression of tgas100 and tgas105 was analysed in 10 :m cross sections of wild type flower buds 5 and 4 mm in length, respectively. Bright field illumination of sections probed with sense (A) and antisense (B) tgas100 RNA probe, and sense (C) antisense (D-E) tgas105 RNA probe.

The untreated *gib-1* mutant initiates flower buds of which the development is arrested prior to meiosis, the moment of sexual cell divisions. Until a bud length of approximately 4 mm the response to GA application is strong and the majority of the treated buds progress their development to normal flowers with fertile pollen and fruit set.

GA application has shown that in anthers mRNAs accumulate which are related to DNA replication (histones H1 and H2B) and cell elongation (expasin and alpha-tubulin). The various cell types in the anthers on the moment of arrest of development showed in longitudinal sections that the elongation of tapetum cells, pollen mother cells and epidermal cells is decreased in camparison with corresponding cells in wild type flower buds.

In cross sections it was found that tapetum and pollen mother cells are significantly smaller than comparable cells in wild type anthers. A number of epidermal cells on the surfaces of wild type anthers differentiate into hair cells (Fig. 2) which eventually interweave with hair cells of neighbouring anthers leading to the tubelike flower development well-known in tomato (Van den Heuvel *et al.*, 1999a).

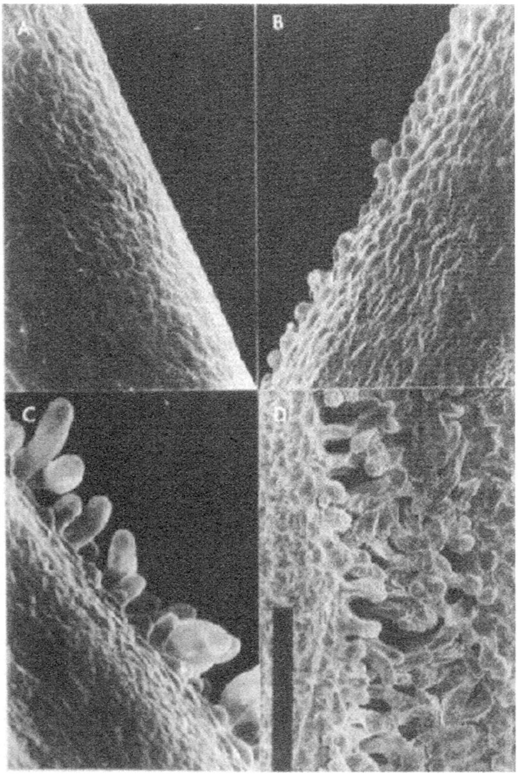

Figure 2. Development of interlocking hairs on the surface of wild type anthers. Analysis using SEM of the middle region of the lateral surface of the anther lobe. Anthers were isolated from 3mm (A), 4mm (B), 5 mm (C) and 6 mm (D) wild type buds.

A study with the scanning electron microscope has revealed that the epidermal cells in the anthers of the *gib-1* mutant which show reduced cell elongation do not develop such hair cells.

However, GA application can induce the formation of these hairs with the same kinetics of development like that in wild type. The biosynthesis of GA may result in various types, whereby it has been shown in our studies that there is a difference in effectiveness between different GAs in overcoming the developmental arrest in the *gib-1* mutant and the GA regulated gene expression. The localisation of transcription has been studied before and after GA application of two marker genes for DNA replication (histon H2B) and cell elongation (expansin) in the anthers of the mutant. It was confirmed that the accumulation is low in the *gib-1* anthers on the moment of the developmental arrest and that the transcription strongly increases after hormone application in those cells which had a reduced cell size (tapetum, pollen mother cells and epidermal cells) (Fig.3).

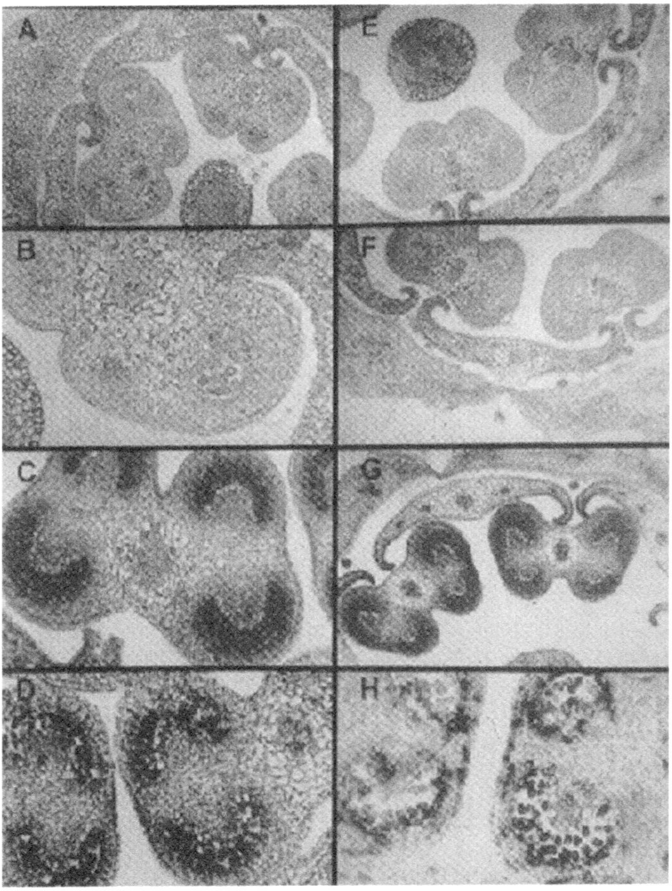

Figure 3. In situ hybridization of cross sections of gib-1 buds with DIG-labelled leH2B and expansin probe 0, 8 and 24 h after GA_3 application. Sections were photographed after hybridization under bright field. A and E developmentally arrested gib-1 bud hybridized with a sense probe of expansin and leH2B

(Figure 3 continued) respectively. B and F section through an untreated gib-1 bud probedwith antisense DIG-labelled expansin and leH2B RNA respectively. C and G mutant bud 8 h after treatment labelled for expansin and leH2B respectively. D and H hybridization of gib-1 buds 24 h after GA_3 application using antisense expansin and leH2B probe respectively.

A study was made whether the hormone required for flower development was synthesized in the flower itself or elsewhere in the plant and consequently transported to the flower bud. Hereto the isolated cDNAs *tgas100* and *tgas118* were used as marker genes. Both genes are expressed upon GA application, however, the kinetics of the response differs and the concentration of hormone required for evoking expression differs for both genes. The *tgas100* mRNA is only expressed when the hormone concentration is sufficient for complete flower development in the mutant, while the *tgas118* mRNA is already expressed at a lower GA concentration. In one experiment paclobutrazol was used to block the GA biosynthesis in the wild type in order to establish whether the expression of both genes could be reduced to a level comparable to that in the *gib-1* mutant. It was shown that the expression was indeed lowered and it was concluded that the rate of hormone translocation from other plant parts to the flower was not sufficient to restore the normal rate of expression. This has been confirmed by grafting experiments, whereby *gib-1* flower buds at the moment of developmental arrest were placed on top of the wild type plant which did not lead to normal flower development. Thus GA transport from elsewhere in the plant does not seem to play a role during the early flower development and the required level of hormone for flowere development is mainly synthesized in the flower itself.

Although it is known that the hormone GA will influence various processes during the early flower development it has been shown in this research that a number of genes show increased expression within hours after GA application in the *gib-1* anthers. Some studied genes appear to be less regulated by this hormone and they show new expression a few days after treatment. A number of these isolated cDNAs codes for proteins which have a possible function related to cell growth and DNA replication. The results obtained indicate that reduced cell elongation and cell division constitute the underlying cellular basis for the nutant phenotype. It appears that the GA deficiency of the *gib-1* mutant influences especially cell elongation and to a lesser extent cell division of several cell types in the developing anthers.

The growth retardation of the *gib-1* anther is most likely the result of smaller cell sizes rather than caused by reduced cell division. Therefore, the increased DNA replication may correlate with increase in endopolyploidy in tapetum and epidermal cells. In tomato it is known that tapetal cells undergo several endomitoses, when meiosis occurs in the pollen mother cells.

In conclusion, in tomato, the expression of several genes is regulated within hours after GA treatment of the developmentally arrested *gib-1* buds and, therefore these genes may have a role during early flower development.

REFERENCES

Barendse, G.W.M., Karssen, C.M. and Koornneef, M.(1991) Role of gibberellins during fruit and seed development, in N. Takahashi, B.O. Phinney and J. MacMillan (eds), *Gibberellins*, Springer-Verlag, N.Y.,pp 179-187.

Jacobsen, S.E. and Olszewski, N.E. (1991) Characterization of the arrest in anther development associated with gibberellin defeciency of the *gib-1* mutant, *Plant Physiol.* **97**, 409-414.

Van den Heuvel, K.J.P.T (2000) *The role of Gibberellin during Anther and Early Flower Development in Tomato: GA Rregulated Gene Expression.* – Thesis (University of Nijmgen), Universal Press, Veenendaal, pp. 1-107.

Van den Heuvel, K.J.P.T., Barendse, G.W.M. and Wullems, G.J. (1999a) GA-regulated development of interlocking hairs on anthers of *Solanum lycopersicon, Flowering Newsletter* **27**, 23-28.

Van den Heuvel, K.J.P.T., Van Esch, R.J., Barendse, G.W.M. and Wullems, G.J.(1999b) Isolation and molecular characterization of gibberellin-regulated H1 and H2B histone cDNAs in the leaf of the gibberellin-deficient tomato, *Plant Mol. Biol.* **39**, 883-890.

Van den Heuvel, K.J.P.T., Van Lipzig, R.H., Barendse, G.W.M. and Wullems G.J. (2002) Regulation of expression of two novel flower-specific genes from tomato (*Solanum lycopersicum*) by gibberellin, *J.Exp.Bot.* **53**, 51-9.

INVOLVEMENT OF GIBBERELLINS IN DEVELOPMENT AND SENESCENCE OF ROSE FLOWERS

N. ZIESLIN AND H. AGBARIA

Department of Horticulture, The Hebrew University of Jerusalem, Faculty of Agriculture, Food and Environmental Science, Rehovot 76100, Israel.

It was previously reported (Goszczynska *et al.*, 1990) that longevity of flowers in certain rose cultivars was improved and petal senescence was postponed following post-harvest treatment of the flowers with GA_3. One of the main findings was a significant decrease in leakage of electrolytes from GA_3-treated petals, although several rose cultivars were insensitive to GA_3 (Sabehat and Zieslin, 1994). High sensitivity of cv. Mercedes flowers to GA_3 treatments and a complete insensitivity of cv. Madelon flowers (Shaul *et al.*, 1995) stimulated investigation of GA_3 effects in petals of these two cultivars. Some of the results of this investigation are described in the present report.

Light-microscopy of the cross-sections showed that the anatomical structure of cv. Mercedes petals treated with GA_3 7 days after harvest was similar to that of fresh harvested petals (day zero), whereas the untreated petals withered and showed all symptoms of senescence. On the other hand, only very small differences were present in the untreated petals of cv. Madelon 7 days after harvest. At day 0, the transducing electron microscopy (TEM) of cv. Mercedes petals showed a normal cytoplasm, with the expected organelles, and normal cell membranes separating the cytoplasm and the vacuole. Seven days after flower harvest, partial destruction of organelles, membranes and formation of myelin bodies were clearly visible in petals of the control treatments. Due to the partial destruction of tonoplast and plasmalemma, the cytoplasm and the vacuole were not separated and vacuole leakage into the cytoplasm occurred, whereas the cell structure of the GA_3-treated petals, unlike the control ones, remained normal and they retained full turgor. In contrast to cv. Mercedes petals, a destruction of internal cell structure was evident 7 days after harvest in GA_3-treated petals of cv. Madelon, although some completely normal cells were also present. We do not have any explanation for this phenomenon. An increase in leakage of electrolytes has been measured two days after flower harvest in the control petals of cv. Mercedes. This increase in leakage has been completely prevented for 10 days by GA_3 treatment. On the other hand, the leakage of electrolytes from cv. Madelon petals was not affected by GA_3 and remained stable during the experimental period (Fig. 1).

I. Macháčková and G.A. Romanov (eds.), Phytohormones in Plant Biotechnology and Agriculture, 49-56

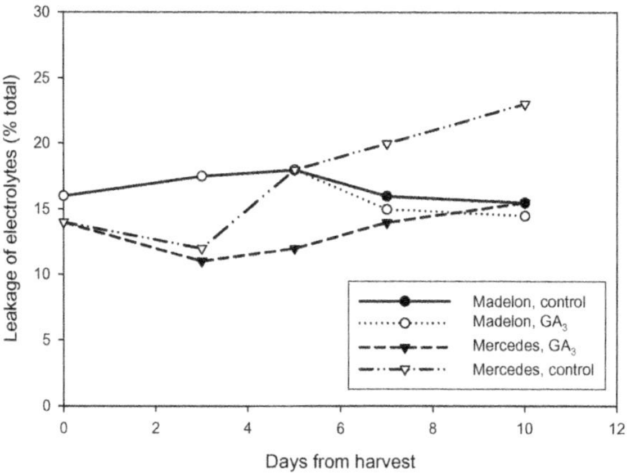

Figure 1. Effect of postharvest petal imbibition in GA_3 on leakage of electrolytes (% of total) from excised petals of rose cvs. Mercedes and Madelon

Effects of GA_3 on leakage of 5 elements: calcium (Ca), sodium (Na), potassium (K), phosphorus (P) and sulfur (S), were investigated. Imbibition of cv. Mercedes petals with the base for 20h in GA_3 solution inhibited leakage of Ca. Only after 7 days in water was a slight increase in Ca leakage detected, whereas the leakage of Ca from untreated petals at termination of the experiment was 50% higher than that from the GA_3-treated petals. In contrast to Mercedes petals, the leakage of Ca from Madelon petals was promoted slightly by the GA_3 treatment during the first week of the experiment, but inhibited during the last three days of the experiment. Ca^{2+} ions are of great importance in regulation of cell functions in plants as well as in improvement of flower longevity (e.g. Hepler and Wayne 1985; Schumaker and Sze, 1985; Eklund and Eliasson, 1990; Ketchum and Poole, 1991; Pantoja *et al.*, 1992). Therefore, a further investigation of the involvement of gibberellins in regulation of cell membrane system in rose flowers is of a great importance. In contrast to Ca^{2+}, the leakage of Na^+ from Mercedes petals was promoted by the GA_3 treatment, while GA_3 application had only a minor effect on leakage of Na^+ from the Madelon petals. An increase in leakage of Na^+ and the inhibition of leakage of Ca^{2+} results in an improvement of cell functions (Cramer *et al.*, 1987) as well as extending rose flower longevity (Zieslin and Abolitz, 1994; Abolitz and Zieslin,1996). The leakage of potassium ions, an important component of rose petal tissue (Acock and Nichols, 1979) was inhibited by the GA_3 treatment in cv. Mercedes petals. The leakage of K^+ was also inhibited by GA_3 treatment in petals of cv. Madelon, although the inhibition effect only became evident on the fifth day after flower harvest. One of the most pronounced effects of GA_3 inhibition of K^+ leakage from cv. Mercedes petals was an increased uptake of water into the petal cells. Leakage of phosphorus (P) from Mercedes petals was inhibited by GA_3 treatment immediately after harvest, while in petals of cv. Madelon the decrease of leakage started only five days after harvest. The possible source of P leaked from the petals is degradation of P-containing compounds such as ATP. The inhibition of Ca leakage by GA_3 may affect the enzymatic activities of phosphatases and phospholipases and thus reduce the leakage of phosphorus. The GA_3-treatment reduced the leakage of sulfur from petals of cv. Mercedes and to a smaller extent from petals of cv. Madelon petals. It has been shown elsewhere (Zieslin *et al.,* 1996) that leakage of amino acids from cv. Mercedes petals is reduced by GA_3 treatments. It is possible therefore, that the decrease of sulfur is due to the decrease in leakage of amino acids containing sulfur.

Proteins known as expansins play an important role in the mechanism of loosening cell walls. The loosening enables stretching of the cell walls and cell growth (McQueen-Mason *et al.*, 1995; Rose *et al.*, 1997). Immunological identification revealed a lower content of expansins at the time of flower harvest in both cultivars. After 3 days the level of expansins increased in cv. Mercedes petals treated with GA_3 while no effect of GA_3 was seen in petals of cv. Madelon. This effect of GA_3 on the level of expansins could be one of the factors responsible for the large increase in size of cv. Mercedes flower buds.

A decrease in fluidity of membranes is a characteristic sign for flower senescence (Borochov et al. 1978, 1982). Imbibition in GA_3 inhibited the decrease of membrane fluidity in petals of cv. Mercedes but not in petals of Madelon (Fig. 2). The degree of membrane fluidity depends on the ratio between sterol and phospholipid components (Shinitzky 1984). An increase in the ratio is one of the main causes of the decreased membrane fluidity in senescing rose flowers (Borochov and Woodson 1989, Itzchaki et al. 1990). A decrease of phospholipids in membranes of Mercedes petals was inhibited by GA_3 treatment, however, GA3 application did not affect the moderate decrease in sterol content (Figs. 3, 4). In contrast to cv. Mercedes, in cv. Madelon petals neither phospholipids nor sterols were affected by GA_3. These data indicate that membrane fluidity is affected by alteration of the phospholipid component of the membranes rather than by the changes in sterol component.

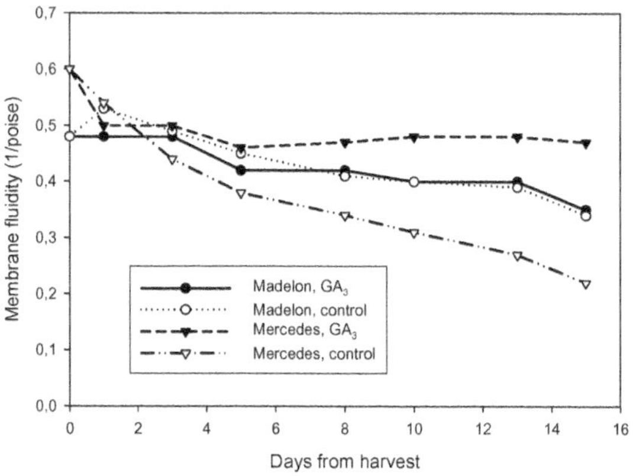

Figure 2. Effect of postharvest petal imbibition in GA_3 on membrane fluidity (1/Poise) in excised petals of rose cvs. Mercedes and Madelon

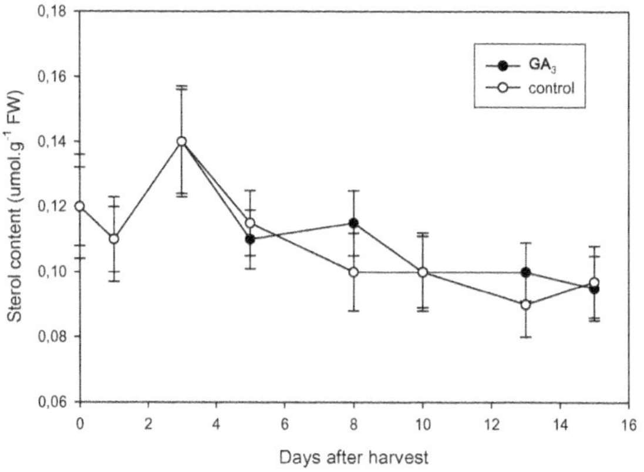

Figure 3. Effect of postharvest petal imbibition in GA_3 on content of sterols in petals of rose cv. Mercedes

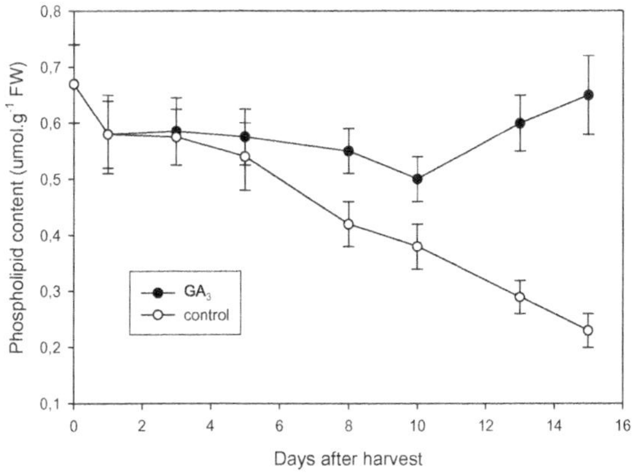

Figure 4. Effect of postharvest petal imbibition in GA_3 on content of phospholipids in petals of rose cv. Mercedes

The activity of sucrose phosphate synthase (SPS), one of the enzymes of sucrose metabolism, was severely inhibited in Mercedes petals by GA_3 treatment (Fig. 5). In spite of the inhibition, the content of sucrose in the petals increased during the experimental period. Hence, the activity of sucrose synthase, an additional enzyme of sucrose metabolism was measured and the results showed that the anabolic activity of sucrose synthase was strongly promoted by GA_3 treatment. Thus, the increase in content of sucrose in petals of cv. Mercedes could be a result of the increased activity of sucrose synthase. On the other hand, an increase in content of sucrose in cv. Madelon petals can be attributed to a slight increase in SPS activity following application of GA_3.

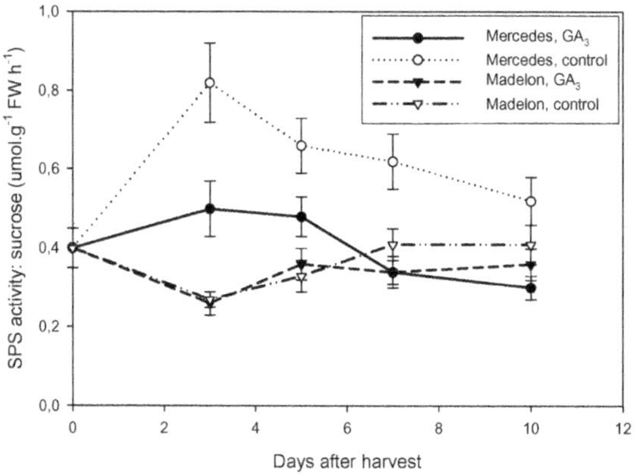

Figure 5. Effect of postharvest petal imbibition in GA_3 on activity of sucrose phosphate synthase (SPS) in excised petals of rose cvs. Mercedes and Madelon

It is difficult to believe that the numerous, various functions of physiological and biochemical cell components in rose petals have been affected by only one gibberellin. Consequently, an analysis of endogenous gibberellins in the petals of both cultivars was carried out. The data showed that the content of GA_1 and GA_3 in petals of cv. Madelon was two and three times respectively higher than the content of these gibberellins in cv. Mercedes petals. The content of GA_{20} was 30% higher than that in Mercedes petals, whereas the content of the three other examined gibberellins (GA_8, GA_{19} and GA_{29}) was lower in the cv. Madelon than in cv. Mercedes petals (Tab. 1). It was logical to assume therefore, that the applied GA_3 has been rapidly altered to other gibberellins, which are able to influence various processes in the plant cells. The content of abscisic acid (ABA) was also measured concurrently with the gibberellins. ABA is known as a compound antagonistic to many activities of plant gibberellins (Ross and O'Neill, 2001). It has also been shown previously that the content of ABA increases during the senescence of rose petals (Borochov *et. al.*, 1976), while treatment with ABA promotes senescence of rose flower buds and of the detached petals (Halevy *et al.* 1974). The analysis of ABA (Tab.2) revealed that in cv. Mercedes petals the content of ABA was two times higher than that in petals of cv. Madelon. Therefore, it is possible, that the increase in gibberellin content and interconversion following GA_3 application could be one of the factors inhibiting the effects of the endogenous ABA in cv. Mercedes petals. Further research into the interactions of gibberellins with various cell components in long and short living rose cultivars could contribute to improvement of rose flower longevity and add to our understanding of the control of gibberellin function in higher plants.

Table 1. Content of endogenous gibberellins (ng.g^{-1} DW) in excised petals of rose cvs. Mercedes and Madelon

Gibberellin	GA content (ng.g^{-1} DW)	
	Mercedes	Madelon
GA_1	3.0 ± 0.3	6.6 ± 1.2
GA_3	13.2 ± 3.7	42.5 ± 8.7
GA_8	2.5 ± 0.4	1.6 ± 0.5
GA_{19}	36.1 ± 6.2	23.5 ± 6.7
GA_{20}	1.3 ± 0.5	1.7 ± 0.7
GA_{29}	3.8 ± 0.4	1.1 ± 0.1
Total	**59.9**	**77.0**

Table 2. Content of endogenous abscisic acid (ABA) (ng.g^{-1} DW) in excised petals of rose cvs. Mercedes and Madelon

Content of ABA (ng.g^{-1} DW)	
Mercedes	Madelon
681 ± 77	330 ± 21

REFERENCES

Abolitz, M. and Zieslin, N. (1996) Effect of pH in root environment on leakage of phenolic compounds and mineral ions from roots of *Rosa indica* major, *Adv. Hortic. Sci.* **10**, 210-214.

Acock, B. and Nichols, R. (1979) Effect of sucrose and water relations in cut senescing carnation flowers, *Ann Bot.* **44**, 221-230.

Borochov, A., Halevy, A.H., Borochov, H. and Shinitsky, M. (1978) Microviscosity of plasmalemmas in rose petals as affected by age and environmental factors, *Plant Physiol.* **61**, 812-815.

Borochov, A., Halevy, A.H., and Shinitsky, M. (1982) Senescence and the fluidity of rose petal membranes, *Plant Physiol.* **69**, 296-299.

Borochov, A., Tirosh, T. and Halevy, A.H. (1976) Abscisic acid content of senescing petals in cut rose flowers as affected by sucrose and water stress, *Plant Physiol.* **58**, 175-178.

Borochov, A. and Woodson, R.W. (1989) Physiology and biochemistry of flower petal senescence, *Hortic. Rev.* **11**, 15-43.

Cramer, G.R., Lynch, J., Läuchli, A. and Epstein, E. (1987) Influx of Na^+, K^+ and Ca^{2+} into roots of salt-stressed coton seedling, *Plant Physiol.* **83**, 510-516.

Eklund, L. and Eliasson, L. (1990) Effect of calcium concentration on cell wall synthesis, *J. Exp. Bot.* **41**, 863-867.

Goszczynska, D.M., Zieslin, N., Mor, Y. and Halevy, A.H. (1990) Improvement of postharvest keeping quality of Mercedes roses by gibberellin, *Plant Growth Regul.* **9**, 293-303.

Halevy, A.H., Mayak, S., Tirosh, T., Spigelstein, H. and Kofranek, A.M. (1974) Opposing effects of abscisic acid on senescence of rose flowers, *Plant Cell Physiol.* **15**, 813-821.

Hepler, P.K. and Wayne, R.O. (1985) Calcium and plant development, *Annu. Rev. Plant Physiol.* **36**, 397-439.

Itzchaki, H., Borochov, A. and Mayak, S. (1990) Age-related changes in petal membranes from attached and detached rose flowers, *Plant Physiol.* **94**,1233-1236.

Ketchum, K.A. and Poole, R.J. (1991) Cytosolic calcium regulates a potassium current in corn *(Zea mays)* protoplasts, *J. Memb. Biol.* **119,** 277-288.

McQueen-Mason, S.J. and Cosgrove, D.J. (1995) Disruption of hydrogen bonding between wall polymers by protein that induce plant wall extension, *Proc. Natl. Acad. Sci. USA* **91**, 6574-6578.

Pantoja,O., Gelli, A. and Blumwald, E. (1992) Voltage-dependent calcium channels in plant vacuoles, *Science* **225**, 1567-1570.

Rose, J.K.C., Lee, H.H. and Benett A.B. (1997) Expression of divergent expansin gene in fruit specific and ripening-regulated, *Proc.Natl. Acad. Sci. USA* **94**, 5955-5960.

Ross, J. and O'Neill, D. (2001) New interactions between classical plant hormones, *Trends Plant Sci.* **6**, 2-4.

Sabehat, A. and Zieslin, N. (1994) GA_3 effects on postharvest alterations in cell membranes of rose (*Rosa hybrida*) petals, *J. Plant Physiol.***144**, 513-517.

Schumaker, K.S. and Sze, H. (1985) A Ca^{2+}/H^+ antiport system driven by the proton electrochemical gradient of tonoplast H+ - ATPase from oat roots, *Plant Physiol.* **79**,1111-1117.

Shaul, O., Elad, Y. and Zieslin, N. (1995) Suppression of *Botrytis* blight in cut rose flowers with gibberellic acid: effect of concentration and mode of application, *Postharvest Biol. Technol.* **6**, 321-330.

Shinitzky, M. (1984) Membrane fluidity and cellular functions, in M. Shinitzky (ed.) *Physiology of Membrane Fluidity*, Vol. 1, CRC Press, Boca Raton, pp. 2-51.

Zieslin,N. and Abolitz, M. (1994) Leakage of phenolic compounds from plant roots: effects of pH, Ca^{2+} and NaCl, *Sci. Hort.* **58**, 303-314.

Zieslin, N., Shaul, O. and Elad, Y. (1996) Suppression of *Botrytis* blight in rose flowers with gibberellic acid. Possible formation of endogenous, inhibitory compounds, *J. Plant Physiol.* **149**, 580-584.

POTATO TUBERIZATION: EVIDENCE FOR A SD-DEPENDENT AND A GIBBERELLIN-DEPENDENT PATHWAY OF INDUCTION

JORDI BOU[1], ARIADNA VIRGÓS, JAIME MARTÍNEZ-GARCÍA AND SALOMÉ PRAT

Dpt. de Genètica Molecular, Institut de Biologia Molecular de Barcelona, Centro de Investigación y Desarrollo-CSIC, Jordi Girona 18-26, 08034 Barcelona, SPAIN.

[1]Present address: Dept Plant Ecophysiology, University Utrecht, Sorbonnelaan 16, 3584 CA Utrecht, The Netherlands

1. INTRODUCTION

Light is an important environmental factor controlling plant growth and development. It is not only used as source of energy in photosynthesis but it also provides information to the plant about the environment which, for example, by sensing variations in the incident light wavelength can recognise the presence of a neighbour competing canopy or, through changes in daylenght are able to perceive the course of the year seasons. Moving from the equator towards the poles, the days become longer in summer and shorter in winter. The rate at which daylength changes varies during the year, with little change from day to day in middle summer or winter, and more rapid changes as days become longer during spring or shorter during fall. In consequence, plants can detect the year season by measuring the relative lengths of day and night, and how they vary in successive days (Salisbury and Ross, 1992). Characteristic seasonal responses are stem elongation, bud dormancy, leaf fall in trees, flowering, storage organ formation and seed germination, with these seasonal responses synchronized by photoperiod.

Potato tubers serve a double function to the plant, as a reserve organ as well as a vegetative propagation system. Under normal conditions, tubers develop at the tip of underground stems of horizontal growth or stolons. Morphologically, they are modified stems with very short swelled internodes and dormant leaf buds or "eyes" which, after an over-winter period of rest or dormancy, become active and sprout to develop a new plant that is genetically identical to the mother plant (Chapman, 1958; Li, 1985). Swelling of stolons is accompanied by major biochemical changes such as accumulation of starch and reserve proteins, which will serve as source of nutrients to the new forming plant (Prat *et al.*, 1990). Low nitrogen supply, low temperatures and short days promote tuberization, with daylength being the most critical and well

I. Macháčková and G.A. Romanov (eds.), Phytohormones in Plant Biotechnology and Agriculture, 57-66

studied environmental factor controlling tuber formation (Krauss, 1985). In general short days (SD) favor tuber formation, with some species such as *Solanum demissum* and some lines of *Solanum tuberosum* ssp *andigena* being strictly dependent on short days (SD, 8h photoperiod) for tuberization (Snyder and Ewing, 1985). These species form tubers in SD conditions, but they are unable to tuberize under long days (LD, 16h photoperiod). Although these plants are called SD plants it is known that is the length of the dark period what is critical for the induction of tuberization. When plants grown under short days are given a 30 min light pulse or night break (NB) in the middle of the dark period (SD+NB), tuberization is inhibited, whereas in plants grown under long days conditions, interruption of the light period does not have any effect on tuber formation (Ewing and Struik, 1992; Jackson, 1999).

2. THE PHOTORECEPTOR PHYB INHIBITS TUBER-FORMATION UNDER NON-INDUCTIVE DAYLENGTH CONDITIONS

Red light is the most effective for a night break and the inhibitory effect of a red light (R) pulse can be reverted if a pulse of far red light (FR) is given immediately afterwards (Batutis and Ewing 1982) . These results implicate the light receptor phytochrome in perception of the relative lengths of day and night and thereby, in photoperiodic control of tuber formation. By transformation of potato *andigena* plants with an antisense construct for the *phyB* phytochrome gene, we could show that this gene plays a key role in daylength perception and, subsequently, in activation of the photoperiod-dependent pathway controlling tuber formation (Jackson *et al.,* 1996). Reduced levels of accumulation of the PHYB photoreceptor lead to altered photoperiodic control of tuber formation, with the antisense lines tuberizing equally well under SD, LD or SD+NB conditions (Fig.1, see color plates). These plants, on the other hand, behave as strongly induced to tuberize, as tuber formation is already observed after 1 month of transfer the plants to soil, under LD greenhouse conditions. These observations are consistent with a negative effect of PHYB on tuber induction, with perception of non-favourable conditions (LD or SD+NB) by this photoreceptor leading to activation of a negative regulatory pathway that blocks tuber formation under non-inductive conditions (Jackson *et al.*, 1996; Jackson *et al.,* 1998).

Besides loss of photoperiodic control of tuberization, the *phyB* lines exhibit an elongated phenotype that reminds that of plants treated with saturating doses of gibberellins (GAs). These plants have paler leaves and very elongated internodes, which suggests that they are altered in GA biosynthesis or in some step of the GA response pathway. Measurement of endogenous levels of GAs showed that GA content is elevated in these lines, with 4 to 6-fold higher levels of GA_1 detected in the leaves of the *phyB* plants as compared to the controls (Martínez-García *et al.,*2002a).

3. REGULATION OF GA BIOSYNTHESIS DURING TUBER INDUCTION

The role of GAs in tuberization has been studied mainly by experiments using GA-biosynthesis inhibitors and application of GAs. These experiments suggested the general idea that GAs are inhibitory for tuber formation (Machackova *et al.*, 1998), but how GAs act in relation to tuber formation is not fully understood. Analysis of GA content in the stolons of plants induced to tuberize showed that the levels of GA_1 strongly decrease in the stolons prior to tuber formation (Xu *et al.*, 1998). Also, alteration of endogenous GA levels by application of inhibitors of GA biosynthesis was found to overcome the strict SD requirement of *andigena* plants for tuberization, thus suggesting that GAs play a role in preventing tuber formation under LDs (Jackson and Prat, 1996).

In order to elucidate the role of GAs in the control of tuberization, we have isolated the genes encoding the two last steps of GA biosynthesis: GA 20-oxidase and GA 3-oxidase (Carrera *et al.*,1999; Bou *et al.*, submitted). These enzymes correspond to key regulatory steps in bioactive GA synthesis and are subjected to negative feed-back regulation by the end-product GA_1. In addition, they have been shown to be controlled by phytochrome or in response daylenght conditions in several plant species (Xu *et al.*, 1997; Wu *et al.* 1996; Toyomasu *et al.*, 1998; Yamaguchi *et al.*, 1998).

We have investigated whether expression of these two GA biosynthetic activities is differentially regulated during transition to tuber formation by analysing gene expression in *andigena* plants induced (SD) and non-induced (SD+NB) to tuberize. Using degenerated primers we isolated three potato GA 20-oxidase cDNAs, designated as *StGA20ox1*, *StGA20ox2* and *StGA20ox3*, with different patterns of tissue-specific expression (Carrera *et al.*, 1999). *StGA20ox1* is abundantly expressed in the shoot tip and the leaves, whereas transcript *StGA20ox2* accumulates in tubers and to a lower level in stolons, fruits and developing seeds. Transcript *StGA20ox3*, in its turn, was found to be expressed to much lower levels than the other two genes, with low levels of mRNA detected in stolon and tubers as well as in stem and roots. Time-course studies in plants entrained to tuber inducing (SD) and non-inducing conditions (LD or SD+NB), did not show significant differences in the levels of accumulation of any of the three potato GA 20-oxidase transcripts. In plants exposed to LD, an extended accumulation of the transcripts was detected during the supplementary hours of light, but in contrast to the studies reported in *Arabidopsis* or spinach (Xu *et al.*, 1997; Wu *et al.* 1996), higher levels of mRNA were not detected in LD or SD+NB conditions as compared to SDs (see schematic outline in Fig. 2).

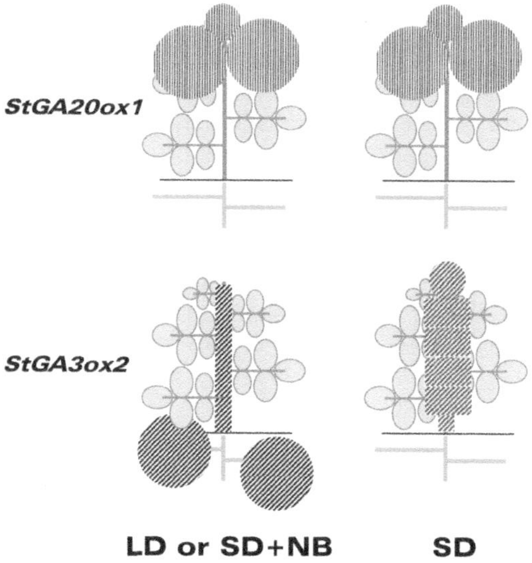

Figure 2. Tissue-specific expression of GA 20-oxidase and GA 3-oxidase in plants grown under LD (non-inducing) and SD (inducing) conditions. GA 20-oxidase is expressed to high levels in the shoot tip and the leaves, with similar levels of transcript observed in plants grown under LD or SD conditions. In plants grown under LD conditions, GA 3-oxidase is abundantly expressed in the stolons and to much lower levels in stem nodes and internodes. Transfer to SD conditions increases levels of expression in the aerial parts of the plant, with high levels of expression observed in nodes and internodes. No expression is detected in induced stolons.

These results would indicate that regulated expression of this biosynthetic activity does not play a main role in the control of GA synthesis during transition to tuber formation. Over-expression of the *StGA20ox1* GA 20-oxidase enzyme in *andigena* plants resulted in taller plants that required of longer exposures to SD conditions to form tubers (Carrera *et al.*, 2000; Fig. 3). Tuber yields in these plants were smaller than those of controls, thus confirming a negative role of GAs in tuber induction.

A

	stem length	tuberization induction
35S-*GA 20ox*	long	late
35S-*GA 3ox*	long	early

B

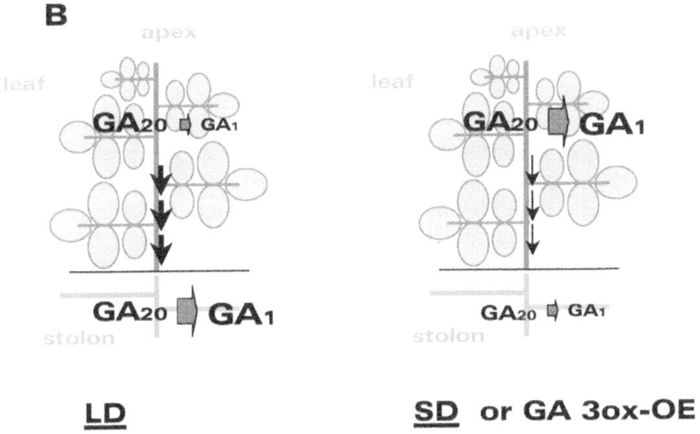

Figure 3. Stem growth and tuberization response of the GA 20-oxidase and GA 3-oxidase over-expresser lines. Hypothetical model based on the preferential transport of GA_{20}. Increased GA_{20} to GA_1 conversion in the leaves reduces GA_{20} transport towards the stolons thereby promoting tuberization.

By using a similar PCR approach, two cDNA clones encoding potato 3 beta-hydroxylase (*StGA3ox1* and *StGA3ox2*) were isolated (Bou et al, submitted for publication). The two cDNAs were expressed in different potato tissues, with transcript *StGA3ox2* being very abundant in the stolons and accumulating to much lower levels in stem nodes and internodes of plants grown in LDs, while the *StGA3ox1* message was mainly detected in flowers. The tissue-specific pattern of expression of gene *StGA3ox2* was found , in addition, to be greatly modified in response to daylength conditions, with high levels of this mRNA observed in the nodes, the shoot tip and the leaves of plants grown under SD inductive conditions but to be down-regulated in SD-induced stolons (see Fig. 2), thereby suggesting an important role of this gene in regulating stolon GA_1 levels in response to daylength conditions (Bou *et al.*, submitted).

4. EARLY TUBERIZATION IN GA 3-OXIDASE OVEREXPRESSERS

We have studied the role of potato GA 3-oxidase in the process of tuberization by over-expressing this enzyme in potato *andigena* plants. Transgenic lines that accumulated higher levels of the GA 3-oxidase enzyme under control of the CaMV 35S promoter (AR lines) or the StLS1 leaf-specific promoter (A6 lines) showed a taller phenotype and increased shoot levels of GAs, with a good correlation observed between levels of expression of the GA 3-oxidase transcript and the taller phenotype of these plants. Quite unexpectedly, GA 3-oxidase over-expression was found to have a positive effect on tuberization, with the GA 3-oxidase lines tuberizing earlier in SD conditions and showing higher tuber yields than the untransformed controls (see Fig. 3). This opposite effect of GA 20-oxidase and GA 3-oxidase over-expression is indicative of different activities of GA_{20} and GA_1on tuber induction; i.e. GA_{20} would be inhibitory while GA_1would promote tuber formation, or alternatively, of a different rate of transport of the GA_{20} precursor and the GA_1 product from the leaves to the stolons; i.e. whereas GA_{20} would be readily transported, GA_1 appears to remain preferentially in the cells where it is produced.

Several lines of experimental evidence are in support of the second of these hypothesis. Moreover, independent evidence for a different mobility of these two GA molecules has also been obtained in pea, using grafting experiments between wild-type plants and the *Le* and *Na* mutants (Proebsting *et al.*, 1992). Constitutive or leaf-specific over-expression of the GA 3-oxidase activity in the AR and A6 over-expresser lines seems to lead to an increase in the rate of GA_{20} to GA_1 conversion in the aerial tissues of these plants and thereby, a reduction in the concentration of GA_{20} available for transport to the stolons, thus mimicking the changes in GA content that occur in SD-conditions (see Fig. 3). A related diversion of the GA_{20} pool into GA_1 may also occur in the *phyB* mutants, in which increased GA_{20} to GA_1 conversion could cause the somehow contradictory phenotype observed in these plants, in which a slender phenotype produced by a 4-6 fold increase in the shoot GA_1 content, paradoxically coexists with a strong induction of tuber formation (Martínez-García *et al.,* 2002). Therefore, contrary to the general idea that GAs inhibit tuberization, our observations indicate that high levels of GAs in the stolons do indeed inhibit tuberization, whereas a high rate of GA_{20} to GA_1 conversion in the shoot would promote tuberization.

5. *phyB* PLANTS DOWN-REGULATED IN GA 20-OXIDASE EXPRESSION TUBERIZE EARLIER THAN *phyB* CONTROLS

To investigate to what extent the slender phenotype of the andigena *phyB* mutants (Jackson *et al.*, 1996) is produced by up-regulated expression of the GA biosynthetic enzymes, we further transformed these lines with an antisense construct for the *StGA20ox1* gene. Several transgenic lines were regenerated which besides showing inhibition of the *phyB* photoreceptor exhibited a strong reduction in the levels of expression of the GA 20-oxidase gene (Figure 4A, see color plates). These plants exhibited reduced stem heights (Figure 4B, see color plates), with internode lengths similar to those observed in the GA 20-oxidase antisense lines (Carrera *et al.*, 1999). The *phyB GA20ox* double mutants, on the other hand, tuberized earlier than *phyB* controls. These plants showed senescence symptoms much earlier than the *phyB* lines, which agrees with an earlier tuber formation response in the double mutants. Both the reduced stem length and early tuberization traits of the double mutants are indicative of a normal response to GAs in the *phyB* lines. High levels of accumulation of GA_1 in the leaves would indeed produce the characteristic slender growth habit observed in these plants, with this elongated phenotype being reverted by a block in GA 20-oxidase expression. Increased GA_{20} to GA_1 conversion, in its turn, would lead to a reduction in GA_{20}ransport towards the stolons, thus resulting in early tuberization. Nonetheless, GAs appear to still exert an inhibitory effect on tuberization transition in the *phyB* mutants, as a block in GA biosynthesis further promotes tuber formation in the *phyB* mutant background.

6. DAYLENGTH CONTROL OF TUBER FORMATION APPEARS TO INVOLVE THE FUNCTION OF A POTATO CONSTANS ORTHOLOGUE

Elegant inter-specific grafting experiments performed by Chailakhyan in early eighties, suggested that flowering and tuberization induction might be controlled by common genetic factors (Chailakhyan *et al.,* 1981). In these experiments, SD, LD, or dayneutral tobacco species for floral induction, were grafted to *andigena* potato plants, and induction of flowering and tuber formation was analyzed in the potato stocks. These studies showed that flowering tobacco scions, regardless of being derived from LD, SD or neutral plants, induced tuberization in the potato stock, thus evidencing that the flowering inducing signal produced in tobacco is able to promote tuberization on potato stocks. This is a fascinating observation, indicative of common signaling steps in the flowering and tuberization inductive pathways. Genetic analyses in the LD plant *Arabidopsis thaliana* has led to the identification of several mutations affecting the flowering time which have been placed in four independent pathways: the autonomous pathway, the vernalization pathway, the GA-dependant pathway, and the LD-dependant pathway (Simpson and Dean, 2002). Detection and transduction of the light signal by phytochromes and cryptochromes, in combination with the circadian clock oscillator, was shown to mediate activation of the daylenght-dependent flowering pathway. The zinc finger regulatory protein CONSTANS (*At*CO) functions as a link between the oscillator and flowering time, playing a key regulatory function in daylength regulation of flowering (Putterill *et al.*, 1995; Suarez-Lopez *et al.*, 2001). Loss-of-function *co* mutants flower late in LD conditions but like wild-type plants in SDs. On the other hand, ectopic over-expression of this gene promotes flowering both in SD and LD conditions.

We have investigated whether a CO-homologue would be involved in photoperiodic control of tuber formation by over-expressing the Arabidopsis

CONSTANS gene in *andigena* plants. Lines over-expressing the *AtCO* gene (pACO lines) were smaller than wild-type plants and in SD conditions tuberized much later than the controls. While control plants started to form tubers after 3 weeks under SD inducing conditions, pACO over-expressers required 16 weeks or longer under SD conditions to tuberize, thereby evidencing a negative effect of *AtCO* on photoperiodic control of tuberization (Martínez-García *et al.*, 2002b).

Analysis of the diurnal rhythm of expression of transcript *StCOL*-1, encoding a zinc finger regulatory protein with homology to CONSTANS, showed that diurnal oscillation of this transcript is not affected in the pACO over-expresser lines, thereby demonstrating that these plants would not be altered in daylength perception (Martínez-García *et al.*, 2002b). This suggests that CONSTANS negative control on tuber formation is exerted through a block in the production or transport of the flowering/tuberization inducing signal, or by alteration in the stolon response to the inductive signal.

Whereas in floral induction the spatial proximity between the site of signal production in the leaves and the evocation site at the shoot apical meristem did not allow to determine where AtCO function is required, potato plants represent a unique experimental system with regard to the spatial arrangement of the induction and evocation sites. Grafting experiments in which wild-type and pACO scions were reciprocally grafted onto wild-type and pACO stocks showed than when wild-type scions were grafted onto pACO stocks, the resulting chimeras tuberized like wild-type plants, whereas when pACO scions were grafted onto wild-type stocks, the resulting chimera tuberized as the pACO plants (Figure 5), thus demonstrating that function of AtCO is required in the leaves.

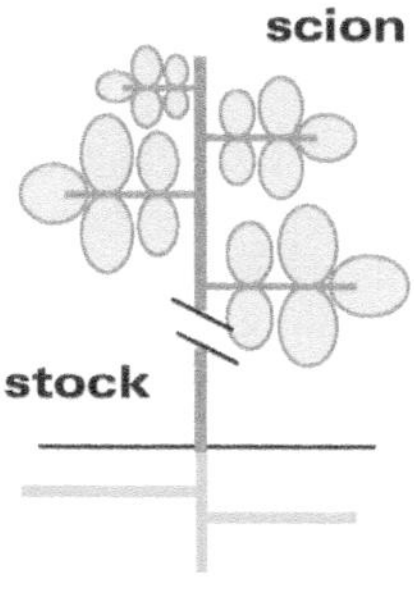

scion / stock	wt	pACO
wt	+	-
pACO	+	-

Figure 5. Results of reciprocal grafting of scions from wild-type andigena and pACO lines onto potato andigena and pACO stocks. The grafted plants were kept for 4 weeks in SD (from Martínez-García et al., 2002b). (+), tuberization ocurred in 100% of the plants; (-), tuberization ocurred in less than 20% of the plants.

Indeed, *AtCO* is likely to act as branch point between the strictly photoperiodic-responsive pathway and the developmental pathways controlling either flowering or tuber formation in Arabidopsis and potato, respectively. Therefore, the genes directly activated by this transcription factor would be different in these two plant species, with the genes directly controlled by *AtCO* in Arabidopsis being involved in flowering control, whereas those in potato involved in tuberization transition. In that respect, while *CO* was found to directly activate expression of the Arabidopsis floral pathway integrator genes *AGL20/SOC1* and *FT* (Samach *et al.*, 2000), preliminary results indicate that in potato *AtCO* over-expression does not affect the expression of a potato *AGL20* homologue (Martínez-García, unpublished results).

7. CONCLUSIONS

Collectively, these results are consistent with two independent regulatory pathways controlling tuber formation in potato: a LD-pathway that inhibits tuber formation in non-favorable daylength conditions and is activated by phyB, and a GA-dependent pathway, which is induced by GA_1 and promotes elongation growth of the stolons, thus preventing tuber formation. In addition, evidence has been obtained for the involvement of a potato *CO* orthologue in tuberization control, which would act as signaling intermediate in the LD-pathway, thereby regulating tuber formation in response to daylength conditions. Function of this regulatory protein is required in the leaves and appears to control generation of the tuberization inducing signal. Defining the primary targets of the CO signaling intermediate should help to elucidate the molecular identity of the tuberization-inducing signal and also to identify the tuberization pathway integrator genes that in response to both endogenous and environmental cues switch the developmental fate of the meristematic cells at the subapical region of the stolon to differentiate into a tuber.

ACKNOWLEDGEMENTS

We thank P. Fontanet, M.C. Badías and A. Sanz for plant care. We are grateful to A. Sánchez for help with the photographic work . This work was supported by grants BIO96-0532-C02-02 and BIO99-0961 from the Spanish Comisión Interministerial de Ciencia y Tecnología (CICYT).

REFERENCES

Batutis, E.J. and Ewing, E.E. (1982) Far-red reversal of red light effect during long-night induction of potato (*Solanum tuberosum* L.) tuberization, *Plant Physiol.* **69**, 672-674.

Carrera, E., Jackson, S.D. and Prat, S. (1999) Feedback control and diurnal regulation of gibberellin 20-oxidase transcript levels in potato, *Plant Physiol.* **119**, 765-74.

Carrera, E., Bou, J., García-Martínez, J.L. and Prat, S. (2000) Changes in GA 20-oxidase gene expression strongly affect stem length, tuber induction and tuber yield of potato plants, *Plant J.* **22**, 247-256.

Chailakhyan, M.Kh., Yanina, L.I., Devedzhyan, A.G. and Lotova, G.N. (1981) Photoperiodism and tuber formation in grafting of tobacco onto potato, *Doklady Akademic Nauk S.S.S.R.* **257**, 1276-1280.

Chapman, H.W. (1958) Tuberization in the potato plant, *Physiol. Plant.* **11**, 215–224.

Ewing, E.E. and Struik, P.C. (1992) Tuber formation in potato: induction, initiation and growth, *Hortic. Rev.* **14,** 89-197.

Jackson, S.D., Heyer, A., Dietze, J. and Prat, S. (1996) Phytochrome B mediates the photoperiodic control of tuber formation in potato, *Plant J.* **9,** 159-166.

Jackson, S.D. and Prat S. (1996) Control of tuberization in potato by gibberellins and phytochrome B, *Physiol. Plant.* **98**, 407-412.

Jackson, S.D. (1999) Multiple signalling pathways control tuber induction in potato, *Plant Physiol.* **119**, 1-8.

Krauss, A. (1985) Interaction of nitrogen nutrition, phytohormones and tuberization, in: P.H. Li (ed.), *Potato Physiology* Academic Press, London, pp 209-231.

Machackova, I., Konstantinova, T.N., Seergeva, L.I., Lozhnikova, V.N.,Golyanovskaya, S.A., Dudko, N.D., Eder, J. and Aksenova, N.P. (1998) Photoperiodic control of growth, development and phytohormone balance in *Solanum tuberosum, Physiol Plant* **102**, 272–278.

Martínez-García, J. , García-Martínez, J.L., Bou, J. and Prat S. (2002a) The interaction of gibberellins and photoperiod in the control of potato tuberization, *J. Plant Growth Regul.* **20**, 377-386.

Martínez-García, J.F., Virgos-Soler, A. and Prat, S. (2002b) Control of photoperiod-regulated tuberization in potato by the *Arabidopsis* flowering-time gene *CONSTANS*m, *Proc. Natl. Acad. Sci. USA* **99**, 15211-15216.

Prat, S., Frommer, W.B., Hofgen, R., Keil, M., Kossmann, J., Koster-Topfer, M., Liu X.J., Muller, B., Penacortes, H., Rochasosa, M., Sanchezserrano, J.J., Sonnewald, U., Willmitzer, L., Kossmann, J. (1990) Gene expression during tuber development in potato plants, *FEBS Lett.* **268**, 334-338.

Proebsting, W.M., Hedden, P., Lewis, M.L., Croker, S.J. and Proebsting, L.N. (1992) Gibberellin concentration and transport in genetic lines of pea, *Plant Physiol.* **100**, 1354-1360.

Putterill, J., Robson, F., Lee, K., Simon, R. and Coupland, G. (1995) The CONSTANS gene of Arabidopsis promotes flowering and encodes a protein showing similarities to zinc finger transcription factors, *Cell* **80**, 847-857.

Salisbury, F. B. and Ross, C. W. (1992) *Plant Physiology*, 4th Edition. Wadsworth Publishing Company, Belmont, California.

Samach, A., Onouchi, H., Gold, S.E., Ditta, G.S., Schwarz-Sommer, Z., Yanofsky, M.F. and Coupland, G. (2000) Distinct roles of CONSTANS target genes in reproductive development of Arabidopsis, *Science* **288**, 1613-1616.

Simpson, G.G. and Dean, C. (2002) *Arabidopsis*, the rosetta stone of flowering time?, *Science* **296,** 285-289.

Snyder, E. and Ewing E.E. (1989) Interactive effects of temperature, photoperiod and cultivar on tuberization of potato cuttings, *Hortic. Sci* **24**, 336–338.

Suarez-Lopez, P., Wheatley, K., Robson, F., Onouchi, H., Valverde, F. and Coupland, G. (2001) CONSTANS mediates between the circadian clock and the control of flowering in Arabidopsis, *Nature* **410**,1116-1120.

Toyomasu, T., Kawaide, H., Mitsuhashi, W., Inoue, Y. and Kamiya, Y. (1998) Phytochrome regulates gibberellin biosynthesis during germination of photoblastic lettuce seeds, *Plant Physiol.* **118,** 1517-1523.

Wu, K., Li, L., Gage, D.A. and Zeevaart, J.A.D. (1996) Molecular cloning and photoperiod-regulated expression of gibberellin 20-oxidase from the long-day plant spinach, *Plant Physiol.* **110,** 547-554.

Xu, Y.L., Gage, D.A. and Zeevaart, J.A.D. (1997) Gibberellins and stem growth in *Arabidopsis thaliana*: effects of photoperiod on expression of the GA4 and GA5 loci, *Plant Physiol.* **114,** 1471-1476.

Xu, X., van Lammeren, A.A.M, Vermeer, E. and Vreughdenhil, D. (1998) The role of gibberellin, abscisic acid and sucrose in the regulation of potato tuber formation *in vitro, Plant Physiol.* **117**, 575-584.

Yamaguchi, S., Smith, M.W., Brown, R.G., Kamiya, Y. and Sun, T. (1998) Phytochrome regulation and differential expression of gibberellin 3beta-hydroxylase genes in germinating Arabidopsis seeds, *Plant Cell* **10**, 2115-2126.

INVESTIGATION ON THE CHEMICAL NATURE OF FLOWER-INDUCING FACTOR(S) IN SHORT-DAY PLANT *PHARBITIS NIL*

[1] A. TRETYN, [2] K. GŁOWACKA, [1] K. KASPRZAK, [3] E. GALOCH,
[3] H. KULIKOWSKA-GULEWSKA, [3] J. KOPCEWICZ

[1]Department of Biotechnology and [2]Department of Plant Physiology and Morphogenesis, Nicholas Copernicus University, 87-100 Toruń, Poland, and [3]Chair of Plant Physiology and Biotechnolog, University of Warmia and Mazury, 10-719 Olsztyn, Poland

1. INTRODUCTION

Plants' flowering precedes production of seeds, which are a basic component of human food and animal fodder. Thus for several decades systematic studies have been conducted to understand physiological and genetic mechanisms underlying this process (Bernier, 1988). One of pioneers in research on flowering was Chailakyhan, who already in 1936 proposed a scientific explanation of the phenomenon of flower induction. In his hypothesis he stipulated that, in response to stimuli that induced flowering, plants generated so-called florigen that moved from leaves to the shoot apex and stimulated morphogenesis of flowers. Attempts to isolate the substance that is able to induce flower formation have been made for more than 60 years (Colasanti and Sundaresan, 2000). Nevertheless, the chemical nature of florigen remains still unknown (Aukerman and Amasino, 1998; Ma, 1999).

One of the most intensively studied plant, in aspect of conditions that influence its flowering, is *Pharbitis nil*, well know as Morning Glory (Vince-Prue and Gressel, 1985). This plant (especially Violet variety) has a number of advantages. It initiates flowers at a very early stage of seedlings growth (4-5 days after germination), when the plant has only well expanded cotyledons. After photoperiodic induction, flower buds develop rapidly, which can be seen two weeks after induction. What is most important, *P. nil* is a very sensitive short-day plant - it initiates flowers in response to a single short day/long night cycle (Kopcewicz and Tretyn, 1998).

2. THE ROLE OF COTYLEDONS IN THE PHOTOPERIODIC FLOWER INDUCTION

It is well documented that in *P. nil* cotyledons are sites of photoperiodic perception. The importance of cotyledons in the flower induction in *P. nil* has been confirmed by microsurgical experiments. They showed that removal of both cotyledons of a seedling at the end of the dark period blocked flowering. The number of flowers produced by the plant increased rapidly with delayed time of defoliation, reaching a

I. Macháčková and G.A. Romanov (eds.), Phytohormones in Plant Biotechnology and Agriculture, 67-78

maximum at h 20. Plants with one cotyledon flowered as intensively as those with both these organs, or even stronger. However, Ogawa (1993a) in his simple, although very elegant, experiments showed that removal of one half or three quarters of an intact cotyledon area at the end of the inductive dark period resulted in lower flower production in comparison with plants with intact cotyledons. Recently we repeated Ogawa studies and our conclusions were similar. In our experiments we removed either a part of a cotyledon above or below main veins or a lateral cotyledon lobe (Fig. 1). It appeared that the flowering response was blocked most efficiently when a lower part of a cotyledon was removed. On this basis we concluded that, although the synthesis of a flowering inductor may happen in the cotyledon, it is its lower part that appears to be more engaged in that process. It is this part of the cotyledon where the highest number of so-called cotyledonary bodies (Tretyn *et al.*, 1996) was observed.

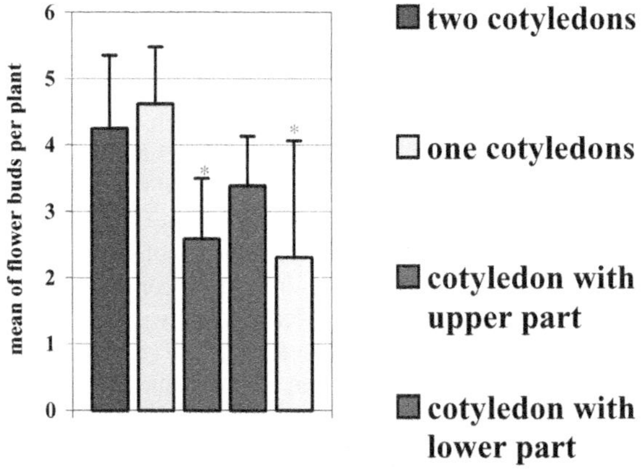

Figure 1. The effect of removal of one cotyledon and different part of this organ Pharbitis nil seedlings, another cotyledon of which was excised before beginning of inductive dark period.

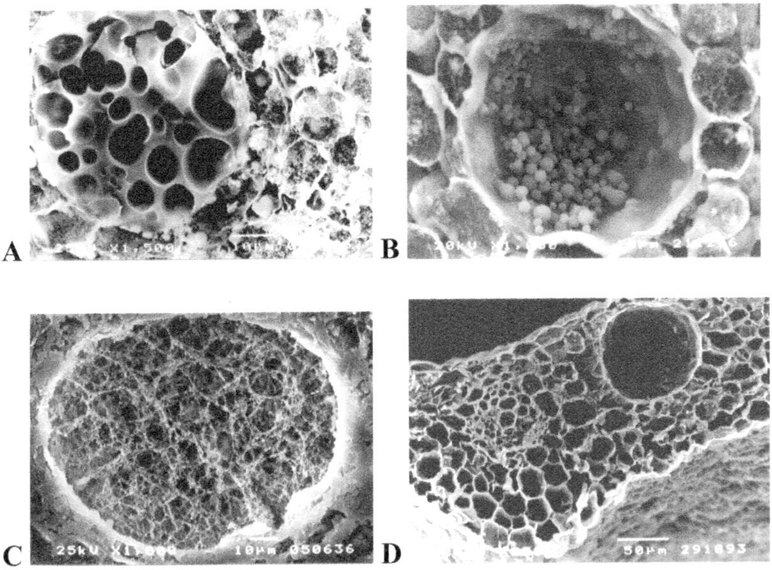

Figure 2. Four types (A-D) of cotyledonary bodies detected in seeds and 5-day-old seedlings of Pharbitis nil. See text for description.

Research done with a scanning electron microscope allowed understanding of the internal structure of cotyledonary bodies (Fig. 2). It appeared that in dry, non-germinating seeds of *P. nil* cotyledonary bodies contain hydrophobic and osmophilic substance of "cheese-like" structure (Fig. 2A). However, during germination and early stages of seedlings growth, cotyledonary bodies appeared filled with small spheres (Fig. 2B), possessing a net-like structure inside (Fig. 2C) or that were completely devoid of hydrophobic and osmophilic substances (Fig. 2D). It appeared that during the inductive night the number of cotyledonary bodies of the first and the second type decreased and the amount of the bodies of the third and the fourth type increased (Tab. 1). It concerned particularly lower part of the cotyledons.

Table 1. The level of glucose, fructose and sucrose (µg/mg fr. wt.) in cotyledons and shoot apices of non-induced (light-grown) and photoperiodically-induced (dark-treated) seedlings of Pharbitis nil.

Sugars	Cotyledons		Shoot apices	
	Non-induced	Induced	Non-induced	Induced
Glucose	3.93	1.02	18.64	2.90
Fructose	5.38	1.71	1.95	0.22
Sucrose	1.12	1.05	2.08	0.18

On this basis we speculate that during the embryogenesis some substances are stored in cotyledonary bodies, which are then used during germination, seedlings growth and transition from a vegetative to a reproductive developmental stage. They can be storage substances. However, it cannot be ruled out that inside the cotyledonary bodies are stored the substances engaged in the process of the photoperiodic flower induction. One of them may be so-called pharbitin, a compound of a complicated chemical structure, which was isolated from cotyledons and cotyledonary bodies by Yokoyama and Wada (1987). These researchers found that pharbitin is synthesised during embryogenesis, and its level in cotyledons decreased rapidly 4 – 5 days after germination. Yokoyama and Wada (1987) found that about 90% of pharbitin was stored inside cotyledonary bodies. However, it was astonishing that they did not link pharbitin presence in cotyledons with the flower induction. Thus, we also undertook that kind of research. From cotyledons and cotyledonary bodies we obtained methanolic extracts, from which we purified pharbitin. Next we examined biological activity of that compound, using shoot apices isolated form non-induced 5-days-old seedlings of *P. nil* as a bio-test. Addition of pharbitin to a medium with isolated shoot apices did not influence formation of flower buds when those buds were kept in non-inductive conditions. However, when pharbitin was added to a medium with isolated shoot apices, that substance efficiently blocked formation of flower buds in *in vitro* conditions, after the isolated shoot apices were treated with three inductive cycles. On this basis we suppose that pharbitin present in cotyledons may block the possibility of receiving an inductive photoperiodic signal at too an early development stage of *P. nil*. Only when the level of this substance in cotyledons decreases, flower inductor formation or release are possible.

3. CHEMICAL NATURE OF FLOWER-INDUCING FACTOR(S)

Beside pharbitin also phenolic compounds, fatty acids, prostaglandins and different mono- and disaccharides were found in methanolic extracts obtained from cotyledonary bodies, using gas chromatography/mass spectrometry (GC/MS). Relatively high amounts of octadecanoic (palmitic), octadecadienoic acid and prostaglandin $F_2\alpha$ were present. Of the phenolic compounds present benzoic acid was the most abundant. It is worth to mention that Ishioka *et al.* (1990) found an effect of benzoic acid on flower induction in cultured apices of *Pharbitis nil*. When the explants were excised from the induced seedlings and cultured under continuous light, only 5% of explants formed flowers. However, nearly 85% of them produced flowers when treated with 10 μms of benzoic acid (Ishioka *et al.* 1990). Up to now, number of substances was identified or isolated from *Pharbitis nil,* which exert inductive or inhibitory influence on flowering of this plant. These substances were isolated either from the whole cotyledons or from phloem exudates. Bernier and co-workers (1981) suggested the involvement of phenolic compounds in the process of flower induction. This opinion was based on some earlier publications. In several short-day (*Nicotiana tabacum* cultivar "Maryland Mammoth", *Xanthium*) and long-day (*Nicotiana sylvestris*, *Lemna gibba* G3) plants, the correlation between flowering and chlorogenic acid (CGA) accumulation in their leaves was observed (Taylor, 1965; Zucker *et al.*, 1965; Umemoto, 1971). The same substance was also identified in cotyledons of *Pharbitis nil* (Shinozaki *et al.*, 1988) induced to flower in continuous light by cultivation under conditions of poor nutrition, high-intensity light and low temperature (Hirai *et al.*, 1988; Shinozaki *et al.*, 1988). Hirai *et al.* (1993) have shown that flowering under poor nutrition conditions is closely correlated with the accumulation of phenylpropanoids. In cotyledons of such plants chlorogenic and p-coumaroylquinic acids, as well as pinoresinol-β-D-glucoside were identified (Hirai *et al.*, 1993). Moreover, the same compounds were found in cotyledons of *Pharbitis nil* stimulated to flower by high-fluence irradiation (Shinozaki *et al.*, 1994). Treating the seedlings with aminooxyacetic acid (AOA), an inhibitor of phenylalanine ammonialyase (PAL; a key enzyme involved in phenylpropanoids synthesis), led to the inhibition of flowering (Amagasa *et al.*, 1992) at the same time as it suppressed the accumulation of chlorogenic acid and pinoresinol-β-D-glucoside (Shinozaki *et al.*, 1994).

Recently Hirai and co-workers (1995) have shown that beside the above-mentioned compounds there is also another substance, which seems to be involved in flower promotion of *Pharbitis nil* seedlings cultivated at low temperature. The analyses of cotyledonary extracts showed the presence of an unstable compound that was identified as ascorbic acid. Endogenous concentration of this substance increased rapidly after the exposure of *Pharbitis nil* seedlings to high-intensity light, i.e. the conditions that promote accumulation of phenylpropanoids inside cotyledons. Hirai *et al.* (1995) have postulated that a major biological function of ascorbic acid could be the protection of plant cells against the toxic hydrogen peroxide released during the treatment of *Pharbitis nil* seedlings with high-intensity light and low temperature. Therefore the changes in ascorbic acid concentration may only reflect the changes in the level of H_2O_2 (Hirai *et al.*, 1995). Phenolic compounds were also found in extracts isolated from cotyledonary bodies (Tretyn *et al.*, 1996). These structures described above were extracted with methanol and the obtained extracts were analysed by gas chromatography/mass spectrometry (GC/MS). The substances present inside the bodies consist of a mixture of fatty acids and phenolic compounds. Relatively high amounts of palmitic, linolenic, linoleic acids as well as p-coumaric, ferulic, and caffeic acids were identified (Tretyn *et al.*, 1996). The most abundant phenolic compound was p-coumaric acid. Its concentration was higher in the extracts obtained from the bodies of dark-induced plants (Tretyn *et al.*, 1996).

During further analysis done with GC/MS method significant differences were found in chemical composition of extracts obtained from cotyledonary bodies isolated from control plants (grown in light) and treated with photoperiodical induction (Fig. 3). It was found that in the extracts of the first kind, besides sugars and fatty acids, the main compound present inside cotyledonary bodies was prostaglandin $F_2\alpha$ (Fig. 3). On the other hand, in the extracts obtained from the cotyledonary bodies purified from the *P. nil* seedlings treated with the inductive darkness, the above-mentioned prostaglandin was their sole component (Fig. 3). The results of these studies are consistent with the data obtained by Groenewald *et al.* (1983), who found that the content of the prostaglandin increased over 20 times in the cotyledons of seedlings treated with an inductive photoperiod. Recently, Groenewald and van der Westhuizen (2001) have proposed model of prostaglandin action in photoperiodic flower induction of *Pharbitis nil*. The results of our own studies seem to confirm that hypothesis. Using isolated shoot apices we have shown that 4 different prostaglandins (E_1, E_2, $F_1\alpha$ and $F_2\alpha$) reveal florigenic activity. All of these substances added to the liquid medium containing shoot apices isolated from non-induced plants stimulated formation of flower buds in continuous white light. Amongst the studied prostaglandins the highest biological activity was shown by prostaglandin $F_2\alpha$, found both by us and by Groenewald *et al.* (1983) in cotyledonary bodies and cotyledons of *P. nil*, respectively. Isolated shoot apices treated with prostaglandin $F_2\alpha$ in concentration of 10^{-7} M, formed 2 flower buds at the average. Other concentrations were less effective or remained without any influence on the studied process (Fig. 4).

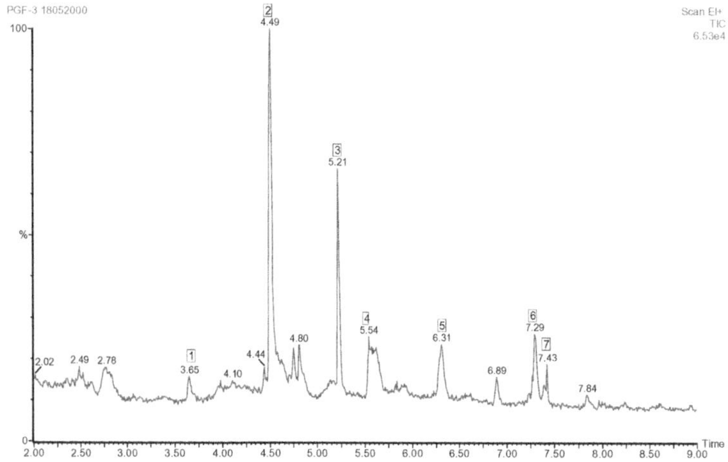

(Figure 3 continued)

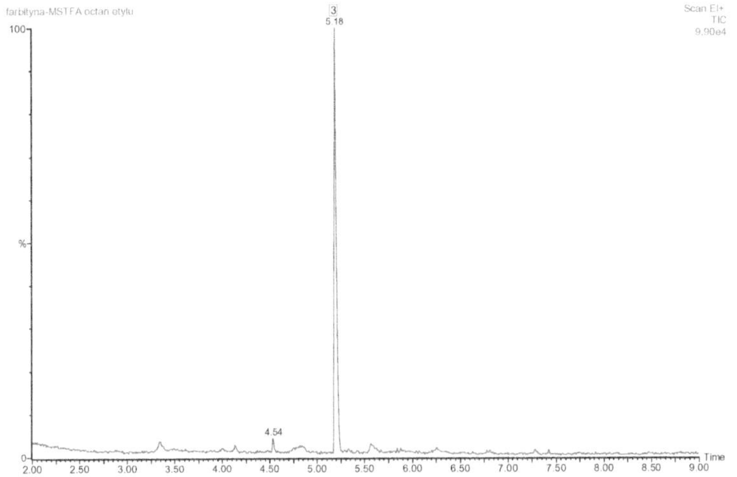

Figure 3. Chromatograms of methanolic extracts of cotyledonary bodies isolated from light-grown (upper panel), and dark-treated (lower panel) seedlings of Pharbitis nil. Compounds detected: ***1*** *- benzeneacetic acid,* ***2*** *– glicerole,* ***3*** *– prostaglandin $F_2\alpha$,* ***4*** *– undecad,* ***5*** *- benzoic acid,* ***6*** *- D-sorbite,* ***7*** *- mirystinic acid.*

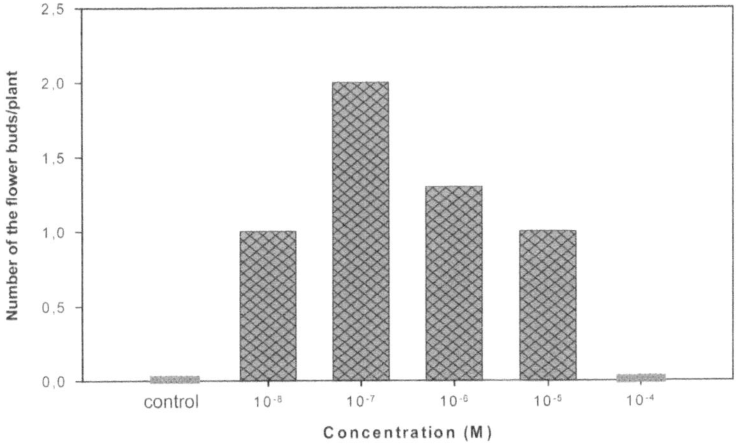

Figure 4. The effect of the prostaglandin $F_2\alpha$ on flower development on shoot apices excised from light-grown Pharbitis nil seedlings under non-inductive conditions (continuous light) and cultured in vitro.

Along with sugars, fatty acids and prostaglandin, benzoic acid was also isolated from the extracts (Fig. 3). It seems an interesting observation because the effect of benzoic acid on flower induction in cultured apices of *Pharbitis nil* seedlings was described by Ishioka *et al.* (1990). When the explants were excised from the induced seedlings and cultured under continuous light, only 5% of explants formed flowers. However, nearly 85% of them produced flowers when treated with 10 μm benzoic acid (Ishioka *et al.,* 1990). Using shoot-tip cultures of *Pharbitis ni,* Nakanishi *et al.* (1995) tested flower-promoting substances isolated from the plant. They have shown that a crude extract from cotyledons excised from 6-day-old seedlings exposed to short-day conditions had no flower promoting activity. However, after its further purification two fractions possessing biological activity were obtained. One of them contained a compound identified as dihydrokaempferol-7-*O*-*β*-D-glucoside. This substance exhibited flower-promoting activity in extremely low concentration: 0.44 nM. However, the maximum effect was observed with the concentration higher by two orders of magnitude. Nevertheless, the above-mentioned compound was not able to induce flowering in a shoot apex kept under not-inductive photoperiod (Nakanishi *et al.,* 1995). There is no doubt that in the case of young *Pharbitis nil* seedlings a flower stimulus is synthesized in cotyledonary cells. However, it is probably accumulated at a very low concentration. Nakanishi *et al.* (1995) used 20.000 cotyledons (1.8 kg fresh weight) to obtain a small amount of flower-promoting dihydrokaempferol which stimulated flowering.

It is generally accepted that a flower inductor may be transported from cotyledons to the shoot apex *via* phloem. This assumption was confirmed both by microsurgery experiments (Ogawa, 1993a) and by studies on the effect of phloem exudates on flowering in cultured apices of *P. nil* seedlings (Ishioka *et al.*, 1990, 1991; Kondoh et al., 1999a,b).

The floral stimulus is transported from the induced cotyledons to the cotyledonary buds and to the plumule no sooner than 14 h after the beginning of the dark period (Ogawa, 1993b). It has been calculated that the floral stimulus moves at a rate of 0.8 cm/h or a little faster in the cotyledon, and then it moves quickly in the petiole (Ogawa, 1993b). It can reach the plumule in 16 to 20 h after the beginning of the dark period (Wada, 1966; Larkin *et al.,* 1990).

The second line of evidence confirming the role of phloem in the transport of flower stimulus was given by Ishioka *et al.* (1990, 1991) and Kondoh *et al.* (1999 a,b). They have shown that phloem exudates obtained from cotyledons of photoperiodically-induced plants stimulated (Ishioka *et al.,* 1990, 1991) and those isolated from non-induced inhibited (Kondoh *et al.,* 1999a,b) flower formation by shoot apices excised from non-induced and induced *P. nil* seedlings, respectively. Flower-inducing activity of those exudates increased as the seedlings were exposed to longer period of darkness. Exudates from cotyledons of seedlings irradiated with white light in the middle of the dark period had a significantly lower activity. On this ground Ishioka *et al.* (1991), as well as Kondoh *et al.* (1999a,b) had pointed out that phloem exudates isolated from *P. nil* possessed an inductive or/and inhibitory activity in flowering of this plant.

To determine the chemical composition of substances transported through veins from cotyledons to a shoot apex, phloem exudates were isolated, which were analysed with the GC/MS method. The results of this study are shown in Fig. 5.

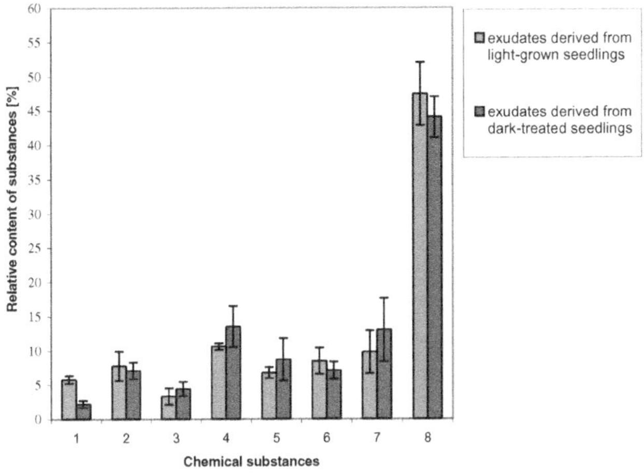

Figure 5. Relative content of chemical compounds present in phloem exudates derived from light-grown and dark-treated seedlings of Pharbitis nil. ***1*** *- mannoic acid,* ***2*** *- β-D-galactofuranoside,* ***3*** *- β-D-galactofuranose,* ***4*** *- mannose,* ***5*** *- glucose,* ***6*** *- hexadecanoic acid, 7- octadecanoic acid,* ***8*** *- D-hexoarabinic acid.*

It was found, that the major fraction of substances with low molecular weight that were present in phloem exudates isolated from control plants and plants subjected to photoperiodic induction, were simple sugars of various types. For many years it has been suggested that those compounds took part in the induction of flowering. Thus, we studied changes in amounts of three sugars: glucose, fructose and sucrose in cotyledons and shoot apices of control and photoperiodically induced plants (Tab. 2). It appeared that before the inductive darkness a dominant sugar in cotyledons and shoot apices was glucose. At the end of the inductive darkness, the level of glucose in cotyledons decreased about 4 times, compared to the control plants, while in the shoot apices that decrease was even about 6 times. When plants were transferred into light, the level of glucose gradually increased, both in cotyledons and in shoot apices. The level of fructose and sucrose in cotyledons was similar to glucose, while in shoot apices the concentration of both sugars was much lower compared to glucose (Tab. 2). At the end of 16-h-long inductive dark period the level of fructose and sucrose was about 10 times lower than in light-grown plants (Tab. 2).

At this point it is worthwhile to recall results of a research done by Durdan *et al.* (2000), concerning the influence of glucose, fructose and sucrose on morphogenesis of flower buds of *P. nil in vitro*. The essence of their research was to study the influence of these sugars added to the medium on morphogenesis of flower buds of isolated shoot apices of *P. nil*. Those authors found that only apices isolated from plants treated with an inductive photoperiod were able to form flower buds. However, the rate of forming of each flower whorl depended on the type of sugar in the medium. If fructose or sucrose was present, first three outer whorls appeared during first 24 hours after mounting the shoot apices on the medium, and the last of them, pistils, about 3 days later. However, if glucose was added to the medium, all of the flower whorls developed during first 24 hours of the experiment. The role of glucose in forming of the flower buds is also supported by the results of Chailakyhan *et al.* (1975). They demonstrated that callus obtained from tobacco was able to form flower buds when mounted on a medium containing glucose. However, flower morphogenesis was not observed when fructose or sucrose was added to the medium.

4. CONCLUSIONS

Photoperiodic flower induction is a multistage process consisting of sequences of events temporally and spatially aligned. These sequential steps have their own specific requirements and are affected differently by environmental and chemical factors. The nature of a transmissible flower inductor is still a controversial issue. Further progress in the studies on the synthesis and chemical nature of flower-promoting substances is required.

We do not have the slightest idea of the chemical structure of this signalling compound. On the basis of the studies done in our laboratory and on those available in literature we believe that flowering of *Pharbitis nil*, either at too an early developmental stage or growing under unfavourable photoperiodic conditions, is blocked by pharbitin or by supraoptimal level of glucose in shoot apices, respectively. Photoperiodic induction of flowering may be accompanied by both an increase in prostaglandin level in cotyledons and a decrease in glucose level in shoot apices. Nevertheless, it seems that prostaglandins and glucose are not the only components of the complex that may influence flowering of *P. nil*. Recently we found that the most active fraction of phloem exudates has had a molecular weight in the range of 1 do 10 kDa (Galoch, unpublished data). We have reasons to speculate that the composition of this fraction may include florigenically active peptides. We hope that in a near future we will be able to isolate these compounds and to determine their detailed chemical structure.

ACKNOWLEDGEMENTS

The research was sponsored by the State Committee for Scientific Research (KBN), Poland, Grant No. 5 P6A 030 16.

REFERENCES

Amagasa T., Ogawa M. and Sugai S. (1992) Effects of aminooxyacetic acid and its derivatives on flowering on *Pharbitis nil, Plant Cell Physiol.* **33**, 1025-1029.

Aukerman M and, Amasino R. (1998) Floral induction and florigen, *Cell* **93**, 491-494.

Bernier G. (1988) The control of floral evocation and morphogenesis, *Annu. Rev. Plant. Physiol. Plant Mol. Biol.* **39**, 175-219.

Bernier G., Havelange A., Houssa C., Petitjean A. and Lejeune P. (1993) Physiological signals that induce flowering, *Plant Cell* **5**, 1147-1155.

Bernier G., Kinet J.M. and Sachs R.M. (1981) *The Physiology of Flowering. Vol II.,* CRC Press, Florida.

Chailakyhan M.K., Aksenova N.P., Konstantinova T.N. and Bavrina T.V. (1975) The callus model of flowering, *Proc. Roy. Soc. Lond., Ser. B* **190**, 330-340.

Colasanti J. and Sundaresan V. (2000) Florigen enters the molecular age: long-distance that cause plant to flower, *Trends Biochem. Sci.* **25**, 236-240.

Durdan S.F., Herbert R.J., Rogers H.J. and Francis D. (2000) The determination time of the carpel whorl is differentialy sensitive to carbohydrate supply in *Pharbitis nil, Plant Physiol.* **123,** 189-200.

Groenewald E., Visser J. and Grobbelaar N. (1983) The occurence of prostaglandin in Pharbitis nil seedlings grown under short day or long day, *S. Afr. Tydskt. Plankt,.* **2**, 82.

Groenewald E. and van der Westhuizen A. (2001) The flowering stimulus and possible involvement of prostaglandins in the flowering of Pharbitis nil, *S. Afr. J. Sci.* **97**, 313-317.

Hirai N., Kojima Y., Koshimizu K., Shinozaki M. and Takimoto A. (1993) Accumulation of phenylpropano-ids in cotyledons of *Pharbitis nil* seedlings during flower induction by poor nutrition, *Plant Cell Physiol* **34**, 1039-1044.

Hirai N., Kojima Y., Shinozaki M., Koshimizu K., Murofushi N. and Takimoto A. (1995) Accumulation of ascorbic acid in the cotyledons of morning glory (*Pharbitis nil*) seedlings during the induction of flowering by low-temperature treatment and the effect of prior exposure to high-intensity light, *Plant Cell Physiol* **36**, 1265-1271.

Ishioka N., Tanimoto S. and Harada H. (1990) Flower-inducing activity of phloem exudate in cultured apieces from *Pharbitis* seedlings, *Plant Cell Physiol.* **31**, 705-709.

Ishioka N., Tanimoto S. and Harada H. (1991) Flower-inducing activity of phloem exudate in from *Pharbitis* cotyledons exposed to various photoperiods, *Plant Cell Physiol.* **32**, 921-924.

Kondoh M., Fukusada Y., Murakami A., Shigeta C.and Tanimoto S. (1999a) Flowering may be controlled by a quantitative balance between flower-inducing and –inhibiting substances in *Pharbitis nil, Plant Growth Regul.* **28**, 101-106.

Kondoh M., Murakami A., Shigeta C.and Takimoto S. (1999b) Flower-inducing and inhibiting activities of phloem exudate from cotyledons of *Pharbitis* seedlings, *Plant Growth Regul.* **28**, 107-116.

Kopcewicz J. and Tretyn A. (1998) Physiological and cytological investigations on photoperiodic flower induction in *Pharbitis nil, Flower. Newslett.* **25**, 26-34.

Larkin J.C., Felsheim R.and Das A. (1990) Floral determination in the terminal bud of the short-day plant *Pharbitis nil, Dev. Biol.* **137**, 434-443.

Ma H. (1999) Flowering time: from photoperiodism to florigen, *Curr Biol.* **8**, 690-692.

Nakanishi F., Kusumi T., Inoue Y. and Fuji T. (1995) Dihyfrokaempferol glucoside from cotyledons promotes flowering in *Pharbitis nil, Plant Cell Physiol.* **36**, 1303-1309.

Ogawa Y (1993a) Effects of cotyledon area and gibberellin A_3 on photoperiodic flower induction in *Pharbitis nil* seedlings, *Bull. Fac. Biores. Mie Univ.* **10**, 195-200.

Ogawa Y. (1993b) Movement of floral stimulus in the cotyledon and petiole of *Pharbitis nil* seedlings, *Bull. Fac. Biores. Mie Univ.* **10**, 201-207.

Shinozaki M., Asada K. and Takimoto A. (1988) Correlation between chlorogenic acid content in cotyledons and flowering in *Pharbitis* seedlings under poor nutrition, *Plant Cell Physiol.* **29**, 605-609.

Shinozaki M., Hirai N., Kojima Y., Koshimizu K.and Takimoto A. (1994) Correlation between level of phe-nylopropanoids in cotyledons and flowering in *Pharbitis nil* seedlings under high-fluence illumination, *Plant Cell Physiol.* **35**, 807-810.

Taylor A.O. (1965) Some effects of photoperiod on the biogenesis of phenylpropane derivatives in *Xanthium*, Plant Physiol. **40**, 273-280.

Tretyn A., Kendrick R.E., Fujioka S. and Sakurai A. (1996) Cytochemical and histohemical characterization of cotyledonary bodies from *Pharbitis nil* seedlings, *Protoplasma* **191**, 205-214.

Umemoto T. (1971) Effect of chlorogenic acid on flower production in long-day duck-weed, *Lemna gibba* G3, *Plant Cell Physiol* **12**, 165-169.

Vince-Prue, D. and Gressel J. (1985) *Pharbitis nil* Chois, in A.H. Halevy (ed.) *Handbook of Flowering, Vol 4*, CRC Press, Boca Raton, pp. 47-81.

Wada K. (1966) Studies on the initiation and development of flower primordia in *Pharbitis nil*, with special reference to the time of arrival of the flowering stimulus at the plumule, *Rap. Fac. Sci. Shizuoka Univ.* **1**, 83-89.

Wada K., Kadota A., Tanihira A. and Suzuki Y. (1991) Giant oil cells in the cotyledons of *Pharbitis nil* and other convolvulacean plants, *Bot. Mag. Tokyo* **94**, 239-247.

Yokoyama R. and Wada K. (1987) Pharbitin content in *Pharbitis nil*, *Rep. Fac. Sci.Shizuoka Univ.* **21**, 77-88.

Zucker M., Nitsch C. and Nitsch J.P. (1965) The induction of flowering in *Nicotiana*. II. Photoperiodic alternation of the chlorogenic acid concentration, *Amer. J. Bot.* **52**, 271-277.

OCCURRENCE AND PUTATIVE ROLE OF MELATONIN IN PLANTS

JAN KOLÁŘ[1,2] AND IVANA MACHÁČKOVÁ[2]
[1] *Department of Plant Physiology, Faculty of Science, Charles University, Viničná 5, 128 44 Praha 2, Czech Republic, e mail: kolar@ueb.cas.cz*
[2] *Institute of Experimental Botany, Academy of Sciences of the Czech Republic, Rozvojová 135, 16502 Praha 6, Czech Republic, e mail: machackova@ueb.cas.cz*

Many processes in living organisms, at both molecular and at higher levels, are rhythmic and many developmental processes (including sexual reproduction, flowering) are regulated photoperiodically. In last two decades much attention has been paid to the mechanisms underlying time measurement - to the biological clock. In both animals and plants, time measurement is based on a network of negative feedbacks in the synthesis of two proteins, which form a dimer (e.g. period and timeless or clock and cycle) (Sassone-Corsi, 1998). Many other components have been identified as well (Krishnan *et al.*, 2001). One of the molecules, which is believed to couple the central oscillator to the measured output rhythms in animals, is the hormone melatonin (see review by Reiter, 1993). Melatonin (5-methoxy-N-acetyltryptamine) regulates daily rhythms and photoperiodic reactions mainly in vertebrates, but it also occurs in invertebrates (Hardeland and Fuhrberg, 1996). It was discovered in 1958 in amphibians, where it regulates accumulation of the skin pigment melanin, from which function its name is derived (Lerner *et al.*, 1958). Since then, melatonin has been found to be ubiquitous in the animal kingdom and has also been shown to be present in some lower and higher plants.

In most species melatonin levels are high at night and much lower during the light period. In vertebrates the duration of elevated melatonin levels is usually proportional to the night length. Melatonin concentration and its daily rhythm can thus inform the organism about time of day and indicate the season (Reiter, 1993). In mammals melatonin is usually synthesized in the pineal gland (Maywood *et al.*, 1993); its synthesis proceeds from serotonin, which is first acetylated to N-acetylserotonin and then methylated to melatonin. Exogenous melatonin can "mimic" darkness and therefore affect circadian rhythms and photoperiodic responses of many animals (Reiter, 1993; Vaněček, 1998; Goldman, 2001). In vertebrates, melatonin usually binds to membrane receptors (Dubocovich *et al.*, 1998). It can also directly interact with Ca^{2+}-calmodulin (Benítez-King and Antón-Tay, 1993; Benítez-King *et al.*, 1996) and possibly also with tubulin (Benítez-King and Antón-Tay, 1993). Another role proposed for melatonin is connected with its

I. Macháčková and G.A. Romanov (eds.), Phytohormones in Plant Biotechnology and Agriculture, 79-87

ability to readily react with free radicals. Thus, Reiter *et al.* (1995) and Reiter (1997) suggested that melatonin may play a role as a radical scavenger.

1. MELATONIN IN LOWER PLANTS

The first report on melatonin in plants described its inductive effect on encystment in the dinoflagellate *Gonyaulax polyedra (Lingulodinium polyedrum)* (Balzer and Hardeland, 1991). When added under non-inductive long photoperiods, melatonin induced full encystment. It was later shown that melatonin is present in this alga and that its level exhibits a daily rhythm similar to that in animals: a high level in darkness and low in light (Poegeller *et al.*, 1991). This rhythm persisted in constant conditions (Balzer et al., 1993). Melatonin was also found in several species of red (Lorenz and Lűning, 1999), brown (Fuhrberg *et al.*, 1996) and green algae (Balzer *et al.*, 1998). High concentrations of melatonin inhibited growth of the brown alga *Pterygophora californica* (Fuhrberg *et al.*, 1996).

2. MELATONIN IN HIGHER PLANTS

The discovery of melatonin in *Lingulodinium polyedrum* (Balzer *et al.*, 1991) prompted a search for melatonin in higher plants and investigations of its possible roles. The first reports came from Kolář and Macháčková (1994) and van Tassel *et al.* (1995), who found evidence for the occurrence of melatonin in *Chenopodium rubrum* and *Pharbitis nil*, respectively. Dubbels *et al.* (1995) and Hattori *et al.* (1995) found melatonin in a number of dicot and monocot species, including tomato, tobacco, cucumber and cereals. Melatonin levels found in these plants range between 10 and 1000 $pg.g^{-1}$ fresh weight, which is comparable to levels found in animals. However, except in the case of tomato, melatonin determinations in these studies were performed using radioimmunoassay (RIA) in more or less crude extracts. This method did not provide reliable quantitative data, because RIA very often highly oversetimated melatonin levels in crude plant extracts. High levels of melatonin (up to 4 $\mu g.\ g^{-1}$ dry weight) were found in flowers of St. John´s wort (*Hypericum perforatum* L.), but also using RIA (Murch *et al.*, 1997). In this plant, conversion of L-tryptophan to melatonin was described (Murch *et al.*, 2000). High melatonin concentrations were also reported in the seeds of some plants (Manchester *et al.*, 2000). Even before this finding it had been suggested that melatonin in seeds may prevent peroxidation of the storage lipids (Hardeland *et al.*, 1995). The only described effect of melatonin application in higher plants was the disruption of the mitotic apparatus in onion root tip cells by high levels of melatonin (8.10^{-4}M). This was probably caused by melatonin interaction with tubulin (Banerjee and Margulis, 1973).

3. MELATONIN IN *CHENOPODIUM RUBRUM*

In our laboratory, studies of the occurrence of melatonin and of its possible function were performed using the short-day plant *Chenopodium rubrum* L. (ecotype 374), which is a model plant in our studies of photoperiodic flower induction (Ullmann *et al.*, 1985). This plant displays many rhythms, e.g. in flowering (King, 1975), in the activities of several enzymes (Frosch and Wagner, 1973), in stem extension rate and leaf movement (Normann *et. al.*, 1997) and in levels of some hormones (Pavlová and Krekule, 1984; Macháčková *et al.*, 1993, 1997). Thus, it seemed to be a convenient plant for melatonin studies. However, melatonin turned out to be unstable in plant extracts and a method for extraction and purification had to be carefully designed (Kolář *et al.*, 1997). Using this method and RIA or liquid chromatography with mass spectrometric detection (LC/MS), the occurrence of melatonin in extracts from shoots of *C. rubrum* was proved (Kolář *et al.*, 1997). Then, the changes of melatonin levels during a daily light/dark cycle were measured. In a cycle of 12h light and 12 h darkness the maximum of melatonin level occurs in the 6th h of darkness and the level started to decrease well before the end of the dark period and stayed very low during the light period (Kolář *et al.*, 1997)(Fig. 1). In animals it is usually the duration of the maximum of melatonin level, which reacts to the length of the darkness; the longer the dark period, the broader the maximum (Reiter, 1993). To check this reaction in *C. rubrum*, melatonin levels were determined in the cycles 8h darkness/16 h light and 16 h darkness/8 h light. It was the position of the maximum of melatonin level that reacted to the duration of the dark period: in 8 h darkness it occurs in the 2nd h and in 16 h darkness in 10th h (Wolf *et al.*, 2001) (Fig. 1). These results were obtained with plants grown for at least 5 cycles before analysis at the respective photoperiodic regime (usually, *C. rubrum* is grown in constant light before flower induction). When plants were grown in constant light and transferred to darkness, Melatonin level started to rise immediately, so that the maximum occurred at about the 1st h, maybe even earlier (unpublished results, Fig. 1). From this part of the work it was concluded that melatonin levels fluctuate rhythmically in *C. rubrum* plants and that the length of the dark period is reflected in the position of the melatonin maximum within the dark period.

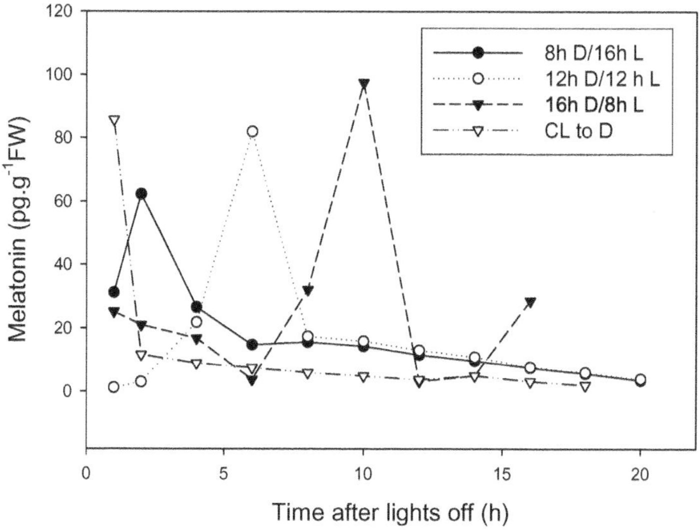

Figure 1.The changes of melatonin content in the shoots of 30-d-old Chenopodium rubrum plants after the transfer from continuous light (CL) to darkness and during photoperiodic regimes of 16 h light/8 h darkness, 12 h light/12 h darkness, and 8 h /16 h darkness. Measured by LC/MS. Daily profiles in each photoperiodic regime were measured at least twice and had the same pattern.

To elucidate the possible role of melatonin in the plant, its effect on photoperiodic flower induction was investigated. In 5-d-old plants or plants older than 14 d, one dark period of about 12-13 h is sufficient to induce flowering. By analogy with *Lingulodinium* (Balzer and Hardeland, 1991) the application of melatonin in subcritical lengths of darkness (6 and 8 h) was expected to promote flowering. But this effect was never observed. On the contrary, an inhibitory effect of Mel was observed when applied to the plumule of 5-d-old *C. rubrum* plants 1 h before the beginning of the inductive dark period of 12 h. Only higher concentrations of melatonin were effective (100 and 500 μM) and the reduction amounted to about 20-40% (Kolář *et al.*, 2003 in press, Fig. 2). Melatonin at this concentration was not toxic and did not influence the vegetative growth of plants. To explain why only such a high concentration of melatonin was effective, uptake of melatonin (500 μM) was measured by means of adding radioactive melatonin (^{3}H, 55 Bq.ml^{-1}). Approximately 60% of applied radioactivity was taken up within 5 h and 80-90 % after 37 h. Taken up Mel was stable and almost no metabolism of it occurred in the first 37 h after application. It was practically immobile - less than 2 % of radioactivity was found in stems or roots after its application to the plumule.

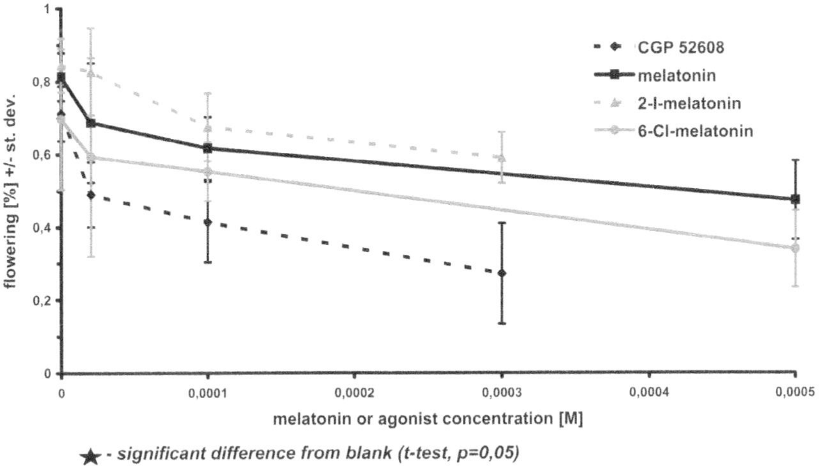

Figure 2. The effect of melatonin and its agonists on flowering in 5-d-old Chenopodium rubrum plants. The substances were applied 1 h before the start of the inductive darkness of 12 h. Flowering was scored under the microscope 7 days after the induction.

Further the dependence of the effect of melatonin on flowering on the time of its application and the effect of melatonin on the rhythm of photoperiodic time measurement in *C. rubrum* was investigated. Melatonin (500 μM) significantly inhibited flowering only when applied at times from 3h before to 6 h after the beginning of a darkness of 12 h. Later applications had no effect (Kolář *et al.*, 2003, in press). When Mel (500 μM) was applied to *C. rubrum* plants 1 h before the beginning of 9h- to 18h-long dark periods (around the first peak of the rhythm of photoperiodic time measurement) or 1 h before 41-h to 47-h-long dark periods (the second peak), it did not affect either the phase or the period of the rhythm. The peaks were at 12 and 45 h of darkness and the only parameter affected by Mel was the amplitude, which was decreased in both peaks (Fig. 3).

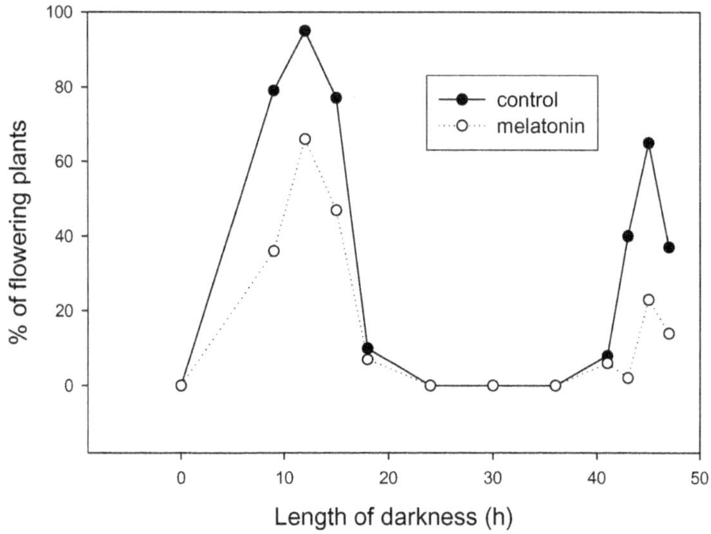

Figure 3. The effect of melatonin on the endogenous rhythm of photoperiodic time measurement in 5-d-old Chenopodium rubrum plants. Control solution or 500 μM melatonin were applied onto the cotyledons and plumule of each plant 1 h before a single dark period of the indicated length. Flowering was scored under the microscope the 7 days after the induction.

To investigate the specificity of Mel effect, several melatonin analogues and agonists known from studies in animals were tested. These were two melatonin analogues which bind to animal membrane receptors for melatonin: 6-Cl-Mel and 2-I-Mel, two indoleamines: 5-methoxytryptamine, which is highly effective in *Lingulodinium*, and serotonin (in animals the precursor of melatonin), and finally a synthetic agonist reported to bind to the putative nuclear receptors of melatonin, CGP 52608 (1-[3-allyl-4-oxo-thiazolidine-2-ylidene]-4-methyl-thiosemicarbazone, see Wiesenberg et al., 1995). All tested substances except 5-methoxytryptamine inhibited flowering when applied 1h before the beginning of 12-h dark period. CGP 52 608 was the most effective; it was inhibitory already in 20μM concentration, 2-I-mel and serotonin in 100 μM and 6-Cl-Mel in 500 μM concentration (Kolář et al., 2003) (Fig. 2).

Thus, melatonin is able to suppress flower induction in *Chenopodium rubrum* without affecting the phase of the rhythm of flowering and this effect is rather specific for melatonin and its derivatives and agonists. The effect of melatonin on flowering in the long-day plant *Arabidopsis thaliana* was also tested. Preliminary results showed that in this plant too, melatonin (5.10^{-4} M) applied to plants grown for the first 20 days in long days (15 h light) and transferred then for 10 days to long days of 18 h, during which melatonin was applied every day 1 h before the "lights off", increased number of leaves produced (both rosette and total leaves). However, the delay in the appearance of the first flower was not statistically significant (unpublished results).

There are a number of possible explanations for the inhibitory effect of melatonin on flowering: 1. melatonin may have an auxin-like effect and it is know that IAA may inhibit flowering in *C. rubrum* (Pavlová and Krekule, 1990); 2. melatonin may interact with Ca-calmodulin system as suggested by Benítez-King and Antón-Tay (1993) and 3. it may induce changes in cytoskeleton (Benítez-King and Antón-Tay, 1993). Melatonin did not have any activity in the wheat coleoptile segment bioassay for auxins (Kolář, unpublished). The effect of melatonin on microtubule arrangement was studied in suspension cultured tobacco cells BY-2 after the incubation of this culture in the medium with melatonin for 20 min or 4 h no changes in tubulin conformation were observed (Petrášek, Černá and Kolář, unpublished results). The mechanism of the inhibitory effect of melatonin on flowering therefore requires further investigations.

ACKNOWLEDGEMENTS

Thanks are due to the Grant Agency of the Czech Republic for financial support for the melatonin research (206/00/1354), to Mrs. B. Čermáková for technical assistance.

REFERENCES

Balzer, I. and Hardeland, R. (1991) Photoperiodism and effects of indoleamines in a unicellular alga, *Gonyaulax polyedra*, *Science* **253**, 795-797.

Balzer, I., Poegeller, B. and Hardeland, R. (1993) Circadian rhythms of indoleamines in a dinoflagellate Gonyaulax polyedra: Persistence of melatonin rhythm in constant darkness and relationship to 5-methoxytryptamine, in: Y. Touitou, J. Arendt and P. Pevet (eds.) *Melatonin and the pineal gland. From basic science to clinical applications, Excerpta Medica,* Amsterdam,pp. 83-86.

Balzer, I., Bartolomaeus, B. and Höcker, B. (1998) Circadian rhythm in melatonin content in *Chlorophyceae,* Abstr. Workshop Eur. Soc. Chronobiol., *News from the Plant Chronobiology Research*, Markgrafenheide.

Banerjee, S. and Margulis, L. (1973) Mitotic arrest by melatonin, *Exp. Cell Res.* **78**, 314-318.

Benítez-King, G. And Antón-Tay, F. (1993) Calmodulin mediates melatonin cytoskeletal effects, Experientia 49, 635-641.

Benítez-King, G., Rios, A., Martinez, A. and Antón-Tay, F. (1996) In vitro inhibition of Ca2+/calmodulin-dependent kinase II activity by melatonin, *Biochim. Biophys. Acta* **1290**, 191-196.

Dubbels, R., Reiter, R.J., Klenke, E., Goebel, A., Schnakenberg, E., Ehlers, C., Schiwara, H.W. and Schloot, W. (1995) Melatonin in edible plants identified by radioimmunoassay and by high performance liquid chromatography-mass spectrometry, *J. Pineal Res.* **18**, 28-31.

Dubocovich, M.L., Cardinalli, D.P., Guardiola-Lemaitre, B., Hagan, R.M., Krause, D.N., Sugden, D., Yocca, F.D. and Vanhoutte, P.M. (1998) Melatonin receptors, in *The IUPHAR Compendium of Receptor Characterization and Classification,* IUPHAR Media, London, pp. 187-193.

Frosch, S., Wagner, E. and Cumming, B.C. (1973) Endogenous rhythmicity in energy transduction. I. Rhythmicity in adenylate kinase, NAD- and NADP-linked glyceraldehydes-3-phosphate dehydrogenase in *Chenopodium rubrum*, *Can. J. Bot.* **51**: 1355-1367.

Fuhrberg, B., Balzer, I., Hardeland, R., Werner, A. and Lünning, K. (1996) The vertebrate pineal hormone melatonin is produced by the brown alga *Pterygophora californica* and mimics dark effects on growth rate in the light, *Planta* **200**, 125-131.

Goldman, B.D. (2001) Mammalian photoperiodic system. Formal properties and neuroendocrine mechanisms of photoperiodic time measurement, *J. Biol. Rhythm* **16**, 283-301.

Hardeland, R., Balzer, I., Poegeller, B., Fuhrberg, B., Uria, H., Behrmann, G., Wolf, R., Meyer, T.J. and Reiter, J. (1995) On the primary functions of melatonin in evolution – mediation of photoperiodic signals in a unicell photooxidation, and scavenging of free radicals, *J. Pineal Res.* **18**: 104-111.

Hardeland, R. and Fuhrberg, B. (1996) Ubiquitous melatonin - presence and effects in unicells, plants and animals, *Trends Comp. Biochem. Physiol.* **2**, 25-45.

Hattori, A., Migitaka, H., Iigo, M., Itoh, M., Yamamoto, K., Ohtanikaneko, R., Hara, M., Suzuki, T. and Reiter, R.J. (1995) Identification of melatonin in plants and its effects on plasma melatonin levels and binding to melatonin receptors in vertebrates, *Biochem. Mol. Biol.* **35**, 627-634.

King, R. W. (1975) Multiple circadian rhythms regulate flowering responses in *Chenopodium rubrum, Can. J. Bot.* **53**: 2631-2638.

Kolář, J. and Macháčková, I. (1994) Melatonin: does it regulate rhythmicity and photoperiodism also in higher plants? *Flower. Newsletter* **17**: 53-54.

Kolář, J., Macháčková, I., Eder, J., Prinsen, E., Van Dongen, W., van Onckelen, H. and Illnerová, H. (1997) Melatonin: occurrence and daily rhythm in *Chenopodium rubrum, Phytochemistry* **44**, 1407-1413.

Kolář, J., Johnson, C.H. and Macháčková, I. (2003) Exogenously apllied melatonin (N-acetyl-5-methoxytryptamine) affects flowering of the short-day plant *Chenopodium rubrum* L., *Physiol. Plant.* **118**, 1-8.

Krishnan, B., Levide, J.D., Lynch, K.S., Dowse, H.B., Funes, P., Hall, J.C., Hardin, P.E. and *Dryer, S.E. (2001) A new role for cryptochrome in a Drosophila circadian oscillator,* Nature **411**, 313-317.

Lerner, A.B., Chase, J.D., Takahashi, Y., Lee, T.H. and Mori, W. (1958) Isolation of melatonin, the pineal factor that lightens melanocytes, *J. Am. Chem. Soc.* **81**, 2587.

Lorenz, M. and Lünning, K. (1999) Detection of endogenous melatonin in the marine red macroalgae *Porphyra umbilicalis* and *Palmaria palmata* by enzyme-linked immunoassay (ELISA) and effects of melatonin administration on algal growth, *Biol. Rhythm Res.* **30**, 243.

Macháčková, I., Krekule, J. Eder, J., Seidlová, F. and Strnad, M. (1993) Cytokinins in photoperiodic induction of flowering in *Chenopodium* species, *Physiol. Plant.* **87**, 160-166.

Macháčková, I., Chauvaux, N., Dewitte, W. and Van Onckelen, H. (1997) Diurnal fluctuations in ethylene formation in *Chenopodium rubrum, Plant Physiol.* **113**, 981-985.

Manchester, L.C., Tan, D.X., Reiter, R.J., Park, W., Monis, K., and Qi, W. (2000) High levels of melatonin in the seeds of edible plants. Possible function in germ tissue protection, *Life Sci.* **67**, 3023-3029.

Maywood, E.S., Hastings, M.H., Max, M., Ampleford, E., Menaker, M. and Loudon, A.S.I. (1993) Circadian and daily rhythms of melatonin in the blood and pineal gland of free-running and entrained Syrian hamster, *J. Endocrinol.* **136**, 65-73.

Murch, S.J. Simmons, C.B. and Saxena, P.K. (1997) Melatonin in feverfew and other medical plants, *Lancet* **350**, 1598-1599.

Murch, S.J., KrishnaRaj S. and Saxena, P.K. (2000) Tryptophan is a precursor for melatonin and serotonin biosynthesis in *in vitro* regenerated St.Johnś wort (*Hypericum perforatum* L. cv. Athos) plants, *Plant Cell Rep.* **19**: 698-704.

Normann J. and Wagner, E. (1997) Specific changes in phase relationship between stem extension rate and leaf movement upon floral induction in *Chenopodium* spec., *Endocytobiosis and Cell Res.* **12**: 117-118.

Pavlová, L. and Krekule, J. (1984) Fluctuation of free IAA under inductive and non-inductive photoperiods in *Chenopodium rubrum, Plant Growth Regul.* **2**: 91-94.

Pavlová, L. and Krekule, J. (1990) The effect of IAA application on endogenous rhythm of flowering in *Chenopodium rubrum* L., *Biol. Plant.* **32**, 277-287.

Poegeller, B., Balzer, I., Hardeland, R. and Lerchl, A. (1991) Pineal hormone melatonin oscillates also in a dinoflagellate *Gonyaulax polyedra, Naturwiss.* **78**, 268-269.

Reiter, R.J. (1993) The melatonin rhythm: both a clock and a calendar, *Experientia* **49**, 654-664.

Reiter, R.J. (1997) Antioxidant actions of melatonin, *Adv. Pharmacol.* **38**, 103-117.

Reiter, R.J., Melchiorri, D., Sewerynek, E., Poegeller, B., Barlow-Walden, L., Chuang, J., Ortiz, G.G. and Acuna-Castroviejo, D. (1995) J. Pineal Res. 18, 1-11.

Sassone-Corsi, P. (1998) Molecular clocks: Mastering time by gene regulation, *Nature* **392**, 871-874.

Ullmann, J., Seidlová, F., Krekule, J. and Pavlová, L. (1985) *Chenopodium rubrum* as a model plant for testing flowering effects of PGRs, *Biol. Plant.* **27**, 367-372.

Vaněček, J. (1998) Cellular mechanisms of melatonin action, *Physiol. Rev.* **78**, 687-721.

Wiesenberg, I., Missbach, M., Kahlem, J.-P., Schräder, M. and Carlberg, C. (1995) Transcriptional activation of the nuclear receptor RZRα by the pineal gland hormone melatonin and identification of CGP 52608 as a synthetic ligand, *Nucl. Acids Res.* **23**, 327-333.

Wolf, K., Kolář, J., Witters, E., van Dongen, W., van Onckelen, H. and Macháčková, I. (2001) Daily profile of melatonin levels in *Chenopodium rubrum* L. depends on photoperiod, *J. Plant Physiol.* **158**, 1491-1493.

INFORMATIONAL SIGNALLING IN STEM APICAL MERISTEMS OF PLANTS DURING TRANSITION TO FLOWERING

E.L. MILYAEVA AND V.U. NIKIFOROVA
Timiryazev Institute of Plant Physiology RAS, Botanicheskaya ul. 35, Moscow 127276 Russia Fax (095) 977-8018, e-mail: e_milyaeva@mail.ru

1. INTRODUCTION

According to Chailakhyan's hormonal theory, during the transition to flowering hormonal signals are produced in leaves and then translocated from leaves to stem apices (Chailakhyan, 1988). Thus, they are long-distance signals. Following arrival of the signal in plant stem apices, morphological modifications take place. Inside the stem apices some short-distance signals exist. These signals may provide the information exchange between cells of the stem apex. These signals may be the basis of the so-called «positional information signalling» and they may determine intra- and intercellular interactions inside the stem apex.

All cells in the stem apical meristem are interconnected with plasmodesmata. Studies on the efficiency of intercellular communications showed that it was proportional to the plasmodesmata frequency in cell walls. A high plasmodesmata frequency was assumed to indicate intense intercellular interactions (Robard and Lucas, 1990).

We tried to study the symplastic way of signalling in stem apices during transition to flowering and elucidate possible ways of signal nature and their transduction. The signals of «positional information» as well as hormonal signals can be transmitted either through the apoplast or the symplast. Earlier we had shown that during transition to flowering of long-day plant coneflower (*Rudbeckia bicolor* Nutt.) after photoperiodic induction by eight long days, frequency of plasmodesmata increased in the first and second cell layers of the central zone and decreased in the cell walls of the medullary zone (Milyaeva and Nikiforova, 1995, 1996). Ormenese *et al.* had shown that the frequency of plasmodesmata increased early in the shoot apical meristem of the long-day plant *Sinapis alba* L. induced to flowering by exposure to a single long day (Ormenese *et al.*, 2000).

In this paper, the frequency of plasmodesmata was determined in the cells of the central and medullar zones of stem apices both of the short–day plant *Perilla nankinensis* and the long-day plant *Rudbeckia bicolor*. The results obtained in stem apices of the plants of different photoperiod - sensitive groups were compared.

I. Macháčková and G.A. Romanov (eds.), Phytohormones in Plant Biotechnology and Agriculture, 89-93

2. MATERIALS AND METHODS

We used the traditional for Chailakhyan's laboratory photoperiodically - sensitive model plants, which possess qualitative photoperiodic reaction: the long-day rosette plant *Rudbeckia bicolor* Nutt. and the short-day plant *Perilla nankinensis* (Lour) Dence. The *Rudbeckia* plants were grown under unfavorable for flowering short days (8 h light+16 h dark) for 3 months and then were induced to flower by 8 long days. The plants of *Perilla nankinensis* were grown under unfavorable for flowering long days (16 h light+8 h dark) for 3 months and then were induced to flowering by 12 short days.

For the transmission electron microscopy the stem apices were prepared from the plants and fixed with 3% glutaraldehyde solution in cold phosphate buffer, pH 7.4, supplemented with 25 mg/ml sucrose. After 18 h fixation the samples were rinsed in cold buffer and postfixed with cold 1% $0s0_4$ for 4 h. After dehydratation in a graded alcohol series, the samples were embedded into Epon resin.

Ultrathin sections of stem apices of induced and control plants were cut with an LKB-4800 ultratome (Sweden), stained first with 1% uranyl acetate in 70% ethanol and then on the grids with lead citrate. The sections were examined under JEM-100 or JEOL-100 (Japan) electron microscopes.

The number of plasmodesmata per 1 μm of cell wall length was counted on electron micrographs using a MOP Videoplan (Reichert-Jung,Austria).

In such experiments, 100 ultrathin sections were examined. The number of plasmodesmata was estimated at a cell-to-cell interface in different zones in both the periclinal and anticlinal cell walls (Fig. 1). The results of plasmodesmata counting were presented as plasmodesmograms, as initially suggested by van Bel *et al.* (van Bel *et al.*, 1988) for convenient visualization of such data. In these plots, plasmodesmata are represented by lines, the number of which is proportional to corresponding plasmodesmata frequencies.

3. RESULTS

The results of our investigations show that in the stem apices of *Perilla* as well as those of *Rudbeckia,* plasmodesmata frequency was different in the central and medullary zones of stem apices and changed after photoperiodic induction of flowering. Fig.2 represent the plasmodesmogram of plasmodesmata frequency in the central and medullary zones of the stem apex of *Rudbeckia bicolor* and Fig.3 shows the same for *Perilla nankinensis.* As can be seen from these figures, the number of plasmodesmata in the central zone of vegetative apices was higher between the first and second cell layers than between the second and the third layers. Floral induction increased the number of plasmodesmata between the cells in all layers. On the other hand, the plasmodesmata frequency between the cells of the medullary zone significantly declined after the floral induction.

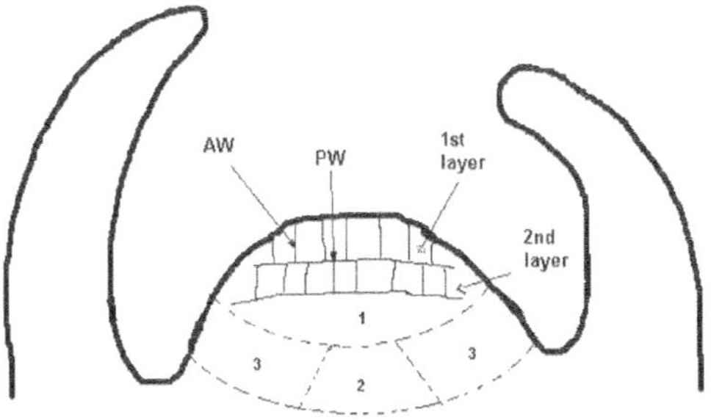

Figure 1. Diagram of longitudial section through a stem apex. 1: central zone 2: medullary zone 3: lateral zone. The arrows show sites in which plasmodesmata frequencies were determined.

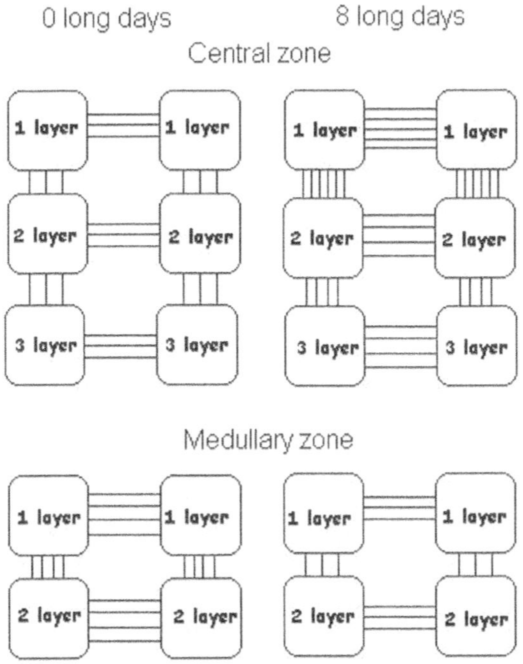

Figure 2. Plasmodesmogram of the plasmodesmata frequency in the cells in stem apical meristem of Rudbeckia

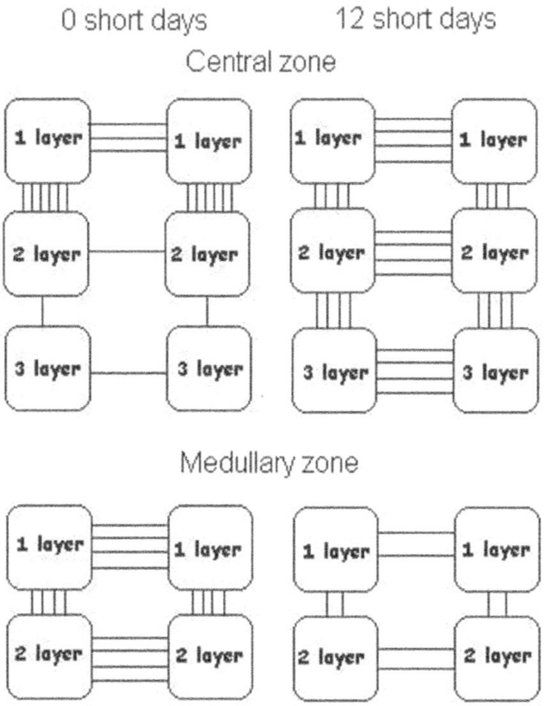

Figure 3. Plasmodesmogram of the plasmodesmata frequency in the cells in stem apical meristem of Perilla

Thus, the results of our work suggest that symplastic barriers exist between different types of cells in *Rudbeckia* and *Perilla* stem apices. This indicates that there exists a certain symplastic compartmentation, which decreased during transition to flowering. We suggest that during transition to flowering, in the reproductive states of stem apices, the symplastic transport was more intensive and signal molecules moved more actively than during the vegetative state.

The formation of «symplastic domains» in the central and medullary zones of the stem apices may be a crucial link in the chain of events from gene expression to initiation of reproductive organogenesis.

REFERENCES

Chailakhyan, M.Kh. (1988) Control on the Higher Plant Flowering, Nauka, Moscow, in Russ., 559 pp.

Milyaeva, E.L. and Nikiforova, V.J. (1995) Symplastic connection in the stem apex of coneflower plants during floral transition. Soviet Plant Physiol. 42, 640-645.

Milyaeva, E.L. and Nikiforova, V.J. (1996) Informational signaling in stem apical meristems during transition to flowering, Flowering Newslett. 22,26-33.

Ormenese, S., Havelange, A., Deltoure, R. and Bernier, G. (2000) The frequency of plasmodesmata increases early in the whole shoot apical meristem of *Sinapis alba L.* during floral transition, *Planta* **211**,370-375.

Robard, A.W. and Lucas, W.J. (1990) Plasmodesmata, *Annu. Rev. Plant Physiol. Plant Mol. Biol.* **41**, 369-381.

van Bel, A.J.E., van Kesteren, W.J.P. and Papenhuijzen, C. (1988) Ultrastructural indications for coexistence of symplastic and apoplastic phloem loading in *Commelia bengalensis* leaves, *Planta* **176**,1591-1599.

CYTOKININS

CYTOKININ AS A REGULATORY FACTOR FOR YIELD AND BIOMASS DISTRIBUTION IN CROP PLANTS

T. SCHMÜLLING, T. WERNER, M. RIEFLER, E. KRUPKOVÁ, I. BARTRINA, Y. MANNS AND H. DORTAY

Institute of Biology, Applied Genetics, FU Berlin, Albrecht-Thaer-Weg 6, D-14195 Berlin, Germany; Email tschmue@zedat.fu-berlin.de

1. INTRODUCTION

In the future, improvement of yield quantity and quality will remain the most important goal in the breeding of crop plants. Classical breeding and hybrid breeding has produced a steady state increase in yield of approximately 2-3% per year over the last 40-50 years. However, yield estimates for the next decades take into account the increasing difficulties encountered by these classical methods and predict a plateau for yield increase in the near future (Reilly and Fuglie, 1998). There are indications that other resources like land, water, chemicals, and mechanisation, are approaching the limits of their capacities to affect significant and sustainable gains in productivity. Gains in productivity will become more dependent on other sources, in particular on an increase of biomass (Lee, 1998). The production of biomass and its distribution within the plant is often seen as the result of a complex interaction of many different factors. However, recent results indicate that these parameters can also be influenced largely by single factors. One such single factor is cytokinin, which regulates the cell cycle and the formation and distribution of primary metabolites.

1.1. Metabolism and signal transduction of cytokinins

Recently, rapid progress has been made in the understanding of cytokinin metabolism and signal transduction (for reviews see Mok and Mok, 2001; Deruere and Kieber, 2002; Schmülling, 2002; Schmülling *et al.*, 2003). Several groups have successfully isolated the genes that code for the proteins catalyzing cytokinin synthesis and breakdown (Houba-Hérin *et al.*, 1999; Morris *et al.*, 1999; Kakimoto, 2001b; Takei *et al.*, 2001). Isopentenyl transferases (IPT) catalyse the first and rate-limiting step of cytokinin synthesis by transferring the isopentenyl moiety of dimethylallyl pyrophosphate to ATP, ADP or AMP. This leads to the formation of isopentenyl-ATP, -ADP and -AMP, which are the precursor molecules of

I. Macháčková and G.A. Romanov (eds.), Phytohormones in Plant Biotechnology and Agriculture, 97-108

biologically active cytokinins. The *IPT* gene family of *Arabidopsis* (*AtIPT*) has seven members. For biotechnological applications of cytokinins it is relevant to note that these genes are expressed in the root and in the shoot (Kakimoto, 2001b,c). This result implies that the classical view of cytokinins, that are synthesized exclusively in the root and govern biological processes in the shoots after being transported there, must be modified. A paracrine model, i.e. a locally restricted function close to their sites of synthesis, is likely for at least some of the activities of cytokinins (Faiss *et al.*, 1997; Schmülling *et al.*, 2002).

Cytokinin oxidases/dehydrogenases (CKX) catalyze the breakdown of cytokinins by oxidative side chain cleavage, producing adenine and an aldehyde as reaction products (Galuszka *et al.*, 2001). The *CKX* gene family has seven members in *Arabidopsis* (*AtCKX1-AtCKX7*) and at least ten members in rice (Bilyeu *et al.*, 2001; Werner *et al.*, 2001; Schmülling *et al.*, 2003). To date 17 *CKX* genes have been fully annotated in databases, 15 genes of plant origin and two procaryotic genes. Sequence comparison shows that large parts of the predicted proteins share relatively few conserved amino acids, usually not more than about 40% identity (Tab. 1). Genes coding for enzymes with a high percentage of amino acid identity originate probably from relative recent gene duplications. This is illustrated by *AtCKX2* and *AtCKX4* of *Arabidopsis*. These genes are located in a duplicated region of the *Arabidopsis* genome (Vision *et al.*, 2000) and share 65.9% amino acid identity (Fig. 1). All CKX enzymes have a predicted molecular weight between 56 kDa and 65 kDa. Sequence analysis identifies N-terminal target peptides for the secretory pathway for the majority of the plant CKX enzymes. It is predicted that three of them are localized to intracellular compartments. The CKX enzymes share a region approximately 170 amino acids long common to FAD-linked oxidases in the N-terminal half of the proteins (Fig. 2) (Schmülling *et al.*, 2003). Phylogenetic analysis shows that several groups of CKX enzymes can be distinguished (Fig. 3). Interestingly, each group is present in both monocots and dicots, suggesting that different CKX enzymes evolved early in plant evolution. The progenitor gene was probably among the genes that were introduced into higher plants from a cyanobacteria-like endosymbiont (Schmülling *et al.*, 2003).

Table 1. Degree of amino acid identity between the CKX proteins

	ZmCKx1	AtCKX1	AtCKX2	AtCKX3	AtCKX4	AtCKX5	AtCKX6	AtCKX7	HvCKX2	DsCKX1	OsCKX1	OsCKX2	OsCKX3	OsCKX4	OsCKX5	RfCKX1	NsCKX1
ZmCKX1	100	34,5	43	43,4	44,9	40,1	40,3	36,5	38	46,6	70,5	50,7	35,7	39,4	42,6	24	18,4
AtCKX1		100	34,3	38,3	35,5	45,2	55,5	36,2	54,7	38,9	35,7	33,6	37,6	54,4	44,3	24,9	18,3
AtCKX2			100	46,8	65,9	41,5	38,8	36,8	37,2	43,3	43	39,5	35,8	38	40,1	27,8	22,4
AtCKX3				100	47	44,1	40,7	37,6	40,8	47	41,2	42,1	38,2	40,3	42,5	26,2	19,8
AtCKX4					100	42,1	38,5	35,5	37,1	43,9	44,3	39	35,8	38,4	41,2	26,7	20,6
AtCKX5						100	47,4	41,9	48,6	44,3	41,8	40,4	39,4	47,5	63	25,1	18,7
AtCKX6							100	39,8	58,4	42,2	40,4	38,1	40,9	57,7	47	27	19,3
AtCKX7								100	39,9	38,5	39,7	37,4	36,8	42,2	44,1	26,8	19,7
HvCKX2									100	43,3	40,2	38,3	40,1	71,8	47,8	26,8	18
DsCKX1										100	45,1	45,4	38,7	42,9	43,9	26,1	19,1
OsCKX1											100	51,2	35,7	41,4	44,5	25,8	18,1
OsCKX2												100	36,4	39,9	42,2	23,8	16,9
OsCKX3													100	39,5	39,1	25,2	16,9
OsCKX4														100	49,8	27,9	18,7
OsCKX5															100	26,3	19,9
RfCKX1																100	24,8
NsCKX1																	100

Values indicate the percentage of identity obtained by pair wise comparisons between protein sequences. Gene names are according to [6]. At, *Arabidopsis thaliana*, Zm, *Zea mays*, Hv, *Hordeum vulgare*, Ds, *Dendrobium sp.*, Os, *Oryza sativa*, Rf, *Rhodococcus fascians*, Ns *Nostoc sp. PCC7120.*

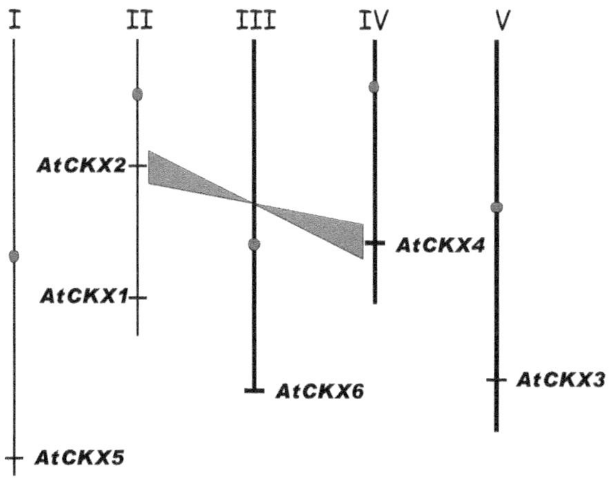

Figure 1. Location of AtCKX genes on the five chromosomes of Arabidopsis. AtCKX2 and AtCKX4 are located in duplicated regions (Vision et al., 2000).

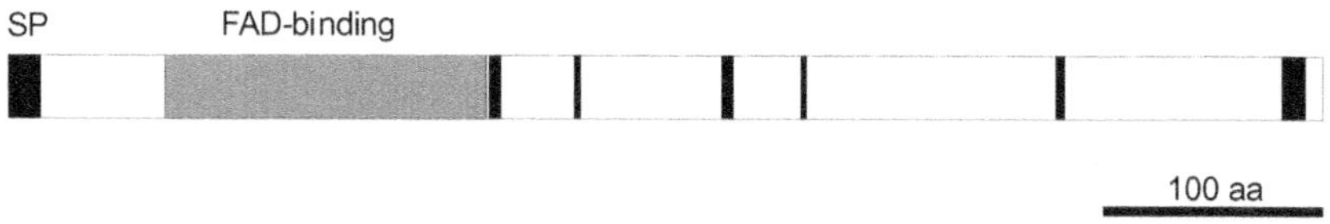

Figure 2. Scheme of CKX protein structure. The N-terminal signal peptide (SP) is shown in black. It varies between different protein family members (see text). The FAD-binding domain is shown in grey. The C-terminal half contains a number of short highly conserved groups of amino acids.

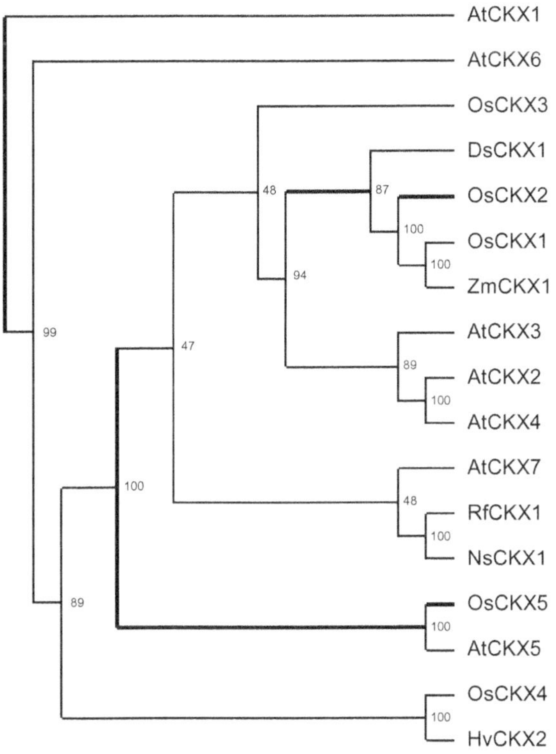

Figure 3. Phylogenetic relation of CKX proteins. Full sequences of AtCKX proteins were used. Programs used were the ClustalW program [60] for protein alignment and the TreeView program for graphical output. The entire amino acid sequences were aligned. Bootstrap values >40% are shown.

Endogenous cytokinin content is also regulated by the reversible or irreversible formation of cytokinin-sugar conjugates. Genes encoding zeatin *O*-glycosyltransferases, zeatin *O*-xylosyltransferase and a *cis*-zeatin *O*-glycosyltransferase were cloned from *Phaseolus vulgaris*, *P. lunatus*, and *Zea mays*, respectively (Martin *et al.*, 1999a, b, 2001).

Parallel to the elucidation of relevant parts of cytokinin metabolism, decisive discoveries that allow a better understanding of cytokinin signalling were made as well. The three known cytokinin receptors (CRE1/WOL/AHK4, AHK2, AHK3) are receptor histidine kinases, which are typical of two component signalling systems (TCS) (Inoue *et al.*, 2001; Suzuki *et al.*, 2001; Yamada *et al.*, 2001). The TCS is common among prokaryotes and lower eukaryotes, among the higher eukaryotes it is unique to plants (Hwang *et al.*, 2001). The available data suggest that cytokinin signals through a typical phosphorelay from the activated receptor to phosphotransfer proteins (AHPs). These transport the signal to the cell nucleus where the phosphoryl group is transmitted to so-called type B response regulators (type B ARRs) (Hwang *et al.*, 2001). Type B response regulators consist of an N-terminal receiver domain and a C-terminal output domain, containing a DNA-recognition motif and transcriptional activator sequences (Lohrmann *et al.*, 1999; Sakai *et al.*, 2001). Activated type B ARRs transcribe primary response genes of cytokinins. Some of the known primary response genes code for type A response regulators (type A ARRs) (D´Agostino *et al.*, 2000). Type A ARRs resemble type B ARRs but lack the C-terminal DNA binding and activation domain. Type A ARRs fulfil at least two different functions. On the one hand they may exert a negative feedback regulation of the cytokinin signalling pathway through protein-protein interaction (Hwang and Sheen, 2001). On the other hand they mediate the cytokinin dependent regulation of other pathways, e.g. light signalling (Sweeree *et al.*, 2001).

1.2. Function of cytokinins in regulating primary metabolism

A wealth of information corroborates the central role that cytokinins play in the regulation of primary metabolism, in particular in sink formation. More specifically, assimilation of inorganic nitrate into organic form, which is essential for plant growth, is stimulated by cytokinin. Nitrate reductase is co-regulated by nitrate, light and cytokinins (Samuelson *et al.*, 1995). An increase in the efficiency with which nitrogen is incorporated into organic form has a marked effect on plant productivity, biomass and crop yield (Oliveira *et al.*, 1997). The link between cytokinins and N-metabolism is underlined by the fact that response regulator genes are primary regulatory targets for both cytokinin and nitrate (Taniguchi *et al.*, 1998). Possibly, cytokinins are signals that are co-transported with nitrate from the roots to the shoot, informing the shoot about the N-status of the root (Sakakibara *et al.*, 1998). Similarly, expression of several genes encoding key enzymes of C-metabolism and transport, such as the extracellular invertase and hexose transporter genes, are controlled by cytokinin (Ehness and Roitsch, 1997). Another important target gene of cytokinins is *CycD3* that regulates the transition of the G1- to S-phase during the cell cycle (Riou-Khamlichi *et al.*, 1999). Regulation of the cell cycle in the growing meristems via *CycD3* could be a central regulatory switchpoint that influences sink strength. Additional roles for cytokinins in determining source-sink relations are their influence on the maturing and activity of plant organelles, both chloroplasts and mitochondria (Letham, 1994; Lohrmann *et al.*, 2001).

1.3. Growth and yield of plants with altered cytokinin content

The *IPT* genes of plants will be used in a similar manner to the earlier known *IPT* gene of the *Agrobacterium tumefaciens* T-DNA to produce plants with enhanced cytokinin content. The phenotypic changes caused in these transgenic plants depend largely on the specific regulatory sequences that drive *IPT* gene expression. They include, for example, delayed leaf senescence or a different branching pattern of the shoot (Gan and Amasino, 1995; Faiss *et al.*, 1997). A very instructive example of the potential of the approach is the expression of the *IPT* gene under the control of the senescence-induced *SAG12* promoter (Gan and Amasino, 1995). *SAG12::IPT* transgenic tobacco plants produce more cytokinins in senescing leaves which leads to retardation of leaf senescence. Seed yield of these plants was increased by approximately 50% (Gan and Amasino, 1995) and several parameters of photosynthesis and nitrogen partitioning were altered (Jordi *et al.*, 2000). The application of this system is being investigated for a number of crop plants. *SAG12::IPT* transgenic lettuce shows a delayed loss of chlorophyll after harvest and 70% lowered nitrate content (McCabe *et al.*, 2001). The transgenic lettuce plants were also more resistant to *Botrytis cinerea*, a fungus that preferentially attacks senescing tissues (McCabe *et al.*, 2001). A similar approach was used to generate transgenic *Lolium* grass, in order to obtain a lawn that stays green longer (www.iger.bbsrc.ac.uk/ierweb/cellbiol/news/abstracts.asp).

Manipulation of yield in cereals by retarding leaf senescence seems to be a more complex matter. In cereals, senescence of the flag leaf determines export of reserve substances to the ripening seed. Therefore, in this case inhibition of leaf senescence may lead to reduced leaf export causing a reduced yield. Notwithstanding these difficulties, the approach to alter yield attributes by altering the cytokinin metabolism remains a promising tool in cereals. As in other cases it seems to be the choice of the right regulatory sequences that determines success.

Other analyses of *IPT* transgenic plants support the idea that endogenously enhanced cytokinin production increases sink strength and biomass formation. Conditional cytokinin-overproducing *Arabidopsis* plants produced a more than 30% increase of fresh weight during the first two weeks following germination. This increase was primarily due to improved leaf growth (higher cell number), the formation of storage cells in the stem and faster growth in the initial growth phase following germination (Rupp *et al.*, 1999). Local expression of the *IPT* gene in axillary buds caused enhanced starch formation in buds, leading to the formation of short and tuber-like lateral branches. These experiments proved that local changes in cytokinin metabolism cause local changes of primary metabolism (Guivarc'h *et al.*, 2002).

Earlier investigations had demonstrated a positive influence of exogenous cytokinins on growth and yield, although the results are not always consistent. Several groups have performed detailed analyses of the influence of cytokinins on yield stability in maize (*Zea mays* L.). During kernel development in maize, peaks in cytokinin levels are apparent between 4 and 12 days after pollination which is the period of endosperm cell divison and amyloplast biogenesis (Jones *et al.*, 1992; Lur and Setter, 1993; Morris *et al.*, 1993; Cheikh and Jones, 1994; Dietrich *et al.*, 1995). High cytokinin content of the kernels was generally correlated with high CKX activity, however not in all kernel tissues (Jones *et al.*, 1992; Dietrich *et al.*, 1995). A strong difference in cytokinin content and CKX activity was noted early during development between kernels that are going to develop normally and those that will abort, the latter having a low cytokinin content and a low CKX activity (Dietrich *et al.*, 1995). Stem infusions of cytokinin significantly reduced the abortion of kernels and increased yield by as much as 30% (Dietrich *et al.*, 1995). The content of cytokinin in maize kernels is lowered by stress and the loss of kernels is enhanced (Cheikh and Jones, 1994). Cytokinin treatment increases thermotolerance and thus protects developing kernels at the ear tip (Cheikh and Jones, 1994). Cytokinin treatment inhibits also the complete loss of ears (Lejeune *et al.*, 1998). Furthermore, in maize cytokinins support the positive effect of increased N-fertilization (Smiciklas and Below, 1992). This is relevant, as the efficiency of the conversion of inorganic nitrogen into organic nitrogen has a strong influence upon plant productivity (Oliveira *et al.*, 1997). Analyses performed in other cereals indicate that these results might be generalized. A positive correlation between cytokinin content and CKX activity was also reported for wheat where cytokinin levels in the developing grain were increasing 4-11 days after anthesis (Kamínek *et al.*, 1994). Barley was found to contain a higher number of fertile flowers after exogenous treatment with cytokinins (Peltonen, 1997). Taken together, the prevailing data indicate that cytokinins have an important function in assuring yield stability in maize and possibly in other cereals.

Oilseed rape is another crop plant that could profit from targeted manipulation of the cytokinin content. One of the undesirable traits of oil seed rape is the high loss of terminal inflorescences which lowers yield significantly (Bouttier and Morgan, 1992). Several parameters that determine yield, like the number of flowers, seeds per silique and seed weight depend on plant nutrition and plant hormones (De Bouille *et al.*, 1990). Morgan *et al.* (1983) increased the number of flowers by treatment with exogenous cytokinin. Roeckel *et al.* (1997a,b) found the number of seeds increased significantly in conditionally or seed specific cytokinin-overproducing transgenic oilseed rape. However, no increase in seed yield was achieved. Cytokinin treatment increased yield in rice and potato plants but a positive effect was not reported in all cases (Pavlista, 1993; Paraye *et al.*, 1995; Caldiz, 1996).

Changing the catabolism of cytokinins is an alternative possibility to endogenously altering the concentration of the hormone. The first cytokinin deficient plants were generated by constitutive overexpression of *AtCKX* genes in transgenic tobacco and *Arabidopsis* plants (Werner *et al.*, 2001; Werner *et al.*, manuscript in preparation). The phenotypic analysis of these plants has yielded information about processes for which cytokinins are limiting and which they might, therefore, regulate. Cytokinin-deficient plants have slower shoot development, leading to dwarfed plants with only 5-15% shoot cells of wild type plants. In contrast, the root system is enlarged, due to faster growth and increased branching. Faster growth is causally linked to an enlarged population of meristematic cells. One function of cytokinin in the root meristem seems to be the regulation of the exit of cells from the meristem. Less cytokinin leads to an increased number of cell divisions before meristematic cells enter the phase of elongation growth. In contrast, in the shoot apical meristem and in the leaves, reduced cytokinin content causes fewer cell divisions. In these tissues, cytokinins are an indispensable promoting factor. Interestingly, the seed size of several lines of cytokinin-deficient plants was increased about twofold compared to wild type, due to a higher number of cells (Werner *et al.*, manuscript in preparation). These results suggest that local interference with cytokinin breakdown is an attractive possibility to achieve improved growth. It will be of particular interest to study the loss of function phenotypes of single *AtCKX* genes and combinations thereof. Currently, T-DNA insertion alleles are available for six of the seven *AtCKX* genes. A selection of these alleles is shown in Fig. 4.

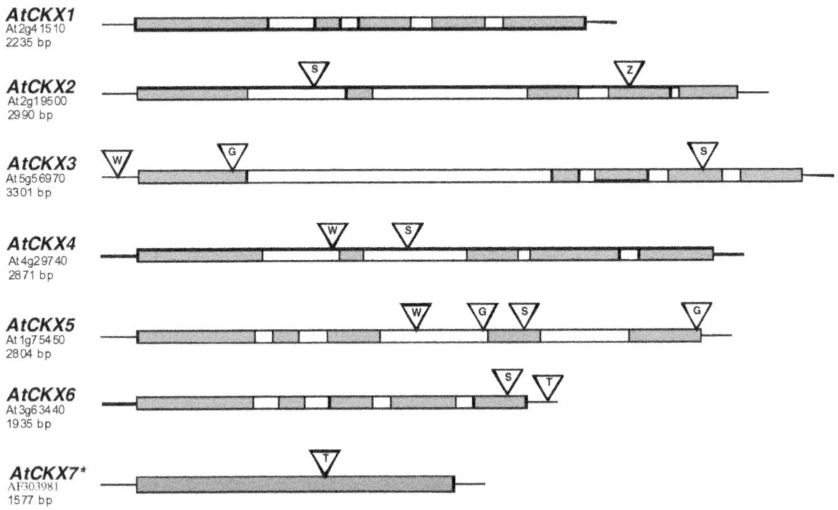

Figure 4.

Genomic structures of *AtCKX* genes and examples of insertion alleles. Exons are shown in gray, introns are white. The exact exon-intron structure of *AtCKX7* is not known. Numbering of genes is according to [6], the MIPS gene code (http://mips.gsf.de/proj/thal/) or the accession number is shown below the gene name. The gene size is in bp. Triangles indicate T-DNA insertion sites. T-DNA insertion lines are available from the collections at T, Torrey Mesa Research Institute (TMRI; http://www.tmri.org), S, Salk Institute Genomic Analysis (http://signal.salk.edu), G, GABI-KAT (www.mpiz-koeln.mpg.de/GABI-Kat/), and Z, ZIGIA (www.mpiz-koeln.mpg.de/~zigia/) respectively.

1.4. Outlook

The possibility of regulating plant growth and yield by modulating the endogenous cytokinin content has a high potential for biotechnological application in agriculture. In particular, altering sink-source relations through manipulation of the cytokinin content is a promising approach to improve yield attributes. For example, an improved root system like that described for cytokinin-deficient plants means improved acquisition of minerals and water, factors which are often limiting for plant growth and yield (Thompson *et al.*, 1994; Lynch, 1995). Currently we are analysing whether tissue-specific *CKX* gene expression will produce plants that have an enhanced root system but lack the detrimental effects on shoot growth. Furthermore, data from maize and oilseed rape indicate that the cytokinin content in reproductive tissues strongly influences seed set and, ultimately, yield. Enhancing the cytokinin content at the right time in the right tissue could be beneficial for these and other species. Although most of the work so far has been carried out by interfering with cytokinin metabolism, the recently discovered signal transduction pathway of cytokinins has similar potential for application in plant breeding. Interfering with signal transduction might permit the fine-tuning of selected downstream pathways.

REFERENCES

Bilyeu, K.D., Cole, J.L., Laskey, J.L., Riekhof, W.R., Esparza, T.J,. Kramer, M.D. and Morris, R.O. (2001) Molecular and biochemical characterization of a cytokinin oxidase from maize, *Plant Physiol.* **125**, 378-386.

Bouttier, C. and Morgan, D.G. (1992) Ovule development and determination of seed number per pod in oilseed rape (*Brassica napus* L.), *J. Exp. Bot.* **43**, 709-714.

Caldiz, D.O. (1996) Seed potato (*Solanum tuberosum* L.) yield and tuber number increase after foliar applications of cytokinins and gibberellic acid under field and glasshouse conditions, *Plant Growth Regul.* **20,** 185-188.

Cheikh, N. and Jones, R.J. (1994) Disruption of maize kernel growth and development by heat stress: Role of cytokinin/abscisic acid balance, *Plant Physiol.* **106**, 45-51.

D'Agostino, I.B., Deruere, J. and Kieber, J.J. (2000) Characterization of the response of the *Arabidopsis* response regulator gene family to cytokinin, *Plant Physiol.* **124**, 1706-1717.

De Bouille, P., Sotta, B., Miginiac, E. and Merrien, M. (1990) Hormones and pod development in oilseed rape (*Brassica napus*), *Plant Physiol.* **90**, 876-880.

Deruere, J.and& Kieber, J.J. (2002) Molecular mechanisms of cytokinin signaling, *J. Plant Growth Regul.* **21,** 32-39.

Dietrich, J.T., Kamínek, M., Blevins, D.G., Reinbott, T.M. and Morris, R.O. (1995) Changes in cytokinins and cytokinin oxidase activity in developing maize kernels and the effects of exogenous cytokinin on kernel development, *Plant Physiol. Biochem.* **33,** 327-336.

Ehneß, R. and Roitsch, T. (1997) Co-ordinated induction of mRNAs for extracellular invertase and a glucose transporter in *Chenopodium rubrum* by cytokinins, *Plant J.* **11**, 539-548.

Faiss, M., Zalubilová, J., Strnad, M. and Schmülling, T. (1997) Conditional transgenic expression of the *ipt* gene indicates a function for cytokinins in paracrine signaling in whole tobacco plants, *Plant J.* **12**, 401-415.

Galuszka, P., Frébort, I., Šebela, M., Sauer, P., Jacobsen, S. and Peč, P. (2001) Cytokinin oxidase or dehydrogenease? Mechanism of cytokinin degradation in cereals, *Eur. J. Biochem.* **268**, 450-461.

Gan, S. and Amasino, R.M. (1995) Inhibition of leaf senescence by autoregulated production of cytokinin, *Science* **270**, 1986-1988.

Guivarc'h, A., Rembur, J., Noin, M., Schmülling, T. and Chriqui, D. (2002) Localized expression of the *ipt* gene in the axillary buds of transgenic tobacco (*Nicotiana tabacum* L. cv. SR1) establishes a role for cytokinins in tuberization, *J. Exp. Bot.* **53,** 621-629.

Houba-Hérin, N., Pethe, C., D'Alayer, J. and Laloue, M. (1999) Cytokinin oxidase from *Zea mays*: Purification, cDNA cloning and expression in moss protoplasts, *Plant J.* **17**, 615-626.

Hwang, I. and Sheen, J. (2001) Two-component circuitry in *Arabidopsis* cytokinin signal transduction, *Nature* **413**, 383-389.

Hwang, I., Chen, H.-C. and Sheen, J. (2002) Two-component signal transduction pathways in *Arabidopsis*, *Plant Physiol.* **129**, 500-515.

Inoue, T., Higuchi, M., Hashimoto, Y., Seki, M., Kobayashi, M., Kato, T., Tabata, S., Shinozaki, K., Jones, R.J., Schreiber, B.M.N., McNeil, K., Brenner, M.L. and Fox, G. (1992) Cytokinin levels and oxidase activity during maize kernel development, in M. Kamínek, D.W.S. Mok and E. Zažímalová (eds.), *Physiology and biochemistry of cytokinins in plants*. SPB Academic Publishing, The Hague, pp. 29-32.

Jordi, W., Schapendonk, C.M., Davelaar, E., Stoopen, G.M., Pot, C.S., de Visser, R., van Rhijn, J.A., Gan, S. and Amasino, R.M. (2000) Increased cytokinin levels in transgenic P_{SAG12}-*IPT* tobacco plants have large direct and indirect effects on leaf senescence, photosynthesis and N partitioning, *Plant Cell Environ.* **23**, 279-289.

Kakimoto, T. (2001a) Identification of CRE1 as a cytokinin receptor from *Arabidopsis*, *Nature* **409**, 1060-1063.

Kakimoto, T. (2001b) Identification of plant cytokinin biosynthetic enzymes as dimethylallyl diphosphate:ATP/ADP isopentenyltransferases, *Plant Cell Physiol.* **42**, 677-685.

Kakimoto, T. (2001c) Biosynthesis and perception of cytokinins, Abstract book, 17th International Conference on Plant Growth Substances, p. 56.

Kamínek M., Trčková M., Motyka, V. and Gaudinová, A. (1994) Role of cytokinins in control of wheat grain development and utilization of nutrients, *Biol. Plant.* **36** (Suppl.), 15.

Lee, M. (1998) Genome projects and gene pools: New germ plasm for plant breeding?, *Proc. Natl. Acad. Sci. USA.* **95**, 2001-2004.

Lejeune, P., Prinsen, E., Van Onckelen, H. and Bernier, G. (1998) Hormonal control of ear abortion in a stress-sensitive maize, *Austr. J. Plant Physiol.* **25**, 481-488.

Letham, D.S. (1994) Cytokinins as phytohormones - sites of biosynthesis, translocation and function of translocated cytokinin, in D.W.S. Mok and M.C. Mok (eds.), *Cytokinins. Chemistry, Activity, and Function*, CRC Press, Boca Raton e.a., pp. 57-80.

Lohrmann, J., Buchholz, G., Keitel, C., Sweere, C., Kircher, S., Bäurle, I., Kudla, J. and Harter, K. (1999) Differentially-expressed and nuclear-localized response regulator-like proteins from *Arabidopsis thaliana* with transcription factor properties, *Plant Biol.* **1**, 495-506.

Lohrmann, J., Sweere, U., Zabaleta, E., Bäuerle, I., Keitel, C., Kozma-Bognar, L., Brennicke, A., Kudla, J., Schäfer, E. and Harter, K. (2001) The response regulator ARR2: A pollen-specific transcription factor involved in the expression of nuclear-encoded mitochondrial complex I genes, *Mol. Gen. Genomics* **265**, 2-13.

Lur, H. and Setter L. (1993) Role of auxin in maize endosperm development, *Plant Physiol.* **103**, 273-280.

Lynch J (1995) Root architecture and plant productivity, *Plant Physiol.* **109**, 7-13.

Martin, R.C., Mok, M.C. and Mok, D.W. (1999a) A gene encoding the cytokinin enzyme zeatin O-xylosyltransferase of *Phaseolus vulgaris*, *Plant Physiol.* **120**, 553-558.

Martin, R.C., Mok, M.C. and Mok, D.W. (1999b) Isolation of a cytokinin gene, *ZOG1*, encoding zeatin O-glucosyltransferase from *Phaseolus lunatus*, *Proc. Natl. Acad. Sci. USA.* **96**, 284-289.

Martin, R.C., Mok, M.C., Habben, J.E. and Mok, D.W. (2001) A maize cytokinin gene encoding an O-glucosyltransferase specific to *cis*-zeatin, Proc. Natl. Acad. Sci. U. S. A. **98**, 5922-5926.

McCabe, M.S., Garratt, L.C., Schepers, F., Jordi, W.J.R.M., Stoopen, G.M., Davelaar, E., van Rhijn, J.H.A., Power, J.B. and Davey, M.R. (2001) Effects of P_{SAG12}-*IPT* gene expression on development and senescence in transgenic lettuce, *Plant Physiol.* **127**, 505-516.

Mok, D.W.S. and Mok, M. (2001) Cytokinin metabolism and action, *Annu. Rev. Plant Physiol. Plant Mol. Biol.* **52**, 89-118.

Morgan, D.G., Keiller, D.R. and Prynne, A.O. (1983) Nitrogen, growth regulators and pod development in oilseed rape, in M.B. Jackson (ed.), *Interactions Between Nitrogen and Growth Regulators in the Control of Plant Development*, British Plant Growth Regulator Group, Monograph 9, pp. 75-86.

Morris, R.O., Blevins, D.G., Dietrich, J.T., Durley, R.C., Gelvin, S.B., Gray, J., Hommes, N.G., Kamínek, M., Matthews, L.J., Meilan, R., Reinbott, T.M. and Sapavedra-Soto, L. (1993) Cytokinins in plant pathogenic bacteria and developing cereal grains, *Austr. J. Plant Physiol.* **20**, 621-637.

Morris, R.O., Bilyeu, K.D., Laskey, J.G. & Cheikh, N.N. (1999) Isolation of a gene encoding a glycosylated cytokinin oxidase from maize, Biochem. Biophys. Res. Comm. **255**, 328-333.

Oliveira, I.C., Lam, H.M., Coschigano, K., Melo-Oliveira, R. & Coruzzi, G. (1997) Molecular genetic dissection of ammonium assimilation in *Arabidopsis thaliana*, Plant Physiol. Biochem. **35**, 185-198.

Paraye, P.M., Kandalkar, V.S., Paliwal, A.K. & Thakur, S.K. (1995) Effect of growth hormones under different fertility levels on grain yield and yield attributes of rice (*Oryza sativa*), Ind. J. Agric. Sci. **65**, 334-336.

Pavlista, A.D. (1993) Morphological changes and yield enhancement or Russet Burbank potatoes by AC 243654, *Am. Potato J.* **70**, 453-462.

Peltonen, S.P. (1997) Nitrogen fertilizer and foliar application of cytokinin affect spikelet and floret set and survival in oat, *Field Crop Res.* **49**, 169-176.

Reilly, J.M. and Fuglie, K.O. (1998) Future yield growth in field crops: What evidence exists?, *Soil Till. Res.* **47**, 275-290.

Riou-Khamlichi, C., Huntley, R., Jacqmard, A. and Murray, J.A. (1999) Cytokinin activation of *Arabidopsis* cell division through a D-type cyclin, *Science* **283**, 1541-1544.

Roeckel, P., Oancia, T. and Drevet, J.R. (1997a) Phenotypic alterations and component analysis of seed yield in transgenic *Brassica napus* expressing the *tzs* gene, *Physiol. Plant.* **102**, 243-249.

Roeckel, P., Oancia, T. & Drevet, J. (1997b) Effects of seed-specific expression of a cytokinin biosynthetic gene on canola and tobacco phenotypes, Transgenic Res. **6**, 133-141.

Rupp, H., Frank, M., Werner, T., Strnad, M. and Schmülling, T. (1999) Increased steady state mRNA levels of the *STM* and *KNAT1* homeobox genes in cytokinin overproducing *Arabidopsis thaliana* indicate a role for cytokinins in the shoot apical meristem, *Plant J.* **18**, 357-363.

Sakai, H., Honma, T., Aoyama, T., Sato, S., Kato, T., Tabata, S. and Oka, A. (2001) ARR1, a transcription factor for genes immediately responsive to cytokinins, *Science* **294**, 1519-1521.

Sakakibara, H., Suzuki, M., Takei, K., A Deji, A., Taniguchi, M. and Sugiyama, T. (1998) A response-regulator homologue possibly involved in nitrogen signal transduction mediated by cytokinin in maize, *Plant J.* **14**, 337-344.

Samuelson, M.E., Campbell, W.H. and Larsson, C.M. (1995) The influence of cytokinins in nitrate regulation of nitrate reductase activity and expression in barley, *Physiol. Plant.* **93**, 533-539.

Schmülling, T. (2002) New insights into the functions of cytokinins in plant development, *J. Plant Growth Regul.* **21**, 40-49.

Schmülling, T., Werner, T., Riefler, M., Krupková, E. and Bartrina y Manns, I. (2003) Structure and function of cytokinin oxidase/dehydrogenase genes of maize, rice, *Arabidopsis* and other species, *J. Plant Res., in press.*

Smiciklas, K.D. and Below, F.E. (1992) Role of cytokinin in enhanced productivity of maize supplied with ammonium and nitrate, *Plant Soil* **142**, 307-313.

Suzuki, T., Miwa, K., Ishikawa, K., Yamada, H., Aiba, H. and Mizuno, T. (2001) The *Arabidopsis* sensor kinase, AHK4, can respond to cytokinin, *Plant Cell Physiol.* **42**, 107-113.

Sweere, U., Eichenberg, K., Lohrmann, J., Mira-Rodado, V., Bäurle, I., Kudla, J., Nagy, F., Schäfer, E. and Harter, K. (2001) Interaction of the response regulator ARR4 with the photoreceptor phytochrome B in modulating red light signalling, *Science* **294**, 1108-1111.

Takei, K., Sakakibara, H. and Sugiyama, T. (2001) Identification of genes encoding adenylate isopentenyltransferase, a cytokinin biosynthesis enzyme, in *Arabidopsis thaliana*, *J. Biol. Chem.* **276**, 26405-26410.

Taniguchi, M., Kiba, T., Sakakibara, H., Ueguchi, C., Mizuno, T. and Sugiyama, T. (1998) Expression of *Arabidopsis* response regulator homologs is induced by cytokinins and nitrate, *FEBS Lett.* **429**, 259-262.

Thompson, J.D., Higgins, D.G. and Gibson, T.J. (1994) CLUSTAL W: improving the sensitivity of progressive multiple sequence alignment through sequence weighting, position-specific gap penalties and weight matrix choice, *Nucleic Acid Res* .**22**, 4673-4680.

Vision, T.J., Brown, D.G. and Tanksley, S.D. (2000) The origins of genomic duplications in *Arabidopsis*, *Science* **290,** 2114-2117.

Werner, T., Motyka, V., Strnad, M. and Schmülling, T. (2001) Regulation of plant growth by cytokinin, *Proc. Natl. Acad. Sci. USA* **98**, 10487-10492.

Yamada, H., Suzuki, T., Terada, K., Takei, K., Ishikawa, K., Miwa, K. and Mizuno, T. (2001) The *Arabidopsis* AHK4 histidine kinase is a cytokinin-binding receptor that transduces cytokinin signals across the membrane, *Plant Cell Physiol.* **42,** 1017-1023.

EXPLOITATION OF MECHANISMS REGULATING CYTOKININ LEVELS TO IMPROVE CEREALS

M. KAMÍNEK[1], B. ŠOLCOVÁ[1], M. TRČKOVÁ[2], V. MOTYKA[1], S. DASKALOVA[3] AND M.C. ELLIOTT[3]
[1]Institute of Experimental Botany, Academy of Sciences of the Czech Republic, Rozvojová 135, Praha 6, Czech Republic
[2]Research Institute of Crop Production, Drnovská 505, Praha 6, Czech Republic
[3]The Norman Borlaug Institute for Plant Science Research, De Montfort University, Scraptoft, Leicester LE7 9SU, U.K.
[*]Corresponding author (kaminek@ueb.cas.cz)

1. INTRODUCTION

Plant growth and development are regulated by phytohormones. Two classes of phytohormones, auxins and cytokinins, play a key role in the regulation of cell division and cell differentiation. Cell cycle progression is regulated at checkpoints, which control the G1/S and G2/M transitions (see Frank and Schmülling, 1999). This enables cell cycle activity to be integrated into complex morphogenetic pathways (Meijer and Murray, 2001). The classical experiments of Skoog and Miller (1957) showed that the concentration ratio of exogenous auxin and cytokinin determines specific morphogenic responses. Such responses can also be achieved by manipulation with intracellular hormone levels in transgenic plants. Hence, changes of auxin/cytokinin ratios and consequent morphogenetic responses can be achieved by expression of genes that encode enzymes involved in auxin and cytokinin metabolism (Akiyoshi *et al.*, 1983; Estruch *et al.*, 1991; Klee and Lanahan, 1995; Faiss *et al.*, 1997, 2001; Gaudin *et al*, 1994). With regard to the current state of knowledge of the biosynthetic and metabolic pathways of the two hormones and mindful of the cloning of the relevant genes (for review see Mok and Mok, 2001; Haberer and Kieber, 2002), transgenic approaches seem to be more readily accessible for manipulation of cytokinin levels in plants than that for the control of auxin levels.

2. MECHANISMS REGULATING CYTOKININ LEVELS

Metabolic regulation of cytokinin levels in plant cells must cause rapid and transient

I. Macháčková and G.A. Romanov (eds.), Phytohormones in Plant Biotechnology and Agriculture, 109-128

changes of hormone concentration in response to environmental and developmental stimuli followed by re-establishment of hormonal homeostasis. While the transient change in cytokinin concentrations in relation to that of auxin is essential for induction of such specific morphogenetic processes, as e.g. adventitious bud formation and tillering, the re-establishment of cytokinin homeostasis is essential for the continuation of initiated events. We have proposed a complex network of mechanisms regulating cytokinin levels, which satisfies these requirements (Kamínek *et al.*, 1997). An updated, advanced version is presented in Fig. 1. Evidence for its operation in plants is presented below.

2.1. Upregulation of cytokinin levels

Habituation of tissues for cytokinins is frequently associated with elevated cytokinin levels which are presumed to be responsible for tumorigenic growth (Ooms *et al.*, 1981). Meins (1989) proposed that the habituated state is maintained by a positive feedback loop in which cytokinins either induce their own biosynthesis or inhibit their own degradation. According to this hypothesis the neoplastic growth of competent cells is maintained as long as the concentration of intracellular cytokinin remains above the threshold level.

The ability of cytokinins to induce their own accumulation in competent cells was demonstrated by application of N^6-benzyladenine (BA) or the synthetic urea-type cytokinin thidiazuron, which resulted in an increased of level of endogenous isoprenoid cytokinins in cytokinin-dependent cells or tissue cultures of soybean (Thomas and Katerman, 1986), tobacco (Hansen *et al*,. 1987; Motyka *et al.*, 2003) and sugar beet (Vaňková *et al.*, 1991). Experiments with immobilized cytokinin-dependent tobacco cells in a column-flow-through system showed that there is a very short lag period between exogenous cytokinin application and increase endogenous cytokinin levels (Vaňková *et al.*, 1987). Cytokinin accumulation may be caused by stimulation of cytokinin biosynthesis, inhibition of cytokinin degradation by cytokinin oxidase/dehydrogenase (CKX) and/or by cytokinin inactivation by N-glycosylation. The recent cloning of higher plant isopentenyl transferase genes (Kakimoto, 2001; Takei *et al.*, 2001a; Sun *et al.*, 2003) may allow testing of involvement of *de novo* cytokinin biosynthesis in this process.

Cytokinin autonomy may be due to cytokinin accumulation but it can also be caused by other mechanisms operating at the cytokinin signal transduction pathway. These include expression of cytokinin receptors with combined sensor and regulator functions (Kakimoto, 1996; Inoue *et al.*, 2001) and cyclinD3, which may mediate the cytokinin signal regulating cell cycle progression (Riou-Khamlichi *et al.*, 1999). The potential for exploitation of these mechanisms for improvement of crop plants deserves further attention.

2.2. Down-regulation of cytokinin levels

2.2.1. Cytokinin degradation by cytokinin oxidase/dehyrogenase

Isoprenoid cytokinin bases and ribosides bearing unsaturated and non-glycosylated side chain serve as substrates of CKX (for review see Armstrong, 1994; Hare and Van Staden, 1994), which actually behaves as dehydrogenase (Galuszka *et al.*, 2001). The enzyme irreversibly degrades cytokinins by removal of the N^6 isoprenoid side chain. There are two molecular forms differing in peptide glycosylation, pH optima and chromatographic properties (Kamínek and Armstrong, 1990; Motyka *et al.*, 2003). The activity of CKX was significantly enhanced *in vivo* following application of exogenous cytokinins (Chatfield and Armstrong, 1986; Kamínek and Armstrong, 1990; Motyka and Kamínek, 1990, 1992; Auer *et al.*, 1999) or following overexpression of the *ipt* gene with a consequent increase of endogenous levels of isoprenoid cytokinins (Motyka *et al.*, 1996, 2003). Both CKX substrate and non-substrate cytokinins as well as cytokinin-active aromatic derivates of urea were capable of increasing the CKX activity *in vivo* within a few hours after their application (Chatfield and Armstrong, 1986; Motyka and Kamínek, 1990, 1992). This indicates that cytokinins enhance CKX activity either by reacting with CKX repressors or by affecting the conformation of CKX as allosteric enzyme. Moreover, isoprenoid cytokinins may act as substrate inducers of the enzyme (see Kamínek *et al.*, 1997). In accordance with the reported presence of a signal sequence in most CKX homologues (Bilyeu *et al.*, 2001), that targets CKX toward excretion most of the increased CKX activity in response to elevated cytokinin levels was associated with the glycosylated form of CKX in tobacco cell suspensions, that was found to be secreted in the cell exterior (Motyka *et al.*, 2003). It seems that CKX can regulate not only intracellular levels of cytokinins but also the access of cytokinins to plant cells from the apoplast. Only CKX substrate cytokinins are subjects of this regulation. This means that cytokinin–*O*-glucosides and cytokinin nucleotides as well as cytokinins bearing saturated and aromatic N^6 side chains (dihydrozeatin- [DHZ]- and BA-type cytokinins, respectively), are resistant to CKX attack. Nevertheless, because the isoprenoid cytokinins bearing an unsaturated and non-glycosylated side chain are very early products of cytokinin biosynthetic pathway(s) CKX evidently plays a crucial role in down-regulation of the broad spectrum of isoprenoid cytokinins in plants. The discovery that overexpression of CKX in transgenic tobacco plants was accompanied by a significant decrease of cytokinin levels, including those which are not substrates of the enzyme (Werner *et al.*, 2001), supports this view.

2.2.2. Cytokinin glycosylation

The physiological activity of cytokinins can be decreased or eliminated by their conjugation with sugar and alanine moieties. Generally, glucosylation of cytokinins on the purine ring at positions 7 and 9, which is catalyzed by more or less specific glucosyltrasferase (Ensch *et al.*, 1980) and alanylation at position 9 catalyzed by cytokinin alanine synthetase (Entsch *et al.*, 1983) results in irreversible inactivation. *N*- glucosylation and alanylation are to date the only ways of inactivation of cytokinins that are resistant to CKX, i.e. DHZ- and BA-type cytokinins. Attachment of the a glucosyl moiety to the purin ring at position 3 and glucose or xylose at the terminal methyl group of the N^6-isopentenyl side chain of Z (*O*-glycosylation) yields storage-type cytokinin conjugates, which can be converted back to the active ompound after their hydrolysis catalyzed by β-glucosidase (for review see Brzobohatý, 1994; Vaňková, 1999; Mok and Mok, 2001).

Conversion of cytokinins to the corresponding *N*- and *O*-glucosides is evidently one of the mechanisms regulating levels of physiologically active cytokinins in plants. This opinion is supported by rapid conversion of [^{3}H] labeled Z, DHZ and BA to the corresponding [^{3}H]7-glucosides in radish hypocotyls and roots and by the predominant conversion of [^{3}H]Z and [^{3}H]ZR to the corresponding *O*-glucosides in poplar leaves (Entsch *et al.*, 1979, 1980; McGaw *et al.*, 1985; Blagoeva *et al.*, 2002). Conversion of Z to ZOG seems to be important for its transport via xylem sap and its role in stimulation of chlorophyll formation in leaves (Kato *et al.*, 2002). An increase in accumulation of Z-*O*- and *N*-glucosides was also recorded in response to application of BA or transient expression of the *ipt* gene in tobacco calli (Zhang *et al.*, 1995; Motyka *et al.*, 1996; Redig *et al.*, 1997) and suspension-cultured cells (Motyka et al., 2003). Similarly Z7G was found to accumulate in tobacco plants after heat-shock induction of *hsp-ipt* expression (Medford *et al.*, 1989). While the cytokinin-*O*-glucosyl- and *O*-xylosyltransferases have a very high affinity for Z-type cytokinins the glucosylation of cytokinins at position 7 catalyzed by 7-glucosyltransferases is much less specific (Letham *et al.*, 1978; Mok and Martin, 1994; Mok and Mok, 2001). Moreover, conversion of ZOG and ZROG to Z and ZR, respectively, was found to be catalyzed by a substrate-specific β-glucosidase in maize roots (Brzobohatý *et al.*, 1993). Interestingly, adenine applied at a in high concentration to suspension cultured coffee cells and to derooted radish seedlings inhibits glucosylation cytokinins at poisons 7 and 9 due to its own rapid glucosylation at position 7 (Baumann *et al.*, 1994; respective Blagoeva *et al.*, 2002). Before cytokinins were discovered Skoog and Tsui (1948) noted, that adenine, applied at a high concentration is capable of inducing cell division and budding in tobacco stem sections. This observation can be explained on the basis of competitive protection of endogenous cytokinins against their *N*-glycosylation. These findings support the idea that cytokinin *O*- and *N*-glucosylation is involved in down-regulation of active cytokinins and maintenance of cytokinin homeostasis in plant cells.

2.2.3. Cytokinin excretion

The uptake and excretion of cytokinins by plant cells represent another mechanism controlling intracellular concentrations of cytokinins. Petrášek *et al.* (2002) reported that changes in actual intracellular levels of different cytokinins during the subculture of cytokinin-autonomous tobacco cells suspension in culture was proportional to their levels in the cultivation medium during the whole subculture period. The physiologically active cytokinin bases are, thanks to their lipophilic character, the main trans-membrane transporting form of cytokinins. Their conversion to inactive and less mobile cytokinin glucosides and ribotides represents a mechanism for their entrapment within the cell and different cell compartments including plastids and vacuoles (Laloue *et al.*, 1981; Kamínek *et al.*, 1997).

2.2.4. Down- regulation of cytokinins by auxin

There are two different types of auxin and cytokinin interaction. Although the two phytohormones act synergistically in supporting cell division they also regulate many other processes associated with organelle/cell/tissue differentiation, such as for example plastid differentiation (Miyazawa *et al.*, 1999), bud formation (Skoog and Miller, 1957), apical dominance (Wickson and Thimann, 1958) and seed development (Morris, 1997), in an antagonistic manner. There are many indications that metabolic down regulation of cytokinins by auxins is involved in the latter type of interactions (for review see Coenen and Lomax, 1997). A dramatic effect of auxin on endogenous cytokinins was found in tobacco tumors initiated with *Agrobacterium* plasmids. Mutation of the auxin biosynthetic genes resulted in a dramatic increase in the cytokinin content of the crown galls that was associated with stimulation of shoot formation (Akiyoshi *et al.*, 1983; Rüdelsheim *et al.*, 1987; Ishikawa *et al.*, 1988; McGaw *et al.*, 1988). Similar, but less pronounced, trends were found in crown galls of carrot induced by a modified Ti plasmid (Ishikawa *et al.*, 1988). Corresponding effects on cytokinin levels were recorded when different auxins were applied to cultured tobacco stem pith and calli (Palni *et al.*, 1988 and Hansen *et al.*, 1987; Zhang *et al.*, 1995, respectively) and cells (Vaňková *et al.*, 1992; Zažímalová *et al.*, 1996). There are conflicting results concerning mechanisms of regulation of cytokinins by auxin. Palni *et al.* (1988) found that the accumulation of degradation products of [^{3}H] ZR in tobacco pith was proportional to the auxin (NAA) concentration in the cultivation medium. They also found moderate increase of CKX activity by NAA when assayed *in vitro* using protein preparations from maize kernels. These observations led the authors to conclude hat auxin effect on cytokinin levels is mediated by enhancement of CKX activity. Similar results with transgenic tobacco tissue expressing *ipt* gene were reported by Zhang *et al.* (1995) from the same laboratory. They observed that exogenously applied NAA enhanced degradation of [^{3}H] ZR to [^{3}H]-labeled adenine derivatives *in vivo* and increased *in vitro* degradation of different isoprenoid cytokinins by partly purified CKX from tobacco tissue cultures. In contrast, we did not find any effect of three different auxins on CKX activity in tobacco either after their application to intact calli *in vivo* or to the enzyme assay solution *in vitro* (Motyka and Kamínek, 1990, 1992). To make the problem more confusing Eklöf *et al.* (1997) reported that transgenic tobacco leaves overexpressing the auxin biosynthetic genes contained reduced levels of cytokinins and exhibited lower CKX activity when assayed *in vitro*. All reported effects of auxins on CKX activity were too moderate to be responsible for the dramatic reduction of cytokinins levels. It is evident that the role of CKX in regulation of cytokinins by auxin needs further research. It seems probable that auxin regulates cytokinins by reducing the rate of their biosynthesis as reported for transgenic tobacco expressing the *ipt* gene where the levels of cytokinins as well as NAA application (Zhang *et al.*, 1996).

2.2.5. Role of cytokinin-binding proteins

Cytokinin-binding proteins (CBPs) have been found in seeds of various cereal species (for review see Brinegar, 1994). CBPs from different cereal species show very different immunochemical cross-reactivities but very similar binding of different adenine- and urea-type cytokinins. Som of their features, for example very high concentrations in embryos (up to 10% of embryo soluble protein) and relatively low binding of isoprenoid cytokinins as compared to cytokinins bearing an N^6 aromatic side chain, suggest that they do not function as cytokinin receptors in the usual sense of that term (Brinegar *et al.*, 1985; Kamínek *et al.*, 2003). Recent identification of a highly active cytokinin, N^6- (3-hydroxybenzyl) adenosine, in wheat grains (Kamínek *et al.*, 2000) indicates that these CBPs may function by regulating the levels of free BA-type cytokinins by their immobilization during the late phases of seed development, thus preventing premature stimulation of embryo cell divisions, and by their release during grain germination.

2.3. Proposed model of regulation of cytokinin content in relation to that of auxin

The proposed network of mechanisms regulating cytokinin levels is summarized in Fig. 1.

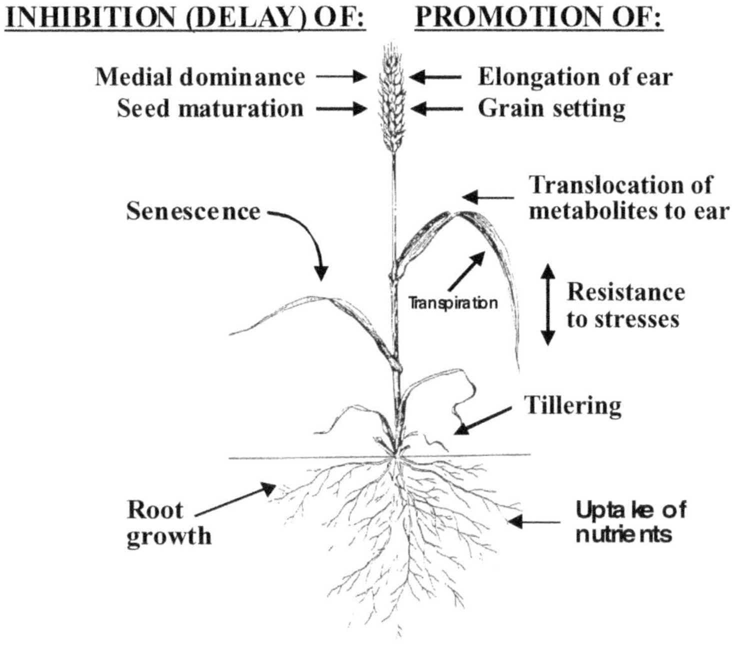

Figure 1.

According to this scheme the accumulation of cytokinins in plant cells can be increased by their uptake from the cell exterior, by induction of their biosynthesis in competent cells or by decrease of their metabolic inactivation/degradation. Once the cytokinin accumulation achieves a threshold level it can be further enhanced in competent cells by a positive feed–back mechanism. In this way the cytokinin/auxin ratio can be increased to the level required for induction of "cytokinin specific" physiological and morphological events as e.g. plastid differentiation, adventitious bud formation, tillering and lateral bud outgrowth. Accumulation of cytokinins activates processes leading to their down-regulation, including intracellular and extracellular degradation by CKX, reversible or irreversible inactivation by *O*- and *N*-glucosylation and (in seeds of cereals) by their reversible immobilization by specific CBPs (Kamínek *et al.*, 2003). Moreover, the access of cytokinins in plant cells and their uptake can be decreased by enhanced secretion of CKX to the cell exterior. In this way the cytokinin/auxin ratio can be decreased to a level corresponding to the hormone homeostasis that is required for further progression of induced physiological and morphological events. According to this scheme the accumulation of auxin either due to the increase of its influx/efflux ratio and intracellular biosynthesis inhibits cytokinin accumulation and decrease the cytokinin/auxin ratio. This may be favorable for induction of "auxin specific" physiological and developmental events, such as for example cell enlargement, root formation and apical dominance. Mechanisms of down regulation of auxin levels are much less understood, however, reduction of auxin influx, formation of inactive IAA conjugates and their translocation to storage compartments may be involved in this process (for review see Zažímalová and Napier, 2003).

3. CEREAL IMPROVEMENT

Cytokinins, in co-operation with auxins, control developmental processes, which are decisive for productivity of crop plants. These processes include tillering, formation and maintenance of photosynthetic apparatus, grain setting and filling as well as acquisition of essential resources, including carbon skeletons and mineral nutrients and their efficient assimilation and allocation. Crop-related physiological and morphogenetic processes regulated by cytokinins in cereals are shown in Fig. 2.

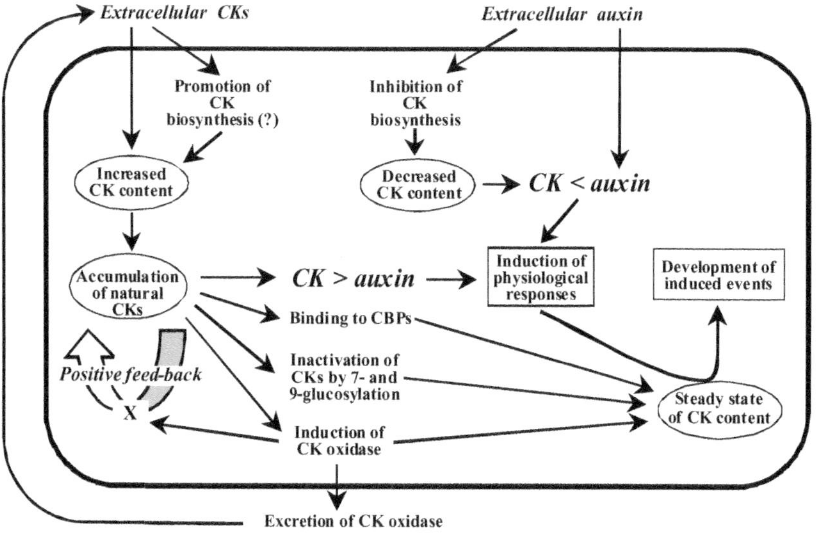

Figure 2. Promotion and inhibition of different physiological process of cereal plants by cytokinins.

3.1. Control of dominance

Auxin/cytokinin concentration ratios play a key role in regulation of various dominance in plants. Cytokinins release lateral buds from dominance of the apical bud and stimulate tillering. This was proved by application of cytokinins to wheat (Shanahan *et al.*, 1985), oat (Harrison and Kaufman, 1980), barley (Ryschka and Leike, 1989; Kalendová *et al.*, 1999) and rice (Buu and Chu, 1983) and it well correlates with the changes in endogenous cytokinins (Harrison and Kaufman, 1983). At the level of the ear, cytokinins suppress the dominance of the early developed grains in the middle part of the ear (so-called"medial", actually "semibasal" dominance) thus enhancing development of grains in the distal parts, which would otherwise be either poorly developed or aborted. This was was reported for barley (Williams and Cartwright, 1980), wheat (Trčková *et al.,* 1992) and maize (Cheikh and Jones, 1994; Dietrich *et al.,* 1995). Cytokinins also suppress the dominance of peripheral grains within the spike by supporting the development of the retarded middle grain. Moreover, application of cytokinins to maize and wheat at pollination and anthesis, respectively, increased the number of fully developed grains per ear by reducing the number of aborting ones. It also ameliorated the detrimental effects of heat and limited nitrogen supply, respectively, on grain abortion (Trčková *et al.*, 1992; Cheikh and Jones, 1994; Dietrich *et al.*, 1995). The effect of an exogenous cytokinin, N^6- (3-hydroxybenzyl) adenosine, on all these characters in wheat is shown in Table 1.

*Tab. 1. Effects of exogenous cytokinin, N^6-(3-hydroxybenzyl)adenosine on grain yield and number of grains and their distribution within the plant, ear and spikelets. Wheat plants (Triticum vulgare L. cv Munk) were grown in hydroponics under 16h{!*h photoperiod }flux density 300µmol m^{-1} sec^{-1}) at 20°C/15°C in nutrient solution containing optimum concentration of nitrate (100 µM NO_3^-). Beginning from day 34th the NO_3^- was maintained at a supraoptimal concentration (70 µM) using a NO_3^- selective electrode controlled supply system. The highly active cytokinin N^6-(3-hydroxybenzyl)adenosine was applied as a spray (3 µM) after the first anther was formed on the main stem, i.e. to plants 66 d old. Grains were collected when plants reached full maturity (98 d after germination).*

Nitrate concentration	3OHBAR (µM)	Spikelet number	Grain number	dry weight (g)	Grains per spikelet	Dry weight mg/1 grain
100 µM:						
Main stem	0	16 ± 1.2	46 ± 2.8	1,37 ± 0.2	2,88 ± 0.1	29,77 ± 0.9
	3	17 ± 1.4	59 ± 3.1	1,95 ± 0.1	3,47 ± 0.3	33,08 ± 2.2
Tillers	0	74 ± 2.1	173 ± 3.1	6,06 ± 0.3	2,32 ± 0.10	35,06 ± 2.2
	3	80 ± 3.0	184 ± 5.9	6,32 ± 0.3	2,30 ± 0,03	34,33 ± 2.4
Whole plant	0	90 ± 2.8	219 ± 7.3	7,43 ± 0.4	2,43 ± 0.2	33,90 ± 1.9
	3	97 ± 3.1	243 ± 8.1	8,27 ± 0.5	2,50 ± 0.1	34,03 ± 2.4
70 µM:						
Main stem	0	13 ± 1.2	24 ± 1.4	1,02 ± 0.1	1,85 ± 0.2	42,32 ± 0.3
	3	16 ± 0.8	34 ± 1.8	1,37 ± 0.2	2,13 ± 0,3	40,22 ± 0.4
Tillers	0	73 ± 1.8	143 ± 3.4	4,95 ± 0.2	1,96 ± 0.1	34,61 ± 0.5
	3	80 ± 2.1	173 ± 4.2	5,24 ± 0.4	2,16 ± 0.1	30,26 ± 0.3
Whole plant	0	86 ± 2.4	167 ± 3.1	6,02 ± 0.4	1,942 ± 0.4	35,63 ± 0.2
	3	96 ± 3.4	207 ± 3.7	6,6 ± 0.3	2,156 ± 0.2	31,90 ± 0.3

(± SE)

3.2. Sink size and sink strength

Stimulation of grain setting by cytokinins is closely related to sink formation. Reproductive sink strength in cereals can be defined as their potential capacity for starch accumulation (Jones *et al.* 1996). At the whole plant level it may be increased by enhancement of the number of amyloplasts per endosperm cell, the number of endosperm cells per seed and the number of seeds per plant. Cytokinins increased the number of seeds per plant by stimulation of formation and survival of secondary fertile tillers and/or grain setting within the ear of barley (Williams and Cartwright, 1980; Mishra and Gaur, 1985), wheat (Trčková *et al.*, 1992) and maize (Cheikh and Jones, 1994; Dietrich *et al.*, 1995). The transient accumulation of cytokinins in young grains soon after pollination found in wheat (Jameson *et al.*, 1982; Kamínek *et al.*, 1999; Banowetz *et al.*, 1999) and maize (Dietrich *et al.*, 1995) correlates with the beginning of cellularization of wheat endosperm (Kamínek *et al.*, 1999) and with the maximum of mitotic activity in maize grains (Dietrich *et al.*, 1995). This indicates that cytokinins are involved in stimulation of cell division at an early stage of endosperm development. Direct evidence supporting the cytokinin effect on the number of endosperm cells is not available but application of cytokinin to cereals at the anthesis slightly increased the grain mass of wheat (Caldiz *et al.*, 1991; Trčková *et al.*, 1992), maize (Cheikh and Jones, 1994; Dietrich *et al.*, 1995) and winter rye (Hradecká and Petr, 1992). Stimulation of amyloplast formation by cytokinins, which favours starch deposition, was reported for some other plant materials (Luštinec *et al.*, 1984). According to Borkovec and Procházka (1989) stimulation of sink formation by cytokinins is associated with an increase of the rate of [^{14}C] sucrose transport from stems to spikes during the filling stage of wheat seeds.

3.3. Chloroplast development and photosynthesis

Fast development of photosynthetic apparatus and its maintenance in active state as long as possible is favorable for enhancement of productivity of annual crop plants (Thomas and Howarth, 2000). There are numerous reports of the co-action of cytokinins and light in plastid multiplication and differentiation (see Reski and Abel, 1992; Reski, 1994, and references therein). By supporting plastid development cytokinins stimulate formation of photosynthetic apparatus and enhance its photosynthetic productivity.

Moreover, cytokinins delay leaf senescence and prolong the period of leaf photosynthetic activity (Van Staden *et al.*, 1988, Singh *et al.*, 1992; Gan and Amasino, 1995). Leaf senescence, which is a type of programmed cell death, is under the control of very complex internal mechanisms involving multiple pathways (Noodén *et al.*, 1997). In cereals and other monocarpic plants the ability of cytokinins to delay senescence seems to be based on their pleiotropic effects on senescence–related processes. These include slowing down the degradation of enzymes involved in photosynthetic metabolism (Wingler et al., 1998), decrease of activities of proteases (Noh and Amasino, 1999; Li *et al.*, 2000), lipase and lipoxygenase and inhibition of respiration (for review see Yoshida, 2003). Cytokinins stimulate stomata opening and enhance transpiration-driven and xylem-mediated nutrient supply to leaves of wheat and oat especially with nitrogen (Badenoch-Jones *et al.*, 1996), which is known to slow down the leaf senescence (Herzog, 1981). Application of cytokinins to intact oat plants after anthesis significantly delayed senescence of the flag leaf (Mishra and Gaur, 1985). Similarly, accumulation of cytokinins in tobacco leaves following expression of a cytokinin biosynthetic *ipt* gene driven by a heat shock or senescence-inducible promoter resulted in very significant delay of leaf senescence (Smart *et al.*, 1991; Gan and Amasino, 1995). Acceleration of *Petunia* corolla senescence by ethylene was found to be preceded by inactivation of isoprenoid cytokinins by their metabolic *O*-glucosylation and degradation (Taverner *et al.*, 1999). Correspondingly, senescence of wheat plants induced by their transfer to darkness was associated with a decline in Z-type cytokinins (Banowetz, 1997).

3.4. Stress tolerance

Cytokinins increase the tolerance of plants to stress, especially that caused by heat, nutrient starvation and salinity. Concentration of cytokinins in xylem exudate was sharply decreased in maize plants exposed to low temperature and well correlated with reduction of shoot growth (Atkins *et al.*, 1973). Pretreatment of wheat root with kinetin solution prevented root damage after their exposure to heat shock (45°C, 2 min) (Skogquist, 1974). Endogenous cytokinin levels in maize kernels cultured *in vitro* sharply declined after their transfer for 4 d to 8 d from 25°C to 35°C. This decline was accompanied by reduction of dry mass accumulation. Stem infusion with BA significantly reduced seed abortion and increased the grain mass of intact heat-stressed maize plants (Cheikh and Jones, 1994).

Salinity stress inhibits transport of cytokinins and nitrate by blocking K-shuttle activity (Cruz *et al.*, 1995). Cytokinin levels were decreased in roots and shoots of salt resistant barley plants shortly after addition of NaCl (65 m*M*) to the nutrient solution. BA applied to the nutrient solution at concentration 10μ*M*, maintained the growth parameters by preventing of decrease of endogenous Z, N^6-isopentenyladenine (iP) and their ribosides. However, the inhibitory effect of salinity on growth of salt-sensitive barley plants preceded the lowering of cytokinin levels in salt-sensitive cultivars. This implies that mediation of growth responses of plants to salinity by cytokinins depends on plant genotype (Kuiper *et al.*, 1989).

3.5. Interactions of nitrate and cytokinins in regulation of cytokinin and nitrate levels

There are several reports about the accumulation of cytokinins in roots and shoots in response to plant re-supply with nitrate, which is the major source of inorganic nitrogen for most of higher plants. Replenishment of nitrogen supply to roots was followed by an increase in the cytokinin levels in birch seedlings (Horgan and Wareing, 1980), *Plantago major* roots and shoots (Kuiper and Staal, 1987), barley roots (Samuelson and Larson, 1993) and *Urtica dioica* shoots (Beck, 1996).

Maize plants responded to re-supply of nitrate to roots in a spatially and time dependent manner. Levels of endogenous cytokinins were sequentially increased in roots, xylem sap and shoots. The first cytokinin accumulated in roots within 1 h after nitrate application was iP ribotide (Takei *et al.*, 2001b, 2002), which is believed to be the primary product of cytokinin biosynthesis (Kakimoto, 2001; Takei *et al.*, 2001a), preceding accumulation of other iP- and Z-type cytokinins. Moreover, the increase of Z and ZR content in roots was associated with a decrease of storage Z-*O*-glucoside indicating not only that the increase of cytokinin biosynthesis but also release of cytokinins from inactive storage conjugates is responsible for the nitrate-induced accumulation of active forms of cytokinins. After cytokinin accumulation in roots an increase in levels of ZR, a transport form of cytokinin, was found in the xylem sap (Takei *et al.*, 2001b).

It has been frequently suggested that cytokinins in addition to nitrate could represent another root-to-shoot signal communicating nitrogen availability (for review see Beck, 1996; Takei *et al.*, 2002b). The physiological significance of such a long-distance signal was demonstrated by Yong *et al.* (2000) who reported that higher nitrogen nutrition increases the delivery of cytokinins per unit of leaf area in cotton. Recent reports on a nitrogen-responsive gene encoding the response regulator ZmPRR1, which is induced in detached maize leaves by Z but not by nitrate (Sakakibara *et al.*, 1998; Takei *et al.*, 2001b), suggests functioning of cytokinins as long-distance signals of nitrogen availability. In this respect it is interesting that in our experiments nitrate alone or in combination with glucose was not able to delay senescence of detached leaves of oat and wheat unless a cytokinin was present (results not shown).

Cytokinins may not only signaling nitrogen availability to the shoots but also enhance nitrate uptake and translocation by activation of meristematic and reproductive sinks. Cytokinin enhanced strength of sinks seems to be a driving force for N uptake, assimilation and allocation within the plant. Indeed, application of a cytokinin to leaves of wheat grown in nutrient solution significantly increased the net nitrate uptake during intensive stem elongation. Moreover, it slowed down the natural decline of these parameters after anthesis. Cytokinin was more efficient in improvement of N economy and grain yield when applied to plants grown under limited nitrate supply (Trčková *et al.*, 1992; Trčková and Kamínek, 2000).

The affect of cytokinins on nitrate uptake is accompanied by stimulation of chloroplast development and suppression of chloroplast ageing that supports supply of photoassimilates and energy required for inorganic nitrogen assimilation. Cytokinins also also induce nitrate reductase (NR) independently of nitrate in many plant species (for review see Gaudinová, 1999). Just as ZR was the most active in induction of the nitrogen-responsive gene *ZmPRR1* (Takei *et al.*, 2001b) also it was most active in induction of NR in *Chenopodium rubrum* cells (Peters *et al.*, 1995) indicating possible role of this transport form of cytokinin in root to shoot signaling in higher plants. There are indications that cytokinins and nitrate affect expression of different genes encoding NR by different mechanisms. In contrast to nitrate cytokinins did not increase nitrite reductase activity and induction of NR by nitrate was not prevented by application of anticytokinins in *Agrostemma githago* embryos (Dilworth and Kende, 1974; Schmerder and Borriss, 1986). Two different genes encoding NR, *Nia1* and *Nia2,* were described in *Arabidopsis thaliana* (Wilkinson and Crawford, 1993; Cheng *et al.*, 1986), However, only expression of the minor *Nia1* was enhanced by cytokinin (Yu *et al.*, 1998).

4. MANUPULATION OF CYTOKININ LEVELS

Most of the physiological processes, which affect productivity and quality of crop plants, are spatially and developmentally determined. This means that the levels of hormones should be changed in the right place (cell, tissue, organ) and at the right stage of plant development - e.g. a transient increase of cytokinin levels in the basal nodes of stems at the beginning of stem elongation and in the flag leaf before decline of its photosynthetic activity may stimulate stem tillering (Harrison and Kaufman, 1983) and delay leaf senescence (Gan and Amasino, 1995), respectively. Application of plant growth regulators by spraying to the whole plant, which is commonly used in the agricultural practice, allows delivering of growth regulators at the appropriate stage of plant development but does not enable their targeting. Interestingly, processes and structures, which are developmentally programmed to respond to a specific plant hormone at certain stage of plant development are generally more sensitive to the particular hormone than the other non-responsive ones and the whole-plant application of corresponding plant regulators is frequently effective. Nevertheless, there is a risk of side effects, such as inhibition of root development when cytokinins are splashed down into the rhizosphere. Moreover, the large-scale application of such chemicals is a costly operation.

Much more sophisticated and potentially efficient is manipulation of cytokinin levels by regulation of expression of genes encoding enzymes involved in cytokinin biosynthesis and metabolism driven by appropriate tissue-specific and/or developmentally inducible promoters.

Site and/or development-specific accumulation of cytokinins can be achieved by coupling of cytokinin biosynthetic genes to promoters of genes which are expressed at early phase of endosperm development, e.g. genes encoding the starch branching enzyme (Zhang *et al.*, 2002), or at the very early stage of leaf senescence, e.g. genes encoding the senescence-associated protein hydrolases (Gan and Amasino, 1995; Weaver *et al.*,1998). In this way the cell division and amyloplast differentiation in endosperm can be achieved together with prolongation of leaf photosynthetic activity. Alternatively, antisense DNA technology can be used to down-regulate CKX activity and preserve endogenous cytokinins. Similarly down-regulation of auxin may reduce dominance of the main stem and support tillering or reduce the medial dominance in the ear to prevent grain abortion at the distal regions of the ear. The fast progress of cloning of the relevant genes, which allows replacement so far used bacterial genes with the higher plant ones, is very promising in this respect. Identification and characterization of suitable promoters will be helpful for full exploitation of this approach. Different cereal species and even different lines of the same species may differ in the source-sink activities and source-sink-communication, which determine the inherited strategy of monocarpic plants for effective use of nutrients and metabolites for seed formation. Our experiments showed that the expression of a P_{SAG}-driven bacterial cytokinin biosynthetic gene (*ipt*) in wheat leaves increased cytokinin levels in the flag leaf, delayed its senescence and increase grain yield (Daskalova *et al.*, 2002). On the other hand, comparison of fast and slowly senescing lines of the New Plant Type cultivar of rice showed that rapid leaf senescence correlates with a high percentage of grain filling and a high grain yield (Rubia *et al.*, 2002). While the grain yield of the transgenic wheat line used in our experiments was increased by prolongation of activity of the source the strategy of the fast-senescing rice lines is based on sudden termination of photosynthetic activity and fast translocation of mobilized metabolites to the developing grain. It is evident that choice of an efficient strategy for improvement of cereals by manipulation with plant hormones must be based on an intimate knowledge of cereal physiology and that ability of plants to retain hormonal homeostasis can be a source of surprises.

LIST OF ABBREVIATIONS

BA = N^6benzyladenine; CBP = cytokinin-binding protein; CKX = cytokinin oxidase(dehydrogenase); iP = N^6-(2-isopentenyl)adenine; iPA = iP riboside; NR = nitrate reductase; Z = *trans*-zeatin, ZR = Z riboside.

ACKNOWLEDGEMENT

This research was supported by the Grant Agency of the Czech Republic (grants No. 522/02/0530 and 206/02/0967).

REFERENCES

Akiyoshi, D.E., Morris, R.O., Hinz, R., Mischke, G.S., Kosuge, T., Garfinkel, D.J., Gordon, M.P. and Nester, E.W. (1983) Cytokinin/auxin balance in crown gall tumors is regulated by specific loci in the T-DNA, *Proc. Natl. Acad. Sci. USA* **80**, 407-411.

Armstrong, D.J. (1994) Cytokinin oxidase and the regulation of cytokinin degradation, in D.W.S. Mok and M.C. Mok (eds.), *Cytokinins: Chemistry, Activity and Function*, CRC Press, Boca Raton, pp. 139-153.

Åstot, C., Doležal, K., Nordstrom, A., Wang, Q., Kunkel, T., Moritz, H., Chua, N.H. and.Sandberg, G. (2000) An alternative cytokinin biosynhetic pathway, *Proc. Natl. Acad. Sci. USA* **97**, 14778-14783

Atkins, R.K., Batton, G.E. and Robinson, D.K. (1973) Effect of room growing temperature on growth substances in xylem exudates of *Zea mays*.

Auer, C.A., Motyka, V., Březinová, A. and Kamínek, M. (1999) Endogenous cytokinin accumulation and cytokinin oxidase activity during shoot organogenesis of *Petunia hybrida, Physiol. Plant* **105**,141-147.

Badenoch-Jones, J., Parker, C.W., Letham, D.S. and Singh, S. (1996) Effect of cytokinins supplied via the xylem at multiplies of endogenous concentrations on transpiration and senescence in derooted seedlings of oat and wheat, *Plant Cell Environ*. **19**,504-516.

Banowetz, G.M. (1997) Cultivars of hexaploid wheat of contrasting stature and chlorophyll retention differ in cytokinin content and responsivness, *Ann. Bot*. **79**,185-190.

Banowetz, G.M., Ammar, K. and Chen, D.D. (1999) Temperature effect on cytokinin accumulation and kernel mass in a dwarf wheat, *Ann. Bot*. **83**, 303-307.

Baumann, T.W., Schulthess, B.H., Linden, A. and Rüedi, P. (1994) Structure and metabolism of 7-β-D-glucopyranosyladenine – the product of cytokinin-sparing reaction?, *Phytochemistry* **36**, 537-542.

Beck, E.H. (1996) Regulation of shoot/root ratio by cytokinins from roots in *Urtica dioica*: Opinion, *Plant and Soil Sci.* **185**, 3-12.

Blagoeva, E., Dobrev, P., Motyka, V. and Vaňková, R. (2002) Effect of exogenous cytokinins, auxins and cdc kinase inhibitors on cytokinin-glucose conjugation in radish, in *Book of Abstracts of 23th Congress of FESPP*, Hersonissos, Heraklion, Crete, 2-6 September 2003, p. 209.

Bilyeu, K., Cole, J.L., Laskey, J.G., Reikhof, W.R., Esparza, T.J., Kramer, M.D. and Morris, R.O. (2001) Molecular and biochemical characterization of a cytokinin oxidase from maize, *Plant Physiol.* **125**, 378-386.

Borkovec, V. and Procházka, S. (1989) The effect of pre-anthesis application of benzyladenine and napthaleneacetic acid on the transport of ^{14}C sucrose in winter wheat, (In Czech), *Rostlinná výroba* **35**, 819-824.

Brinegar A.C. (1994) Cytokinin binding proteins and receptors, in D.W.S. Mok and M.C. Mok (eds.), *Cytokinins: Chemistry, Activity, and Function*, CRC Press, Boca Raton, pp. 217-232.

Brinegar A.C., Stevens A. and Fox J.E. (1985) Biosynthesis and degradation of a wheat embryo cytokinin-binding protein during embryogenesis and germination, *Plant Physiol* **79**, 706-710.

Brzobohatý, B., Moore, I. and Palme, K. (1994a) Cytokinin metabolism: implications for regulation of plant growth and development, *Plant Mol. Biol.* **26**, 1483-1497.

Brzobohatý, B., Moore, I., Kristoffersen, P., Bako, L., Campos, N., Schell, J. and Palme, K. (1993b) Release of active cytokinin by a β-glucosidase localized to the maize root meristems, *Science* **262**, 1051-1054.

Buu, R-H. and Chu, C. (1983) Effects of temperature on the tillering and the endogenous cytokinin activities in rice roots, *J. Agricult. Assoc. China, New Series* **123**, 12-21.

Cadiz, D.O., Beltrano, J., Fernandez, L.V., Saradon, S.J. and Favoretti, C. (1991) Effects of foliar applied benzyladenine on grain yield and grain protein in wheat (*Triticum aestivum* L.), *Plant Growth Regul.* **10**, 197-204.

Chatfield, J.M., Armstrong, D.J. (1986) Regulation of cytokinin oxidase activity in callus cultures of *Phaseolus vulgaris* L. cv Great Northern, *Plant Physiol.* **80**, 493-499.

Cheikh, N. and Jones, R.J. (1994) Disruption of maize kernel growth and development by heat stress, *Plant Physiol.* **106**, 45-51.

Cheng, C.L., Dewdney, J., Kleinhofs, A. and Goodman, H.M. (1986) Cloning and nitrate induction of nitrate reductase, *Proc. Natl. Acad. Sci. USA* **83**, 6825-6828.

Coenen C. and Lomax, T.R. (1997) Auxin-cytokinin interactions in higher plants" old problems and new tools, *Trends Plant Sci.* **2**, 351-356.

Cruz, C., Lips, S.H. and Martins-Loucao, M.A. (1995) Uptake regions of inorganic nitrogen in roots of carob seedlings, *Physiol. Plant.* **95**, 167-175.

Daskalova, S., Sorokin, A., Scott, N.W., Slater, A., Fowler, M.R., Kamínek, M., Gaudinová, A., Trčková, M. and Elliott, M.C. (2002) Enhancement of wheat grain yields by delaying leaf senescence. *10th International Association for Plant Tissue Culture and Biotechnology Congress, Orlando, Florida, 23-28 June 2002*, p.1082, p 53-A.

Dietrich, J.T., Kamínek, M., Blevins, D.G., Reinbott, T.M. and Morris, R.O. (1995) Changes in cytokinins and cytokinin oxidase activity in developing maize kernels and the effects of exogenous cytokinin on kernel development, *Plant Physiol. Biochem.* **33**, 327-336.

Dilworth, M.F. and Kende, H. (1974) Comparative studies on nitrate reductase in *Agrostemma githago* induced by nitrate and bezyladenine, *Plant Physiol.* **54**, 821-825.

Eklöf, S., Åstot, C., Moritz, T., Olson, O. and Sandberg, G. (1997) Auxin-cytokinin interactions in wild-type and transgenic tobacco, *Plant Cell Physiol.* **38**, 225-235.

Entsch, B., Letham, D.S., Parker, C.W., Summons, R.E. and Gollnow, B.I. (1979) Metabolites of cytokinins, in F. Skoog (ed.) Plant Growth Substances 1979, Springer-Verlag, Berlin, pp. 109-118.

Entsch, B., Parker, C.W. and Letham, D.S. (1983a) An enzyme from lupin seeds forming alanine derivatives of cytokinins, *Phytochemistry* **22**, 375-381.

Entsch, B., Parker, C.W., Letham, D.S. and Summons R.E. (1979b) Preparation and characterization using high-performance liquid chromatograpy of an enzyme forming glucosides from cytokinins, *Biochim. Biophys. Acta* **570**, 124-139.

Estruch, J.J., Prinsen, E., Van Onckelen, H.A., Schell, J. and Spena, A. (1991) Viviparous leaves produced by somatic activation of an inactive cytokinin-synthesizing gene, *Science* **254**, 1364-1367.

Faiss, M., Zalubilová, J., Strnad, M. and Schmülling, T. (1997) Conditional transgenic expression of the *ipt* gene indicates a function for cytokinins in paracrine signaling in whole tobacco plants, *Plant J.* **12**, 401-415.

Frank, A. and Schmülling, T. (1999) Cytokinin cycles cells, *Trends Plant Res.* **4**, 234-244.

Galuszka, P., Frébort, I., Šebela, M., Sauer, P., Jacobsen, S. and Peč, P. (2001) Cytokinin oxidase or dehydrogenase? Mechanism of cytokinin degradation in cereals, *Eur. J. Biochem.* **268**, 450-461.

Gan, S. and Amasino, R.M. (1995) Inhibition of leaf senescence by auto-regulated production of cytokinin, *Science* **270**, 1986-1988.

Gaudin, V., Vrain, T. and Jouanin, L. (1994) Bacterial genes modifying hormonal balances in plants, *Plant Physiol. Biochem.* **32**, 11-29.

Gaudinová, A. (1999) Effects of phytohormones on nitrate reductase, in M. Strnad, P. Peč and E. Beck (eds) *Advances in Regulation of Plant Growth and Development*, Peres Publishers, Prague, pp. 59-66.

Haberer, G. and Kieber, J.J. (2002) Cytokinins. New insights into a classic phytohormone, *Plant Physiol.* **128**, 354-362.

Hansen, C., Meins, F. and Aebi, R. (1987) Hormonal regulation of zeatin-riboside accumulation in cultured tobacco cells, *Planta* **172**, 520-525.

Hare, P.D., van Staden, J. (1994) Cytokinin oxidase: Biochemical features and physiological significance, *Physiol. Plant.* **91**, 128-136.

Harrison, M.A. and Kaufman, P.B. (1983) Estimation of free and bound indole-3-acetic acid and zeatin levels in relation to regulation of apical dominance and tiller release in oat shoots, *J. Plant Growth Regul.* **2**, 215-223.

Herzog, H. (1981) Einfluss von Stickstoff und Cytokiningaben auf die Fahnenblattes und der Karyopsen von Weizen, *Zeitschr. Pflanzenernährung und Bodenkunde* **144**, 16-29.

Horgan, J. M. and Wareing P F (1980) Cytokinins and the growth responses of seedlings of *Betula pendula* Roth. and *Acer pseudoplatanus* L. to nitrogen and phosphorus deficiency, *J. Exp. Bot.* **31**, 525-532.

Hradecká, D. and Petr, J. (1992) Effects of cytokinins on some cereals, in M. Kamínek, D.W.S. Mok and E. Zažímalová (eds.) *Physiology and Biochemistry of Cytokinins in Plants*, Academic Publishing bv., The Hague, The Netherlands, pp. 245-247.

Inoue, T., Higuchi, M., Hashimoto, Z., Seki, M., Kobayashi, M., Kato, T., Tabata, S., Shinozaki, K. and

Ishikawa, K., Kamada, H., Yamaguchi, I., Takahashi, N. and Harada, H. (1988) Morphology and hormone levels of tobacco and carrot tissues transformed by *Agrobacterium tumefaciens*. I. Auxin and cytokinin contents of cultured tissues transformed with wild-type and mutant Ti plasmids, *Plant Cell Physiol.* **29**, 461-466.

Jameson, P.E., McWha, J.A. and Wright, G.J. (1982) Cytokinins and changes in their activity during the development of wheat (*Triticum aestivum* L.), *Z. Pflanzenphysiol.* **106**, 27-36.

Jones, R.J., Schreiber, B.M.N. and Roessler J.A. (1996) Kernel sink capacity in maize: genotypic and maternal regulation, *Crop. Sci.* **36**, 301-306.

Kakimoto, T. (1996) CKI1, a histidine kinase homolog implicated in cytokinin signal transduction, *Science* **274**, 982-985.Kakimoto, T. (1998) Genes involved in cytokinin signal transduction, *J. Plant. Res.* **111**, 261-265.

Kakimoto, T. (2001) Identification of plant biosynthetic enzymes as dimethylallyl diphosphate:ATP/ADP isopentenyltransferases, *Plant Cell Physiol.* **42**, 677-685.

Kakimoto, T. (2001) Identification of CRE1 as cytokinin receptor from *Arabidopsis*, *Nature* **409**, 1060-1063.

Kalendová, A.M., Novák, V. and Kamínek, M. (1999) Improvement of yield formation and seed germination of spring barley by exogenous cytokinins, *Biol. Plant.* **42**(suppl.), p. 118.

Kamínek, M. and Armstrong, D.J. (1990) Genotypic variation in cytokinin oxidase from *Phasoleus* callus cultures, *Plant Physiol.* **93**, 1530-1538.

Kamínek, M., Dobrev, P., Gaudinová, A., Motyka, V., Malbeck, J., Trávníčková, A. and Trčková, M. (2000) Potential physiological function of cytokinin binding proteins in seeds of cereals, *Plant Physiol. Biochem.* **38**,79.

Kamínek, M., Motyka, V. and Vaňková, R. (1997) Regulation of cytokinin content in plant cells, *Physiol. Plant* **101**, 689-700.

Kamínek, M., Trčková, M., Gaudinová, A., Dobrev, P., Motyka, V., Albrechtová, J. and Votrubová, O. (1999) Regulation of cytokinin and auxin levels in developing wheat grains and its use for crop improvement, *Biol. Plant.* **42**(suppl.), 85.

Kamínek, M., Trčková, M., Fox, J.E. and, Gaudinová, A. (2003) Comparison of cytokinin-binding proteins from wheat and oat grains, *Physiol. Plant.* **117**, 1-6.

Kato, C., Kato, H., Asami, T., Yoshida, S., Noda, H., Kamada, H. and Satoh, S. (2002) Involvement of xylem sap zeatin-O-glucoside in cucumber shoot greening, *Plant Physiol. Biochem.* **40**, 949-954.

Klee, H.J. and Lanahan M.B. (1995) Transgenic plants in hormone biology, in P.J. Davies (ed.) *Plant Hormones: Physiology, Biochemistry and Molecular Biology*, Kluwer Acad. Publ., Dordrecht, The Netherlands, pp. 340-353.

Kuiper, D., Kuiper, P.J.C., Lambers, H., Schuit, J. and Staal, M. (1989) Cytokinin concentration in relation to mineral nutrition and benzyladenine treatment in *Plantago major* ssp. *Pleiosperma*, *Physiol. Plant.* **75**, 511-517.

Kuiper, D., Schuit, J. and Kuiper, J.C. (1990) Actual cytokinin concentration in plant tissue as indicator for salt resistance in cereals, *Plant Soil* **123**, 243-250.

Kuiper, D. and Staal, M. (1987) The effect of exogenously applied plant growth substances on the physiological plasticity in *Plantago major* ssp. *pleiosperma*: Responses to growth, shoot to root ratio and respiration, *Physiol. Plant* **69**, 651-658.

Laloue, M., Pethe-Terrine, C. and Guern, J. (1981) Uptake and metabolism of cytokinins in tobacco cells: Studies in relation to the expression of their biological activities, in J. Guern and C. Péaud-Lenoël (eds.) *Metabolism and Molecular Activities of Cytokinins*, Springer-Verlag, Berlin, pp. 80-96.

Letham, D.S., Summons, R.E., Entsch, B., Gollnow, B., Parker, C.W. and MacLeod, J.K. (1978) Glucosylation of cytokinin analogues, *Phytochemistry* **17**, 2053-2057.

Luštinec, J., Kamínek, M., Beneš, K. and Conrad, K. (1984) Hormone-like effect of vascular tissue on starch accumulation in stem explants of kale, *Brassica oleracea*, *Physiol. Plant.* **61**, 224-230.

Li, Q., Bettany, J.E., Donnison, I., Griffiths, C.M., Thomas, H. and Scott, I.M. (2000) Characterization of a cysteine protease cDNA from *Lolium multiflorum* leaves and its expression during senescence and cytokinin treatment, *Biochim. Biophys. Acta* **1492**, 233-236.

McGaw, B.A., Horgan, R. and Heald, J.K. (1985) Cytokinin metabolism and the modulation of cytokinin activity in radish, *Phytochemistry* **24**, 9-13.

McGaw, B.A., Horgan, R., Heald, J.K., Wullems, G.J. and Schilperoort, R.A. (1988) Mass-spectrometric quantitation of cytokinins in tobacco crown-gall tumors induced by mutated octopine Ti plasmids of *Agrobacterium tumefaciens*, *Planta* **176**, 230-234.

Medford, J.I., Horgan, R., El-Sawi, Z. and Klee, H.J. (1989) Alternation of endogenous cytokinins on transgenic plants using chimeric isopentenyl transferase gene, *Plant Cell* **1**, 403-413.

Meijer, M. and Murray, J.A.H. (2001) Cell cycle controls and development of plant form, *Curr. Opinion Plant Biol.* **4**, 44-49.

Meins, F., Jr. (1989) Habituation: Heritable variation in the requirement of cultured plant cells for hormones, *Annu. Rev. Genet.* **23**, 395-408.

Mishra, S.D. and Gaur, B.K. (1985) Modification of flag leaf senescence and yield characters in barley (*Hordeum vulgare* L.) by gibberellic acid and kinetin, *J. Plant Growth Regul.* **4**, 63-70.

Miyazawa, Z, Sakai, A., Miyagishima, S., Tokano, H., Kawano, S. and Kuroiwa, T. (1999) Auxin and cytokinin have opposite effects on amyloplast development and expression of starch synthesis genes in cultured Bright Yellow- tobacco cells, *Plant Physiol.* **212**, 461-70.

Mok, D.W.S. and Martin, R.C. (1994) Cytokinin metabolic enzymes, in D.W.S. Mok and M.C. Mok (eds.) *Cytokinins: Chemistry, Activity and Function*, CRC Press, Boca Raton, pp. 129-137.

Mok, D.W. and Mok, M.C. (2001) Cytokinin metabolism and action, *Annu. Rev. Plant Physiol. Plant Mol. Biol.* **52**, 89-118.

Morris, R. (1997) Hormonal regulation of seed development, in B. Larkins and I.K. Vasil (eds.) *Cellular and Molecular Biology of Plant Seed Development*, Kluwer Academic Publishers, Dordrecht, The Netherlands, pp. 117-149.

Motyka, V., Feiss, M., Strnad, M., Kamínek, M. and, Schmülling, T. (1996) Changes in cytokinin content and cytokinin oxidase activity in response to derepression of *ipt* gene transcription in transgenic tobacco calli and plants, *Plant Physiol.* **112**, 1035-1043.

Motyka, V. and Kamínek, M. (1990) Regulation of cytokinin catabolism in tobacco callus cultures, in H.J.J. Nijkamp, L.H.W. Van Der Plas and J. Van Aartrijk (eds.) *Progress in Plant Cellular and Molecular Biology*, Kluwer Academic Publishers, Dordrecht, The Netherlands, pp. 492-497.

Motyka, V. and Kamínek, M. (1994) Cytokinin oxidase from auxin-and cytokinin-dependent callus cultures of tobacco (*Nicotiana tabacum* L.), *J. Plant Growth Regul.* **13**, 1-9.

Motyka, V. and Kamínek, M. (1992) Cytokinin oxidase and degradative metabolism of cytokinins, in M. Kamínek, D.W.S. Mok and E. Zažímalová (eds.) *Physiology and Biochemistry of Cytokinins in Plants*, Kluwer Academic Publishing, The Hague, The Netherlands, pp. 29-39.

Motyka, V., Vaňková, R., Čapková, V., Petrášek, J., Kamínek, M. and Schmülling, T. (2003) Cytokinin-induced upregulation of cytokinin oxidase activity in tobacco includes changes in enzyme glycosylation and secretion, *Physiol. Plant.* **117**, 11-21.

Noodén, L.D., Guiamét. J.J. and John, I. (1997) Senescence mechanisms, *Physiol. Plant.* **101**, 746-753.

Noh, Y-S. and Amasino, R.M. (1999) Regulation of developmental senescence is conserved between *Arabidopsis* and *Brassoca napus*, *Plant Mol. Biol.* **41**, 195-206.

Ooms, G., Hooykaas, P.J., Moleman, G. and Schilperoort, R.A. (1981) Crown plant tumors of abnormal morphology induced by *Agrobacterium tumefaciens* carrying mutated octopine Ti-plasmids; analysis of T-DNA functions, *Gene* **14**, 33-50.

Palni, L.M.S., Burch, L. and Horgan, R. (1988) The effect of auxin concentration on cytokinin stability and metabolism, *Planta* **174**, 231-234.

Peters, W., Füchtenbauer, B. and Beck, E. (1995) Nitrate reductase activity is endogenously induced by zeatin riboside in habituated suspension cultured *Chenopodium rubrum* cells, *J. Plant Physiol.* **147**, 401-407.

Petrášek, J., Březinová, A., Holík, J. and Zažímalová, E. (2002) Excretion of cytokinins into the cultivation medium by suspension-cultured tobacco cells, *Plant Cell Rep.* **21**, 97-104.

Redig, P., Motyka, V., Van Onckelen, H.A. and Kamínek, M. (1997) Regulation of cytokinin oxidase activity in tobacco callus expressing the T-DNA *ipt* gene, *Physiol. Plant.***99**, 89-96.

Reski, R. (1994) Plastid genes and chloroplast biogenesis, in D.W.S. Mok and M.C. Mok (eds.) Cytokinins: Chemistry, Activity and Function, CRC Press, Boca Raton, pp. 139-153.

Reski, R. and Abel, W.O. (1992) Analyzing cytokinins and plastid differentiation at the molecular level: Interference with genome plastome-interaction, light quality, developmental state and endogenous oscillators. (Mini-review), in M. Kamínek, D.W.S. Mok and E. Zažímalová (eds.) Physiology and Biochemistry of Cytokinins in Plants, Kluwer Academic Publishing, The Hague, The Netherlands, pp. 179-195.

Riou-Khamlichi, C., Huntley, R., Jacqmard, A. and Murray, J.A.H. (1999) Cytokinin activation of *Arabidopsis* cell division through a D-type cyclin, *Science* **283**, 1441-1544.

Rubia, L.G., Fowler, M.R., Dobrev, P., Ghareyazie, B., Visperas, R., Peng, S., Khush, G., Slater, A., Scott, N.W., Kamínek, M., Bennett, J. and Elliott, M.C. (2002) Leaf senescence and grain filling in New Plant Rice lines, in *International Rice Congress 2002*, Beijing, China 16 – 20 September, Book of Abstracts p 335.

Rüdelsheim, P., Prinsen, E., van Lijsebettens, M. Inze, D., van Montagu, M., De Greef, J. and Van Onckelen, H. (1987) The effects of mutations in the T-DNA encoded auxin pathway on the endogenous phytohormones content in cloned *Nicotiana tabacum* crown gall tissues, *Plant Cell Physiol.* **28**, 475-484.

Ryschka, S.and Leike, R.H. (1989) Effects of 6-benzylaminopurine and Camposan on lateral buds of cereals cultured *in vitro, Biochem. Physiol. Pflanzen* **185**,109-122.

Sa, G., Mi, M., He-Chun, Ye., Nen-Ye, L., Guo-Feng, L and Kang, C. (2001) Effetcs of *ipt* gene expression on the physiological and chemical characteristics of *Arthemisia annua* L., *Plant Sci.* **160**, 691-698.

Sakakibara, H., Suzuki, M., Takei, K., Deji, A., Taniguchi, M. and Sugiyama, T. (1998) A response-regulator homologue possibly involved in nitrogen signal transduction mediated by cytokinin in maize, *Plant J.* **14**, 337-344.

Samuelson M. E. and Larson C-M (1993) Nitrate regulation of zeatin riboside levels in barley roots: effect of inhibitors on N assimilation and comparison with ammonium, *Plant Sci.* **93**, 77-84.

Schmerder, B. and Borriss, H. (1986) Induction of nitrate reductase by cytokinin and ethylene in *Agrostemma githago* L. emryos, *Planta* **169**, 589-593.

Shanahan, J.F., Donnely, K.J., Smith, D.H. and Smika, D.E. (1985) Shoot developmental properties associated with grain yield in winter wheat, *Crop. Sci.* **25**, 770-775.

Singh, S., Palni, L.M. and Letham, D.S. (1992) Cytokinin biochemistry in relation to leaf senescence. V. Endogenous cytokinin levels and metabolism of zeatin riboside in leaf discs from green and senescing tobacco (*Nicotiana rustica*) leaves, *J. Plant Physiol.* **139**, 279-283.

Skogquist, I. (1974) Induced heat sensitivity of wheat roots and protecting effect of ethanol and kinetin, *Physiol. Plant.* **32**, 166-169.

Skoog, F. and Miller, C.O.: (1957) Chemical regulation of growth and organ formation in plant tissue cultures cultured *in vitro*, *Symp. Soc. Exp. Biol.* **11**, 118-131.

Skoog, F. and Tsui, C. (1948) Chemical control of growth and bud formation in tobacco stem segments and callus cultured *in vitro*, *Amer. J. Bot.* **35**, 782-787.

Smart, C.M., Scifield, S.R., Bevan, M.V. and Dyer, T.A. (1991) Delayed leaf senescence in tobacco plants transformed with *tmr*, a gene for cytokinin production in *Agrobacterium, Plant Cell* **3**, 647-656.

Sun, J., Niu, Q-W., Tarkowski, P., Zheng, B., Tarkowska, D., Sandberg, G., Chua, N-H. and Zuo, J. (2003) The *Arabidopsis AtIPT8/PGA22* gene encodes an isopentenyl transferase that is involved in *de novo* cytokinin biosynthesis, *Plant Physiol.* **131**,167-176.

Takei, K., Sakakibara, H. and Sugiyama, T. (2001a) Identification of genes encoding adenylate isopentenyltransferase, a cytokinin biosynthesis enzyme, in *Arabidopsis thaliana*, *J. Biol. Chem.* **276**, 26405-26410.

Takei, K., Sakakibara, H., Taniguchi, M. and Sugiyama, T. (2001b) Nitrogen-dependent accumulation of cytokinins in root and the translocation to leaf: Implication of cytokinin species that induces gene expression of maize response regulator, *Plant Cell Physiol.* **42**, 85-93.

Takei, K., Takahashi, T., Sugiyama, T., Yamaya, T. and Sakakibara, H. (2002) Multiple routes communicating nitrogen availability from roots to shoots: a signal transduction pathway mediated by cytokinin, *J. Exp. Bot.* **53**, 971-977.

Taverner, E., Letham, D.S., Wang, J., Cornish, E. and Willcocks (1999) Influence of ethylene on cytokinin metabolism in relation to *Petunia* corolla senescence, *Phytochemistry* **51**, 341-347.

Thomas, H. and Howarth, C. (2000) Five ways to stay green, *J. Exp. Bot.* **51**, 329-337.

Thomas, J.C. and Katterman, F.R. (1986) Cytokinin activity induced by thidiazuron, *Plant Physiol.* **81**: 681-683.

Trčková, M. and Kamínek, M. (2000) Nitrate uptake and nitrogen allocation in wheat as affected by exogenous cytokinins,.in M.A. Martins-Loucão and S.H. Lips (eds.) *Nitrogen in a Sustainable Ecosystem: From the Cell to the Plant*, pp. 261-268, Bachyus Publishers, Leiden, The Netherlands, pp. 261-268.

Trčková,M., Kamínek, M. and Zmrhal, Z. (1992) Grain formation and distribution of nutrients in wheat plants after application of synthetic cytokinin N^6-(*meta*-hydroxybenzyl)adenosine, in M.-M. Kamínek, D.W.S. Mok and E. Zažímalová (eds.) Physiology and Biochemistry of Cytokinins in Plants, Kluwer Academic Publishing, The Hague, The Netherlands, pp. 241-244.

Van Staden, J., Cook, E.L. and Noodén, L.D. (1988) Cytokinins in senescence, in L.D. Noodén and A.C. Leopold (eds.) *Senescence and Aging in Plants*, Academic Press, San Diego, pp. 281-328.

Vaňková, R. (1999) Cytokinin conjugates – distribution, metabolism and function, in M. Strnad, P. Peč and E. Beck (eds.) *Advances in Regulation of Plant Growth and Development*, Peres Publishers, Prague, Czech Republic, pp. 67-78.

Vaňková, R., Gaudinová, A., Kamínek, M. and Eder, J. (1992) The effect of interaction of synthetic cytokinin and auxin on production of natural cytokinins by immobilized tobacco cells, in M. Kamínek, W.S. Mok and E. Zažímalová (eds.) *Physiology and Biochemistry of Cytokinins in Plants*, pp 47-51, Academic Publishing bv., The Hague, The Netherlands, pp. 47-51.

Vaňková, R., Hsiao, K.C., Bornman, C.H. and Gaudinová, A. (1991) Effects of synthetic cytokinins on levels of endogenous cytokinins

Vaňková, R., Kamínek, M. and Eder, J. (1992) Dynamics of production of *trans*-zeatin and *trans*-zeatin riboside by immobilized cytokinin-autonomous and cytokinin-dependent tobacco cells, *J. Plant Growth Regul.* **6**, 147-157.

Weaver, L., Gan, S., Quirino, B. and Amasino, R.M. (1998) A comparison of the expression of several senescence-associated genes in response to stress and hormone treatment, *Plant Mol. Biol.* **37**, 455-469.

Werner, T., Motyka, V., Strnad, M. and Schmülling (2001) Regulation of plant growth by cytokinins, *Proc. Nat. Acad. Sci. USA* **18**, 10487-10492.

Wickson, M. and Thimann, K.V. (1958) The antagonism of auxin and kinetin in apical dominance, *Physiol. Plant.* **11**, 62-74.

Wilkinson, J.Q. and Crawford, N.M. (1993) Identification and characterization of a chlorate resistant mutant of *Arabidopsis* mutations in both *Nia1*and *Nia2* nitrate reductase structural gene, *Mol. Gen. Genet.* **239**, 289-297.

Williams, R.H. and Cartwright, P.M. (1980) The effect of applications of a synthetic cytokinin on shoot dominance and grain yield in spring barley, *Ann. Bot.* **46**, 445-452.

Wingler, A., von Schaewen, A., Leegood, R.C., Lea , P. and Quick, W.P. (1998) Regulation of leaf senescence by cytokinin, sugars and light, *Plant Physiol.* **116**, 329-335.

Yoshida, Y. (2003) Molecular regulation of leaf senescence, *Curr. Opinion Plant Biol.* **6**, 79-84.

Yong, J.W.H., Wong, S.C., Letham, D.S., Hocart, C.H. and Farquhar, G.D. (2000) Effects of elevated [CO_2] and nitrogen nutrition on cytokinins in the xylem sap and leaves of cotton, *Plant Physiol.* **124**, 767-780.

Yu, X., Sukumaran, S. and Márton, S., (1998) Differential expression of the *Arabidopsis Nia1* and *Nia2* genes, *Plant Physiol.* **166**, 1091-1096.

Zažímalová, E., Březinová, A., Holík, J. and Opatrný, Z. (1996) Partial auxin deprivation affects endogenous cytokinins in an auxin-dependent, cytokinin.-independent tobacco cell strain, *Plant Cell Rep.* **16**,76-79.

Zažímalová, E. and Napier, R.M. (2003) Points of regulation of auxin action, *Plant Cell Reports* (in press).

Zažímalová, E., Kamínek, M., Březinová, A. and Motyka V. (1999) Control of cytokinin biosynthesis and metabolism, in P.J.J. Hooykaas, M.A.Hall and K.R. Libbenga (eds.) *Biochemistry and Molecular Biology of Plant Hormones*, Elsevier, Amsterdam, The Netherlands, pp. 141-160.

Zhang, X.D., Letham, D.S., Zhang, R. and Higgins T.J.V. (1996) Expression of the isopentenyl transferase gene is regulated by auxin in transgenic tobacco callus, *Transgen Res.* **5**, 57-65.

Zhang, C.L., Daskalova, S., Jiang, X.-C., Rubia, L.G., Rangan, L., Fowler, M.R., Scott, N.W., Slater, A., Bennett, J., Peng, S., Khush, G.S., Kamínek, M. and Elliott, M.C. (2002) Transformation of rice (*Oryza sativa* L.) with putative endosperm-specific Promoter::*ipt* constructs, in *International Rice Research Congress 2002, Beijing, China*, 16-20 September 2002, p. 500.

Zhang, R., Zhang, X., Wang, J., Letham, D.S., McKinney, S.A. and Higgins, T.J. (1995) The effect of auxin on cytokinin levels and metabolism in transgenic tobacco tissue expressing an *ipt* gene, *Planta* **196**, 84-94.

PRIMARY ALCOHOLS, SUBSTRATES FOR PHOSPHOLIPASE D-CATALYZED TRANSPHOSPHATIDYLATION, SUPPRESS THE CYTOKININ ACTION

G.A. ROMANOV[1], I.A. GETMAN[1], YU. P. BOLYAKINA[1], N.YU. RAKOVA[1], J.J. KIEBER[2] AND T. SCHMÜLLING[3]

[1]*Institute of Plant Physiology, Russian Academy of Sciences, Botanicheskaya 35, 127276 Moscow, Russia;* [2]*University of North Carolina, Biology Department, Chapel Hill, North Carolina 27599-3280, USA;* [3]*Freie Universität Berlin, Institute of Biology/Applied Genetics, Albrecht-Thaer-Weg 6, D-14195 Berlin, Germany*

1. INTRODUCTION

Cytokinins have been recognised as potent plant growth regulators since 1955 (Skoog, 1994). However, it is only recently that elucidation of the molecular mechanism of their perception and signal transduction has come about (Schmülling, 2001; Hutchinson and Kieber, 2002; Romanov, 2002). Cytokinin receptors discovered so far are integral transmembrane proteins belonging to the family of receptor histidine kinases. The downstream signaling obviously occurs *via* the phosphorelay pathway including phosphotransmitters (HP) and response regulator (RR) proteins (Schmülling, 2001; Hutchinson and Kieber, 2002; Romanov, 2002). A parallel set of data obtained by means of pharmacological analysis indicated the possible involvement of other signaling components, in particular phospholipase D (PLD), in the early events triggered by cytokinins (Romanov *et al.*, 2000; Romanov *et al.*, 2002). Taking into account that the transduction of a defined signal inside the cell is a complex process usually comprising crosstalk among different signaling elements, both these sets of data are not necessarily contradictory. For example, subtly tuned hormone signaling pathways comprising Ca^{2+}, PLD, cyclic GMP, protein kinase(s), protein phosphatase(s), calmodulin, and G-proteins constitute an integrated network in the aleurone and stomatal guard cells (Ritchie *et al.*, 2002).

PLD cleaves the terminal phosphodiesteric bond of phospholipids to phosphatidic acid (PdtOH) and water-soluble free head groups such as choline. Both components, especially some kinds of PdtOH, can serve as second messengers (Chapman, 1998; Munnik, 2001; Wang, 2001; Wang *et al.*, 2002). The genomic organization of plant PLDs is much more diverse than that of animals and microorganisms. To date, in the small *Arabidopsis* genome 12 PLD genes belonging

I. Macháčková and G.A. Romanov (eds.), Phytohormones in Plant Biotechnology and Agriculture, 129-139

to five subfamilies (α, β, γ, δ and ζ) have been identified, but only two PLD genes have been found in mammals and one in yeast (Wang *et al.*, 2002). This comparison suggests an important role for PLDs in plants. Indeed, some PLDs (especially α-type) play pivotal roles in plant response to different stresses, such as freezing, drought, wounding, pathogen infection, etc. Also there is strong evidence for the role of PLD in the ABA (Ritchie *et al.*, 2002; Wang, 2001) and Nod factor (Munnik, 2001) signaling pathways. It is therefore not excluded that different PLDs can take part in the mechanism of action of some other phytohormones as well.

Almost all known PLDs can catalyze a transphosphatidylation reaction utilizing short-chain primary alcohols as acceptors of phosphatidyl residues (Yu *et al.*, 1996; Chapman, 1998; Munnik, 2001). This reaction is a unique property of PLD, and results in the production of phosphatidylalcohols instead of normally produced PtdOH. Therefore primary alcohols suppressing the production of PtdOH by PLDs are often used to confirm the involvement of PLD activity in a studied process. Typical characteristics of the transphosphatidylation reaction driven by plant PLDs (Munnik *et al.*, 1995; Yu *et al.*, 1996; Ella *et al.*, 1997; Ritchie and Gilroy, 1998) are listed below:

a) PLDs use only primary alcohols as cosubstrate for transphosphatidylation, not secondary or tertiary alcohols;

b) the length of the alcohol hydrocarbon chain can vary from C1 to C8, however, C4 alcohols (1-butanol) are usually the most efficient cosubstrates;

c) the alcohol concentrations that efficiently inhibit PtdOH formation by PLDs *in vivo* or *in vitro* are rather small, ranging from 0.1 to 1%.

The present research was undertaken to study in more detail the effect of primary alcohols on the *in vivo* primary cytokinin responses based on cytokinin-induced gene expression. Two different plant models were used. Etiolated *Amaranthus* seedlings respond to cytokinins by a rapid accumulation of the red pigment, betacyanin, which can be measured quantitatively (Köhler and Conrad, 1968; Biddington and Thomas, 1973). Our *Arabidopsis* model is based on plants harboring the reporter gene *GUS* under control of the promoter of *ARR5* gene (Romanov *et al.*, 2002), the latter being a cytokinin primary response gene (D´Agostino *et al.*, 2000).

2. MATERIAL AND METHODS

2. 1. Plant Material and Bioassays

Seeds of *Amaranthus caudatus* L were germinated in distilled water in the dark at 24-25°C for 3 days. Bioassays were performed according to a modified procedure (Romanov *et al.*, 2000) using derooted seedlings, i.e. cotyledons and apical parts of the hypocotyls. Cytokinin and light effects were quantified by measuring betacyanin accumulation spectrophotometrically at 540 nm.

Seeds of transgenic *Arabidopsis* plants (*Arabidopsis thaliana* L.) harboring the *GUS* reporter gene fused to 1.6 kb of the *ARR5* (P_{ARR5}:*GUS*) gene promoter (D'Agostino *et al.*, 2000; Zvereva and Romanov, 2000) were germinated for 3-4 days at 24°C. Measurement of GUS activity was performed by a standard method using 4-methylumbelliferyl glucuronide as a substrate (Zvereva and Romanov, 2000). For more experimental details, see (Romanov *et al.*, 2002).

2. 2. *RNA blot analysis*

Total RNA was extracted from *Arabidopsis* plants according to (Puissant and Houdebine, 1990) or by using the RNeasy kit from Qiagen (Germany). 25 µg RNA was separated in a denaturing 1.5% agarose-formaldehyde gel, transferred to nylon membranes and hybridized with radioactive DNA probes labelled with ^{32}P using Random Primer Labelling kit (Stratagene, USA). The lowest stringency wash was performed in 0.2 x SSC, 0.1 % SDS at 65° C. The *ARR5* gene-specific probe was a full length cDNA of *A. thaliana* (accession Col-0). As a control for loading, the blot was rehybridized with an *actin 2* probe. This experiment was independently repeated twice.

3. RESULTS

Four liquid alcohols differing in carbon chain length or structure were used in this study: the primary alcohols methanol, ethanol and 1-butanol and the secondary alcohol 2-butanol. Only primary, but not secondary or tertiary alcohols have been shown (Munnik *et al.*, 1995; Yu *et al.*, 1996; Ella *et al.*, 1997; Ritchie and Gilroy, 1998) to serve as cosubstrates for PLD-catalyzed transphosphatidylation. On the other hand, all the alcohols used, especially 1- and 2- butanols, possess similar physico-chemical properties. The only pronounced difference is the degree of hydrophobicity, which becomes stronger with the longer carbon backbone of the molecule. Correspondingly, methanol and ethanol are fully soluble in water while the solubility of 1- and 2-butanols is limited to approx. 10% content in water. In our experiments the concentration of alcohols never exceeded 1%. That means that like other alcohols butanols were fully solved at concentrations far from saturation.

In accordance with previous data (Romanov *et al.*, 2000), in etiolated *Amaranthus* seedlings, cytokinin (5 µM benzyladenine) induced rapid betacyanin accumulation, which was measured quantitatively after 8 hours of treatment (Fig.1). The application of primary alcohols such as methanol, ethanol or 1-butanol together with cytokinin in the assay medium strongly reduced the cytokinin effect. The inhibition was dose-dependent, and any alcohol at concentrations of 1% almost totally suppressed the response to cytokinin. At low concentrations (0.2%) only 1-butanol exerted marked inhibition of cytokinin action, while methanol and ethanol produced much smaller effect (Fig. 1). It should be noted that the same % (v/v) concentration of the alcohols correspond to different molar concentrations as their molecular masses are different. For example, 1% (v/v) of butanol, ethanol and methanol is equal to 11, 17 and 25 mM, respectively. After recalculation in molar values, the range of inhibiting activity 1-butanol > ethanol > methanol became evident.

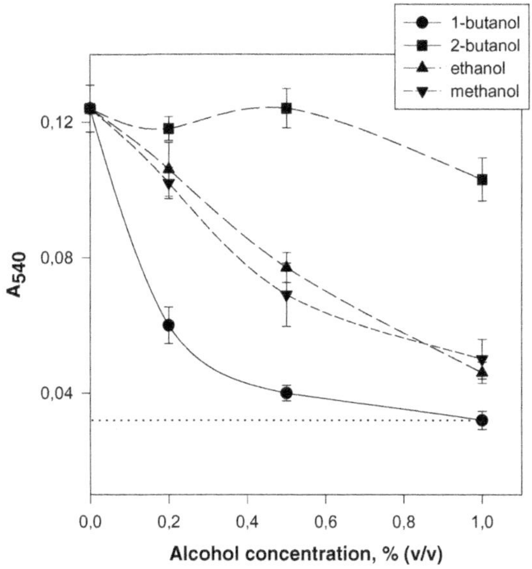

Figure 1. Dose-response curves of primary alcohol effect on cytokinin-induced betacyanin accumulation in Amaranthus seedlings. 3-day-old etiolated seedlings were treated with 5 μM 6-benzyladenine (BA) and with or without alcohols, for 8 h in darkness at 24-25 °C. Betacyanin was extracted and measured as described earlier [5]. Betacyanin content in non-induced seedlings (H_2O control) is indicated by the dotted line.

The comparison of action of primary alcohols with the corresponding secondary or tertiary alcohols is a very important criterion of the specificity of the alcohol effect. As secondary isomers do not exist for methanol and ethanol, we have used 1- and 2-butanols for this purpose. The general properties of 1 and 2-butanol are similar, so non-specific damages that they may cause are not expected to differ significantly. However, the two isomers differ profoundly in their interactions with PLD. The latter result was obtained in *Amaranthus* seedlings induced with cytokinin in the presence of 1- or 2-butanols (Fig. 1). In contrast to total inhibition by 1-butanol, 2-butanol showed almost no effect at 0.2-0.5% and rather small (evidently non-specific) inhibition (20-25%) at the highest concentration (1%).

To investigate the time period during which the inhibitors exert their action, special kinetic experiments were carried out. For this purpose BA was added simultaneously to all assay plants and then 1-butanol was added to different samples at different timepoints. Similar experiments were made on this model with other inhibitors of known cellular targets. These inhibitors included actinomycin D (inhibitor of transcription), cycloheximide (inhibitor of translation) and 1,10-phenanthroline (inhibitor of oxygenases and some other enzymes). After subtracting the background values, the resulting curves were expressed as percent of the maximal cytokinin effect and plotted on the same graphics together with the curve of the time course of betacyanin accumulation (Fig. 2).

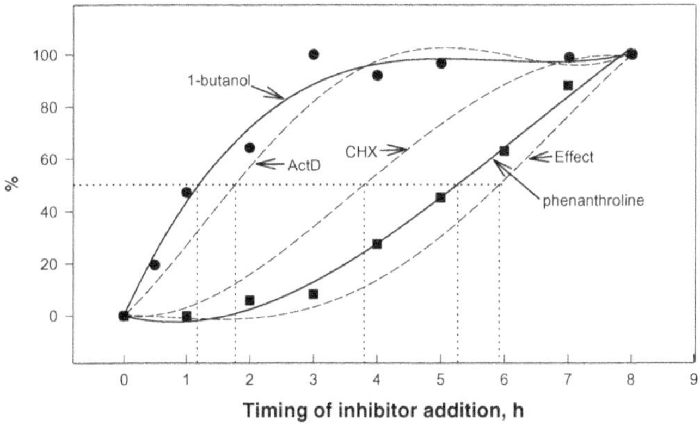

Figure 2 Time course of action of different compounds interfering with the cytokinin effect in Amaranthus seedlings. BA was added at the start time to all seedlings, other chemicals were added at various timepoints as indicated on abscissa. Data are expressed as percent of maximal values after background subtraction. The inhibition curves for actinomycin D and cycloheximide as well as the curve for induced amaranthin accumulation are indicated by intermittent lines. Half-maxima of inhibition and betacyanin accumulation curves and corresponding timepoints on abscissa are indicated by dotted lines.

Results showed that actinomycin D inhibited betacyanin accumulation only during the first 3-4 h, while the inhibition curve for cycloheximide was shifted to the later part of the 8 h-incubation period. This reflected the activation of transcription and subsequent translation of newly synthesized mRNAs under influence of cytokinin. As was expected, the main action of phenanthroline, the enzyme activity inhibitor, occurred at the late stage of the assay, close to the onset of betacyanin accumulation. By contrast, 1-butanol inhibited cytokinin effects only if added at the very early stage of the assay, even earlier than actinomycin D. When added 3 h after cytokinin or later, 1-butanol had no effect (Fig. 2). These data provided evidence that 1-butanol operated on an early event of the chain of intracellular events evoked by cytokinin, while phenathroline acted on a later link of the same chain.

Another means of assessing the specificity of inhibitor action is based on the unique property of the *Amaranthus* model to respond to different factors - cytokinin or light - with the same process of betacyanin accumulation. The two types of stimulus act independently (Kochbar *et al.*, 1981), evidently using their own signal transduction pathways, though subsequent stages, including transcription, new protein and then betacyanin biosynthesis, seem to proceed uniformly. Therefore the inhibitor effectiveness is expected to be the same in both assays in the case of the suppression of later steps leading to pigment accumulation. Only inhibitors that act early, at the perception/signal transduction stage, display a clear difference in their effect upon their target: phytohormone or light signaling. In keeping with this prediction, the enzymatic inhibitor phenanthroline was equally effective in both cytokinin- and light-induced bioassays. Unlike phenanthroline, 1-butanol demonstrated a clear difference in effects in parallel assays: the cytokinin effect was almost fully suppressed while the light effect was suppressed by 25% (Fig. 3). The latter value is close to that shown in Fig. 1 for non-specific inhibition produced by similar concentration of 2-butanol.

Transgenic *Arabidopsis* plants expressing the reporter gene *GUS* under control of the cytokinin-dependent promoter of the *ARR5* gene, represent another convenient plant model for the study of primary cytokinin action. This model is simpler than that of *Amaranthus* and relies on the expression of a defined known gene. Therefore the cytokinin action can be monitored, not only at the level of GUS activity, but also by measuring steady state mRNA concentrations.

Cytokinin was shown to specifically induce GUS activity in P_{ARR5}:*GUS* *Arabidopsis* (D´Agostinjo *et al.*, 2000; Romanov *et al.*, 2002). The main features of the cytokinin-induced response were similar to those described for *Amaranthus*. *Arabidopsis* seedlings were also sensitive to primary alcohols: 1-butanol strongly repressed the cytokinin-induced GUS activity while 2-butanol had almost no effect (Fig. 4). Similar reactions occurred when 1- and 2-propanol were used (not shown). These results provide strong evidence for the involvement of PLD in the primary cytokinin effects in *Arabidopsis* (as well as *Amaranthus*).

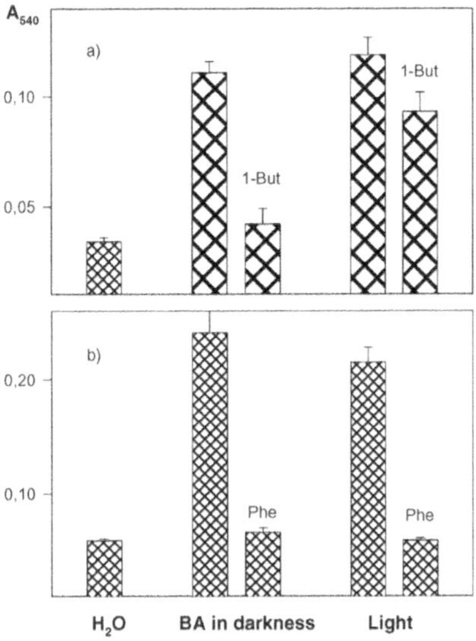

Figure 3. Effect of 1-butanol or 1,10-phenanthroline on cytokinin- or light-induced betacyanin accumulation in Amaranthus plantlets. Etiolated seedlings were divided into two sets of equivalent samples, one set was induced by BA in darkness, the second set by white light (40 Wt/m^2), both for 8 h at 24 °C, with or without 0.75% 1-butanol (a) or 5 mM 1,10-phenanthroline (b).

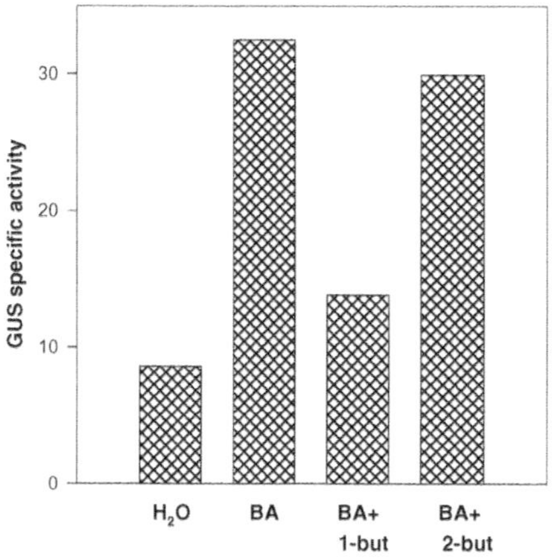

Figure 4. Effect of 1- or 2-butanols on cytokinin-induced GUS activity in P_{ARR5}:GUS transgenic arabidopsis seedlings. 4-day-old seedlings were treated with 5 μM BA, with or without 1% alcohol, for 15.5 h in darkness at 24-25 °C. GUS activity was determined by spectrofluorometry and plotted in graphics in arbitrary units.

To determine the stage at which primary alcohols exert their action, northern blots were made using RNA isolated 35-40 min after cytokinin administration [6]. Results on the determination of GUS specific activity and *ARR5* mRNA content are presented in Fig. 5. Primary butanol significantly reduced not only the activity of the reporter gene, but also mRNA content, both to a similar degree. Similar results were obtained using the *GUS* gene as a labeled hybridization probe to monitor the transcription activation of the reporter gene (not shown). This indicates that 1-butanol operated at the stage prior the onset of transcription activation, i.e. the stage of signal perception and/or transduction.

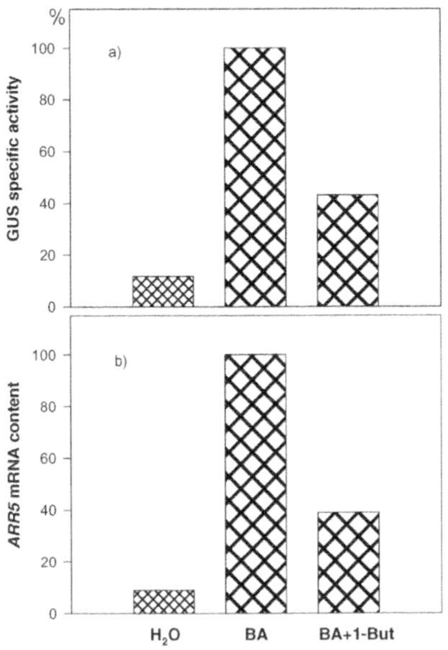

Figure 5. Effect of 1-butanol on GUS activity and ARR5 mRNA content in P_{ARR5}:GUS transgenic arabidopsis seedlings. 3-day-old seedlings were treated with 5 μM BA, with or without 1% butanol for either 7 h for GUS activity determination (a) or 35 min for RNA extraction (b).

4. DISCUSSION

In the present report we show the effect of different alcohols on primary cytokinin effects in two distinct plant species: *Amaranthus* and *Arabidopsis*. The results are very similar irrespective of the plant model: all primary alcohols strongly inhibited the cytokinin response while secondary alcohols had little or no effect. Curiously, data on the much stronger inhibitory activity of 1-propanol versus 2-propanol in the *Amaranthus* bioassay had already been published long ago (in 1968, see Köhler and Conrad, 1968), before the «era of PLD». Data presented here are in good agreement with typical characteristics of transphosphatidylation reaction driven by plant PLDs (see Munnik *et al.*, 1995; Yu *et al.*, 1996; Ella *et al.*, 1997; Ritchie and Gilroy, 1998) and «Introduction»): (i) primary alcohols were always much more efficient inhibitors than secondary ones; (ii) 1-butanol seemed to be a stronger inhibitor than shorter alcohols, and (iii) primary alcohols were effective at the concentration range of 0.2-1.0%. Hence these results represent a strong argument in favor of the participation of PLD in the early cytokinin effects.

Additional experiments were performed to determine at which level in the chain of cytokinin-induced events PLD exerted its action. In *Amaranthus*, results of kinetics and cytokinin/light experiments were in good accord, providing evidence that PLD operated at the earliest stage of cytokinin action, namely at the stage of cytokinin perception and/or signal transduction. More direct proof for this conclusion was obtained using the *Arabidopsis* model, where 1-butanol was shown to prevent the accumulation of cytokinin-induced mRNAs. Alcohol inhibition of cytokinin-responsive gene transcription seems to be specific as some other inhibitors of the final cytokinin effect were not effective at the mRNA level (Romanov *et al.*, 2002). These results correspond well with our knowledge (Chapman, 1998; Munnik, 2001; Wang, 2001; Wang *et al.*, 2002) about the mode of PLD participation in different signaling processes.

At present it is not possible to ascertain the precise role of PLD in the molecular mechanism of cytokinin action. It seems plausible to suggest that PdtOHs produced by PLD act as second messengers in parallel with the two-component system in the process of cytokinin signaling. On the other hand, some cellular roles of PdtOH may result from its effect on membrane properties and configuration as well. Also, PdtOH could create docking sites to recruit enzymes from the cytosol to the specific membrane domain (Munnik, 2001; Wang, 2001). Further investigations will show the precise role PLD and liberated PdtOHs play in the early events triggered by cytokinins.

ACKNOWLEDGEMENT

This work was supported by grants of the Deutsche Forschungsgemeinschaft (Schm 814/13-1), the Russian Foundation of Basic Research (N 01-04-04002) and the INTAS (N 01-0602).

REFERENCES

Biddington N.L. and Thomas T.H. (1973) A modified *Amaranthus* betacyanin bioassay for the rapid determination of cytokinins in plant extracts, *Planta* **111**, 183-186.

Chapman, K.D. (1998) Phospholipase activity during plant growth and development and in response to environmental stress, *Trends Plant Sci.* **11**, 419-426.

D'Agostino, I.B., Deruere, J. and Kieber, J.J. (2000) Characterization of the response of the *Arabidopsis* response regulator gene family to cytokinin, *Plant Physiol.* **124**, 1706-1717.

Ella, K.M., Meier, K.E., Kumar, A., Zhang, Y. and Meier, G.P. (1997) Utilization of alcohols by plant and mammalian phospholipase D, *Biochem. Mol. Biol. Int.* **41**, 715-724.

Hutchison, C.E. and Kieber, J.J. (2002) Cytokinin signaling in *Arabidopsis, The Plant Cell*, Supplement: S47-S59..

Kochhar, V.K., Kochhar, S. and Mohr, H. (1981) Action of light and kinetin on betalain synthesis in seedlings of *Amaranthus caudatus*: a two-factor analysis, *Ber. Deutsch. Bot. Ges.* **94**, 27-34.

Kohler, K.H. and Conrad, K. (1968) Zur Spezifitat des *Amaranthus*-Cytokinintest. III. Benzimidazolderivate und andere Verbindungen, *Flora* **159**, 293-298.

Munnik, T. (2001) Phosphatidic acid: an emerging plant lipid second messenger, *Trends Plant Sci.* **6**, 227-233.

Munnik, T., Arisz, S.A., de Vrije, T. and Musgrave, A. (1995) G protein activation stimulates phospholipase D signalling in plants, *Plant Cell* **7,** 2197-2210.

Puissant, C. and Houdebine, L.-M. (1990) An improvement of the single-step method of RNA isolation by acid guanidium thiocyanate-phenol-chloroform extraction, *Biotechniques* **8**, 148-149.

Ritchie, S. and Gilroy, S. (1998) Abscisic acid signal t ransduction in the barley aleurone is mediated by phospholipase D activity, *Proc. Nat. Acad. Sci. USA* **95**, 2697-2702.

Ritchie, S.M., Swanson, S.J. and Gilroy, S. (2002) From common signalling components to cell specific responses: insights from cereal aleurone, *Physiol. Plant.* **115**, 342-351.

Romanov, G.A. (2002) The phytohormone receptors, *Russian J. Plant Physiol.* (English translation) **49**: 552-560.

Romanov, G.A., Getman, I.A. and Schmülling, T. (2000) Investigation of early cytokinin effects in a rapid *Amaranthus* seedling test, *Plant Growth Regul.* **32**, 337-344.

Romanov, G.A., Kieber, J.J. and Schmülling, T. (2002) A rapid cytokinin response assay in *Arabidopsis* indicates a role for phospholipase D in cytokinin signalling, *FEBS Lett.* **515**, 39-43.

Schmülling, T. (2001) CREam of cytokinin signalling: receptor identified, *Trends Plant Sci.* **6**, 281-284.

Skoog F. (1994) A personal history of cytokinin and plant hormone research, in D.W.S. Mok and M.C. Mok (eds.), *Cytokinins. Chemistry, Activity, and Function.* CRC Press, Boca Raton., pp 1-14.

Wang, H. (2001) Plant phospholipases, *Annu. Rev. Plant Physiol. Plant Mol. Biol.* **52**, 211-231.

Wang, X., Wang, C., Sang, Y., Qin, C. and Welti, R.(2002) Networking of phospholipases in plant signal transduction, *Physiol. Plant.* **115**, 331-335.

Yu, C.H., Liu, S.Y. and Panagia, V. (1996) The transphosphatidylation activity of phospholipase D, *Mol. Cell. Biochem.* **157**, 101-105.

Zvereva, S.D. and Romanov G.A. (2000) Reporter genes for genetic engineering: characteristics and detection, *Russian J. Plant Physiol.* (English translation) **47**, 424-432.

AUXINS

PHYSIOLOGICAL AND MOLECULAR GENETIC ASPECTS OF AUXIN TRANSPORT: RECENT DEVELOPMENTS

DAVID A. MORRIS[1, 2] AND EVA ZAŽÍMALOVÁ
Institute of Experimental Botany, Academy of Sciences of the Czech Republic, Rozvojova 135, 165 02 Prague 6, Czech Republic.
[1]Permanent address: Division of Cell Sciences, School of Biological Sciences, University of Southampton, Biomedical Sciences Building, Bassett Crescent East, Southampton, SO16 7PX, UK. [2]Author for correspondence; e-mail: dam3@soton.ac.uk

1. INTRODUCTION

Auxins, of which indole-3-acetic acid (IAA) is the commonest naturally occurring example, are a group of important signal molecules that play crucial roles in the regulation of spatial aspects of plant growth and development. As well as being required for normal division, enlargement and differentiation of individual plant cells, auxins are vital signals between cells, tissues and organs, and thereby contribute to the coordination of growth and development and to physiological responses to environmental cues (for examples, see Davies, 1995).

Auxins owe their importance as key regulatory signals in part to their pleiotropic effects and in part to the two very different mechanisms by which they are transported through the plant. The first of these is by mass transfer in the xylem and phloem translocation streams. IAA is a constituent of xylem and phloem saps, and may enter these translocation pathways following synthesis in leaves (for discussion see Baker, 2000). The transport of IAA in these translocation systems contributes to its role as a signal in the coordination of growth and development of widely separated organs and tissues. In particular, through its effects on cellular activity in growth sinks where, *inter alia*, it stimulates the synthesis of the sucrose hydrolysing enzyme, acid invertase (Morris and Arthur, 1984, 1985), IAA may regulate sink activity, phloem unloading and source/sink relationships in the plant (reviewed by Morris, 1996).

Of more significance for the regulation of development is the strictly polar (directional) long- and short-distance cell-to-cell transport of auxins through parenchymatous cells (including those of the vascular cambium; Morris and Thomas, 1978, and references therein). This unique polar transport of auxin, unconfirmed for other groups of plant hormones, confers directional properties (axial and radial) on an auxin signal. As a result, auxin transport is necessary for the establishment and

I. Macháčková and G.A. Romanov (eds.), Phytohormones in Plant Biotechnology and Agriculture, 143-155

maintenance of axiality and polarity in plant cells, tissues and organs, and plays a key role in pattern development and organogenesis in plants (Berleth *et al.*, 2000; Berleth and Sachs, 2001). Although occurring by very different mechanisms, the transport of auxin in the vascular and polar pathways is functionally linked. It has been found that auxin moving through the phloem can enter the root-directed (basipetal) polar transport stream at the shoot apex (Cambridge and Morris, 1996). Therefore, in addition to IAA synthesised in apical tissues and leaf primordia, auxin produced in older leaves also may act as a source of auxins.

Several comprehensive reviews covering aspects of the mechanism, regulation and significance in development of polar auxin transport have appeared in the last few years (e.g. Lomax *et al.,* 1995; Bennett *et al.,*1998; Morris,2000; Muday and DeLong, 2001; Muday and Murphy, 2002; Friml and Palme, 2002; Friml, 2003; Zažímalová and Napier, 2003). Consequently no attempt will be made here to cover the whole topic of auxin transport. Rather this paper will concentrate on some of the new discoveries that have forced us to re-examine and modify some of our previous notions of the mechanism and regulation of polar auxin transport. These discoveries have revealed that polar auxin transport is a far more complex and dynamic process than we once thought.

2. CELLULAR TRANSPORT OF AUXIN – THE BASIC MECHANISM

A wealth of experimental evidence now exists to support the original proposal by Rubery and Sheldrake (1974) and Raven (1975) that the polarity of auxin transport results from the differential permeabilities of each end of transporting cells to auxin anions (IAA^-) and undissociated auxin molecules (IAA; reviewed by Goldsmith, 1977). Being a relatively lipophilic a weak organic acid, IAA readily enters cells by diffusion across the plasma membrane (PM) from the more acidic extracellular space. In contrast, the auxin anion (IAA^-), being more hydrophilic does not easily diffuse across membranes. Accordingly, IAA^- tends to accumulate in the cell by a process of "anion trapping", and can do so to a total concentration ($IAA + IAA^-$) that greatly exceeds that in the external medium (for discussion see Goldsmith, 1977). To exit the cell, IAA^- requires the intervention of specific transmembrane auxin anion efflux carriers. A substantial body of physiological, biochemical and molecular genetic evidence supports the view that the polarity of auxin transport results from an asymmetric (polar) distribution of auxin uptake and efflux carriers in the PM (Goldsmith, 1977; Lomax *et al.*, 1995; Bennett *et al.,* 1998; Morris, 2000; Swarup *et al.,* 2001; Friml and Palme, 2002; Grebe *et al.,* 2002). Several genes that encode putative auxin influx and efflux carriers have now been identified from *Arabidopsis* and other species (reviewed by Morris, 2000; Muday and DeLong, 2001; Friml and Palme, 2002). Efflux carrier proteins, encoded by members of the *PIN-FORMED* (*PIN*) gene family and related genes are targeted to specific regions of the PM in auxin transporting cells, including basal, apical and lateral regions of the membrane (Gälweiler *et al.,* 1998; Müller *et al.,* 1998; Friml *et al.,* 2002; reviewed by Friml and Palme, 2002; Friml, 2003). More recently, influx carriers (encoded by *AUXIN1* [*AUX1*] and related genes; Bennett *et al.,* 1996) also have been shown to be distributed asymmetrically in *Arabidopsis* root apical and epidermal cells (Swarup *et al.*, 2001) and may also contribute to the polarity of auxin transport (Grebe *et al.*, 2002).

The development in the last 4-5 years of new biochemical approaches and the rapid advances that have been made in molecular genetic techniques have allowed aspects of auxin transport to be studied in far greater detail than ever before. Some of these discoveries resulting from these studies are described and discussed in the succeeding sections.

3. GENES, PROTEINS AND TRANSPORT

Characterization of auxin transport mutants has provided a wealth of information about the identity and sequence of genes coding essential components of the auxin transport machinery, including putative auxin uptake and efflux catalysts and proteins that are essential for their targeting to the PM and for the regulation of their activity. A number of genes, belonging to several major gene families, have now been identified as encoding putative trans-membrane auxin carriers. These include members of the *AUX1* and *LIKE-AUX1* [*LAX1*] families, which encode amino acid carriers, including putative auxin uptake carriers (e.g. Bennett *et al.*, 1996); and members of the *PIN* gene family, which encode putative auxin efflux carriers (e.g. Chen *et al.*, 1998; Luschnig *et al.*, 1998; Gälweiler *et al.*, 1998; reviewed by Friml and Palme, 2002; Friml, 2003). Although the evidence for an auxin transport role for the proteins encoded by these genes appears overwhelming, it nevertheless remains largely indirect (Friml and Palme, 2002).

The large family of PIN proteins is of particular interest. It is now well established that the direction of polar auxin transport correlates well with the region of the PM to which individual PIN-proteins are targeted. Thus, acropetal transport of NAA in seedling *Arabidopsis* roots coincides with the apical location of PIN1 in transporting root cells (Steinmann *et al.*, 1999); the lateral redistribution of auxin in similar roots following gravistimulation correlates closely with redistribution of PIN3 to lateral membranes of root cells (Friml *et al.*, 2002). Furthermore, the existence of many of *PIN* genes, which are differentially expressed in different cells and tissues, and the expression of which may be sensitively modified by a variety of plant and environmental factors (Morris, 2000; Friml, 2003), endows the auxin transport system with a considerable degree of flexibility.

Despite the key role played by the AUX and PIN proteins in polar auxin transport, recent studies have revealed that other proteins may also play a part in the distribution of auxin through the plant. Perhaps the most surprising candidates for a role in auxin transport are members of the multidrug resistance (MDR) sub-group of ABC transporters (Noh *et al.*, 2001; Gaedeke *et al.*, 2001). Mutants in several *Arabidopsis* members of this group, including *AtMDR1, APGP1* (Noh *et al.*, 2001) and *AtMRP5* (Gaedeke *et al.*, 2001), exhibit characteristics typical of plants in which auxin efflux is inhibited (i.e. auxin accumulates as a consequence of impaired transpot;Noh *et al.*, 2001). AtMDR1 and APGP1 were shown by Noh *et al.* (2001) to specifically bind the auxin efflux inhibitor, 1-*N*-naphthylphthalamic acid (NPA).

Even so, considerable binding of NPA also occurred in microsomal membranes prepared from *Atmdr1* mutants (more than 60% of wild type binding), and NPA remained able to reduce auxin transport to background levels in *Atmdr1* tissues (Noh *et al.,* 2001). This indicates that in addition to MDRs, other NPA-binding proteins (NBPs) are present in *Arabidopsis* tissues (for a detailed discussion see Muday and Murphy, 2002). Whilst together these observations strongly suggest a potential role for MDRs in auxin transport, it remains unclear exactly how, and under what conditions these wide-specificity transporters function in this role. Among the possibilities suggested are that MDRs act in auxin homeostasis, by removing auxin ions to the vacuole or to the exterior under conditions of high internal auxin concentration; alternatively been suggested that the NPA-binding MDRs may act to regulate the activity of PIN-family auxin efflux carrier (Noh *et al.,* 2001; for discussion see Luschnig, 2002). Whatever their mechanism of action in auxin transport, it is clear that a very precise regulation of these multifunctional carrier systems is likely. For a recent detailed review of plant ABC transporters, the reader is referred to Martinoia *et al.* (2002).

Recently a new class of amino acid transporters has been identified from *Arabidopsis* (Chen *et al.,* 2001). By expressing novel cDNAs with sequence similarity to the amino acid transporter gene, *AAP1*, in a yeast amino acid transport-deficient mutant, a group of aromatic and neutral transporters (ANTs) was identified. Among these, ANT1 was found to transport both IAA and 2,4-dichlorophenoxyacetic acid (2,4-D; Chen *et al.,* 2001). Thus the AUX/LAX proteins may not be the only transporters in plant cells mediating auxin influx. However, as in the case of the MDRs (see above), the conditions under which ANT1 protein might function in auxin transport remain unclear. Attempts to integrate these new discoveries into a single scheme describing fluxes of auxin into and out of cells, and between compartments within cells, thereby contributing to auxin homeostasis, have recently been presented by Luschnig (2002) and Zažímalová and Napier (2003). Given the current lack of detailed information about the auxin transport functions of MDRs and ANTs, however, for the moment such schemes must remain speculative.

4. THE ROLE OF THE CYTOSKELETON IN AUXIN CARRIER TRAFFIC AND CYCLING

The actin cytoskeleton plays an important role in directing vesicular traffic to delivery sites at the PM and to other membrane-enclosed compartments in the cell. It has been known for several years that the NPA-binding component of the efflux carrier complex is closely associated with actin filaments (Cox and Muday, 1994; Butler *et al.,* 1998; Hu *et al.*, 2000). The application of the actin-depolymerising drug, cytochalasin, reduced polar auxin transport in *Zea mays* L. coleoptiles (Cande *et al.*, 1973) and zucchini hypocotyl segments (Butler *et al.*, 1998). Interestingly, it has now been convincingly demonstrated that treatment of tissues with such drugs also perturbs the normal polar distribution of auxin carriers in the PM, including that of PIN1. Geldner *et al.* (2001) found that cytochalasin D and latrunculin B (another actin-depolymerising drug) had relatively little effect on the polar accumulation of PIN1 at the PM when applied alone. However, both drugs strongly inhibited not only the internalisation of PIN1 caused by treatment with brefeldin A (BFA; - a drug that strongly inhibits the Golgi-mediated trafficking of auxin carrier proteins to the PM; see below), but also the recovery of the polar localization of PIN1 following BFA washout. These results indicate that the BFA-sensitive traffic of PIN1 to and from the PM involves actin-dependent processes. The precise mechanism by which PIN-carrying vesicles are sorted and directed to specific sites at the PM along actin filament pathways, however, are not known. The demonstrated link between the high affinity NBP and actin microfilaments suggests that NBPs may play a crucial role in this process. This possibility has been strengthened by the identification of *TIR3/BIG* as a member of the *CALOSSIN/PUSHOVER (CAL/O)* gene family (Gil *et al.,* 2001; see discussion above, and Luschnig, 2001). In *Drosophila*, CAL/O functions in the cycling of synaptic vesicles (Richards *et al.,* 1996), and an attractive possibility is that the NPA-binding TIR3/BIG protein may play a similar role in *Arabidopsis* and function to regulate PIN traffic to appropriate locations in the PM by an actin-dependent mechanism (for detailed discussions, see Gil *et al.,* 2001; Luschnig 2001; Muday and Murphy, 2002). Association of NPA or other phytotropins with NBPs conceivably might interfere with this function, although the mechanism remains unknown. Whatever the mechanism of action of NPA, it does not involve changes to the structure of the actin cytoskeleton; we have recently shown that NPA has no effect on the arrangement of either microtubules or actin filaments in BY-2 tobacco cells, even at concentrations well above those that saturate the inhibition of auxin efflux (Petrášek *et al.,* 2003).

5. AUXIN EFFLUX CARRIERS ARE MULTI-COMPONENT SYSTEMS

An early and important discovery that prompted further research and provided a framework around which subsequent results could be interpreted, was that auxin efflux carriers were multi-component systems consisting of at least two, but possibly more individual protein components, each with a different function (Morris *et al.,* 1991). Specific inhibitors of polar auxin transport have been known for some time and they have played a major role in investigations of the polar auxin transport machinery and the way in which it is regulated. NPA, which is a well-characterized member of a group of inhibitors known as phytotropins (Katekar and Geissler, 1980; Rubery, 1990), is the most widely employed of such inhibitors. NPA application to various plant tissues strongly inhibits auxin efflux carrier activity and in doing so, increases auxin accumulation in cells (reviewed by Morris, 2000). Although the mechanism of NPA action on polar auxin transport is still unknown, available evidence suggests that it is mediated by a specific, high affinity, NPA-binding protein (NBP; Sussman and Gardner, 1980; Rubery, 1990). Protein synthesis inhibitors such as cycloheximide (CH) rapidly uncouple carrier-mediated auxin efflux and the inhibition of efflux by NPA; however, CH does not affect auxin efflux itself, or the level of saturable NPA binding to microsomal membranes (Morris *et al.,* 1991). This suggests that the NBP and the efflux catalyst are separate proteins that may interact through a third, rapidly turned over transducing protein (Morris *et al.,* 1991; discussed in Morris, 2000; Luschnig, 2001). The NBP itself is probably a peripheral membrane protein located on the cytoplasmic face of the PM and associated with the actin cytoskeleton (Cox and Muday, 1994; Dixon *et al.*, 1996; Butler *et al.*, 1998; but cf. Bernasconi *et al.*, 1996). Although the identity of the NBP and its mechanism of action on auxin efflux carriers remain uncertain, recent evidence suggests that the NBP may be essential to direct the traffic of auxin efflux catalysts to the correct location at the PM (Gil *et al.*, 2001). An *Arabidopsis* mutant, *transport inhibitor response 3* (*tir3*), exhibits a reduced number of NPA binding sites and a reduction in polar auxin transport (Ruegger *et al.*, 1997). Therefore *TIR3* (renamed *BIG* by Gil *et al.*, 2001, to reflect the unusually large size of the protein it encodes – 566 kD), may encode an NBP or may be required for NBP expression, localization or stabilization (for discussion see Gil *et al.*, 2001; Luschnig, 2001). Phosphorylation/dephosphorylation reactions might be expected to play a key role in auxin transport and in the interactions between the individual components of such complex carrier systems. Indeed, efflux carrier activity in suspension-cultured tobacco cells (*N. tabacum* L. cv. Xanthi XFHD8) has been shown to be inhibited by treatment with the kinase inhibitors staurosporine and K252a (Delbarre *et al.*, 1998). The *A. thaliana* gene *AtRCN1* (*ROOTS CURL IN NPA*) has been shown to encode the regulatory A subunit of protein phosphatase 2A; mutation of this gene (*Atrcn1*) results in enhanced sensitivity to NPA and causes phenotypic characteristics consistent with reduced auxin efflux (Garbers *et al.*, 1996). Furthermore, treatment of wild type seedlings with the phosphatase inhibitor cantharidin phenocopies the *Atrcn1* mutant (Druère *et al.*, 1999). Although associated protein kinases have not yet been unequivocally identified, a possible candidate kinase is encoded by the *Arabidopsis PINOID* (*AtPID*) gene (Benjamins *et al.*, 2001), which has been shown to enhance polar transport and to encode a protein kinase. For further discussion of this topic, see Muday and Murphy (2002) and Zažímalová and Napier (2003). The latter authors have speculated that RCN1 and PID might be candidate proteins for the hypothetical labile protein coupling the NBP and the auxin efflux catalyst (see Morris *et al.*, 1991 and above).

6. EFFLUX CATALYSTS TURN OVER RAPIDLY IN THE PM

Another significant recent advance was the discovery that auxin efflux catalysts turn over very rapidly in the PM. This was first revealed by investigations of the effect of the inhibitors of Golgi-mediated vesicle traffic to the PM, monensin and brefeldin A (BFA), on the accumulation of IAA by zucchini (*Cucurbita pepo* L.) hypocotyl segments (Wilkinson and Morris, 1994; Morris and Robinson, 1998; Robinson *et al.*, 2000), and the effect of BFA on NAA accumulation by suspension-cultured tobacco cells (*Nicotiana tabacum* L.; Delbarre *et al.*, 1998). Both monensin and BFA very rapidly stimulated IAA or NAA accumulation (depending on system), but in both systems had no effect on the accumulation of 2,4-D (Delbarre *et al.*, 1998; Morris and Robinson, 1998). 2,4-D is a substrate for auxin uptake carriers but, in most species, is not a substrate for auxin efflux carriers (Delbarre *et al.*, 1996). Together with an observed reduction in the efflux of NAA or IAA (but not of 2,4-D) from pre-loaded cells and tissue segments in the presence of BFA (Delbarre *et al.*, 1998;

Morris and Robinson, 1998), the lack of effect of BFA on 2,4-D accumulation indicated that the target for BFA action was the efflux carrier system. Consistent with this conclusion, treatment of long (30 mm) pea (*Pisum sativum* L.) and zucchini stem segments with BFA abolished polar (basipetal) auxin transport (Robinson *et al.*, 1999). Interestingly, BFA treatment had no effect on NPA binding to microsomal preparations from zucchini, providing additional evidence that the NBP and the auxin efflux catalyst are different proteins. Careful comparisons (Delbarre *et al.*, 1998; Robinson *et al.*, 1999) demonstrated that the lag for BFA action was extremely short (minutes or less) compared to the lag for responses to the protein synthesis inhibitor CH (up to 2 h in zucchini). Taken together, these observations provided strong indirect evidence that an essential component of the auxin efflux carrier system (possibly the efflux catalyst itself, but not the NBP) is targeted to the PM through the BFA-sensitive Golgi-mediated secretory system, and that this component turns over very rapidly at the PM without the need for concurrent protein synthesis (Delbarre *et al.*, 1998; Morris and Robinson, 1998; Robinson *et al.*, 1999). Subsequently, more direct evidence for a rapid turnover of the efflux catalyst came from studies by Steinman *et al.* (1999) of the effect of BFA on the distribution of the putative efflux carrier protein PIN1 in cells of *Arabidopsis* seedling roots. In these, BFA treatment was found to abolish the polar localization of PIN1 in the PM and to result in its accumulation in intracellular compartments. PIN1 mislocalization was also found to occur in the *gnom* mutant of *Arabidopsis*, which also exhibits severe disruption of apical/basal polarity in the developing embryo (Mayer *et al.*, 1993; Geldner *et al.*, 2000). The *GNOM* gene encodes a BFA-sensitive regulator of vesicle trafficking - a guanine nucleotide exchange factor on an ARF-type small GTP-binding protein (ARF-GEF), which is essential for both normal establishment of polarity in the embryo and for polar localization of PIN1 (Steinman *et al.*, 1999).

7. EFFLUX CARRIERS CYCLE RAPIDLY

It has been argued that since the rate of turnover of the efflux carrier proteins is extremely rapid and that the proposed intracellular pools of carrier proteins are likely to be of finite capacity, the rate of efflux carrier turnover observed by Delbarre *et al.* (1998) and Morris and Robinson (1998) could only be sustained if a proportion of the carriers continuously cycled between the PM and the putative internal pool of carrier protein (Morris and Robinson, 1998; Robinson *et al.*, 1999). Support for such a possibility has come from studies of the effects of 2,3,5-triiodobenzoic acid (TIBA), an inhibitor of polar auxin transport, and BFA on the behaviour of PIN1 proteins in *Arabidopsis* root cells (Geldner *et al.*, 2001). It was demonstrated that the BFA-induced internalisation of PIN1 into large and as yet unidentified intracellular compartments (of which most cells appear to contain two), was completely and very rapidly reversed when BFA was washed out. As shown previously in physiological studies (Robinson *et al.*, 1999; see above), treatment of roots with CH for up to 2 hours had no detectable effect on the density of PIN1 in the PM. Furthermore, CH had no effect on the internalisation of PIN1 in response to BFA treatment, or on its re-localisation when BFA was washed out (Geldner *et al.*, 2001). These findings supported the conclusions reached earlier by Robinson *et al.* (1999) and the authors concluded that the BFA response resulted from the inhibition of the PM-directed traffic of PIN1 proteins that normally cycle rapidly between the PM and as yet unidentified endosomal pools (Geldner *et al.*, 2001).

There are several previously reported precedents from animal systems for the rapid cycling of PM located carriers and receptors. Examples of particular interest inrelation to the auxin efflux carriers are the Menkes copper-transporting ATPase (MNK; Petris *et al.*, 1996) and the insulin-dependent GLUT4 glucose transporter (e.g. see Baumann and Saltiel, 2001; Simpson *et al.*, 2001). Similarities between the cycling of GLUT4 in insulin-sensitive mammalian cells and the putative cycling of PIN1 in plant cells have been discussed in detail by Muday and Murphy (2002). Cycling of GLUT4 involves a completely endosomal pathway, whilst MNK cycles between the PM and the *trans* Golgi network (TGN). The steady-state distribution of the MNK protein between the PM and the TGN changes in response to variation in substrate (Cu) levels in the cell (Petris *et al.*, 1996). This property is of particular interest because indirect evidence suggests that the targeting of auxin carriers to the PM also may be stimulated by the carrier substrate, namely auxin itself. This evidence, which has been discussed in detail by Morris (2000), includes observations that auxin is required for the induction of the new pathways of polar auxin transport necessary to induce the formation and axial development of new vascular tissues (eg. Sachs, 1981; Gersani and Sachs, 1985; Berleth and Sachs, 2001); and the well-known requirement for auxin to maintain its own polar transport (Morris and Johnson, 1990; and references therein). Withdrawal of an auxin source leads to a reversible loss of the ability of tissue to sustain polar transport that may be caused by a re-distribution of auxin efflux carriers (Morris and Johnson, 1990). The mechanism by which auxin gradients themselves might establish (or re-establish) new polar transport pathways or change the direction of polar transport in response to environmental or internal cues remains unknown. One possibility arising from the observations discussed above is that auxin gradients in a cell may regulate the direction of Golgi-mediated vesicle traffic to the PM (discussed in Morris, 2000). Auxins are known to stimulate Golgi-mediated vesicle traffic (Hager *et al.*, 1991; Hawes and Satiat-Jeunemaitre, 1996) and by doing so to rapidly increase the level of some PM components, including a PM H^+-ATPase (Hager *et al.*, 1991). Furthermore, as discussed above, *GNOM* (which encodes a BFA-sensitive ARF-GEF) is essential for proper localization of the putative auxin efflux carrier PIN1 (Steinmann *et al.*, 1999).

8. DO POLAR AUXIN TRANSPORT INHIBITORS ACT BY INTERFERING WITH AUXIN CARRIER TARGETING?

Recent observations by several groups have raised some important questions about the interactions between phytotropin binding proteins, auxin efflux carrier trafficking, and the role of the actin cytoskeleton in carrier protein cycling. In an interesting and intriguing development, Geldner *et al.* (2001) reported that TIBA, an inhibitor of polar auxin transport (but not a phytotropin; Katekar and Geissler, 1980; Rubery 1990), blocked the actin-dependent cycling of PIN1. In the absence of BFA, TIBA (25 μM) alone only slightly affected the polar localization of PIN1. However, TIBA prevented the BFA-induced internalisation of PIN1. Moreover, when included in the washout medium, TIBA also blocked the relocalisation of PIN1 to the PM following BFA washout (Geldner *et al.*, 2001). The authors observed similar effects of TIBA on the cycling of other PM-located proteins (a PM H^+-ATPase and the syntaxin KNOLLE) and reported that other polar auxin transport inhibitors behaved similarly (although no supporting data for the latter observation was presented in their paper). On the basis of their observations Geldner *et al.* (2001) suggested that polar auxin transport inhibitors may exert their action through a general effect on the cycling of PM-located proteins – they may, in effect, be inhibitors of membrane trafficking rather than specific non-competitive inhibitors of auxin efflux carriers. Such a possibility might explain why NPA and BFA appear to mimic each other's action. However, for a number of reasons discussed in detail elsewhere (Petrášek *et al.*, 2003), such a suggestion may be hasty. On structural grounds, TIBA cannot be classified as a phytotropin and it behaves very differently from them (Katekar and Geissler, 1980; Rubery, 1990). Depta *et al.* (1983) found that the high affinity binding of TIBA to *Zea mays* L. microsomal membrane preparations was only partially displaced by NPA, suggesting that the two auxin transport inhibitors probably have different loci of action. Indeed, unlike NPA, TIBA undergoes polar transport on carriers that can be competed by IAA, 2,4-D and NAA (Depta and Rubery, 1984), consistent with its own action as a weak auxin. Furthermore, it has not yet been excluded that at the relatively high concentration of 25 μM used by Geldner *et al.* (2001), TIBA does not affect PM protein cycling as a result of non-specific side effects. For example, unlike NPA and IAA, at 25 μM TIBA causes substantial cytoplasmic acidification (see Depta and Rubery, 1984), which in turn would be expected to affect many cellular processes, possibly including vesicle traffic and protein cycling. The concentrations of TIBA (25 μM) and NPA (200 μM) required to inhibit BFA-sensitive PIN1 cycling (Geldner *et al.*, 2001) are both considerably greater than those required to saturate auxin efflux inhibition. TIBA-induced stimulation of NAA accumulation (2 nM) in BY-2 tobacco cells was found to be saturated by as little as 1 μM TIBA (Slizowska, Elčkner and Zažímalová, unpublished data), and from ca. 0.3 to 3 μM NPA (Petrášek *et al.*, 2002, 2003).

Thus, despite the interesting suggestion by Geldner *et al.* (2001), we believe that the mechanism of inhibition of auxin efflux carrier activity by NPA remains an open question.

9. CONCLUSIONS AND FUTURE LINES OF ENQUIRY

It is apparent that the polar auxin transport system is much more complex than was originally believed. There is much evidence that the auxin efflux carrier is a multi-component system consisting of separately targeted transport catalysts and at least one, possibly more, associated regulatory proteins. The transport catalysts are encoded by a large family of related genes that are highly conserved throughout the

Plant Kingdom. The expression of these genes is cell and tissue specific, and their products are targeted to different positions at the PM by mechanisms involving the actin cytoskeleton. At least some efflux catalysts turn over very rapidly in the PM and are probably recycled from internal stores without the need for concurrent protein synthesis, probably via an endosomal pathway, although other possible pathways involving, for example the TGN, have not yet been excluded. The transport catalysts themselves interact with a variety of regulatory proteins, including phytotropin-binding proteins, and it is possible that protein phosphorylation may be important in these interactions.

Such a complex transport system seems to offer considerable scope for regulation at a variety of levels, enabling both the direction and flux of auxin transport to respond rapidly and subtly to changing internal and external conditions. The case of PIN3 is a prime example of such a response. A lateral relocation of this efflux carrier takes place in the *Arabidopsis* seedling hypocotyl in response to gravistimulation. This in turn allows a lateral redistribution of auxin in the hypocotyl and root and an appropriate growth response to the gravistimulus (Friml *et al.,* 2002).

There are clearly many challenges for future research on the polar auxin transport system and many fundamental questions remain to be answered. A mong these are: Firstly, what are the genetic and physiological mechanisms that regulate the differential expression of members of the PIN gene family; and to what extent is PIN gene expression modified by auxin itself, by other internal signals and by external cues? Secondly, what are the targeting and vesicle-docking mechanisms that result in different PIN proteins being targeted to different locations at the PM; are specific targeting and docking signal sequences involved? And precisely how do ARF-GEF proteins like the *GNOM* gene product operate in this process? Thirdly, if as seems likely, at least some of the auxin efflux catalysts (eg. PIN1) can cycle rapidly between internal compartments and the PM, what regulates the steady-state distribution of these proteins? We have seen that the distribution can readily and reversibly be changed using drugs like BFA. Can the distribution between the compartments change in response to physiological signals and/or environmental cues? The existence of mechanisms that would rapidly target pre-existing auxin carriers from internal compartments to specific PM domains in response to internal or environmental signals might facilitate rapid appropriate growth responses to stimuli. There are many more such questions remaining unanswered. Nevertheless, thanks to the rapid advances in molecular biology and cytology techniques, the answers to many of them are no longer beyond our grasp.

ACKNOWLEDGEMENTS

DAM is grateful to Professor Ivana Macháčková, Director, for her generous invitation to undertake research at the Institute of Experimental Botany, Academy of Sciences of the Czech Republic, Prague; and to the UK Royal Society and the Academy of Sciences of the Czech Republic for the award of grants under the European Science Exchange Scheme which made his visits possible. Our joint research was supported in part by grants to EZ from the EU (INCO Copernicus project no. ERBIC 15 CT98 0118) and from the Ministry of Education, Youth and Sports of the Czech Republic (project no. LN00A081), receipt of which is gratefully acknowledged.

REFERENCES

Baker, D.A. (2000) Vascular transport of auxins and cytokinins in *Ricinus*, *Plant Growth Regul.* **32**, 157-60.

Baumannn, C.A. and Saltiel, A.R (2001) Spatial compartmentalization of signal transduction in insulin action, *Bioessays* **23,** 215-222.

Benjamins, R., Quint, A., Weijers, D., Hooykaas, P. and Offringa, R. (2001) The PINOID protein kinase regulates organ development in *Arabidopsis* by enhancing polar auxin transport, *Development* **128,** 4057-4067.

Bennett, M.J., Marchant, A., Green, H.G., May, S.T., Ward, S.P., Millner, P.A., Walker, A.R., Schulz, B. and Feldmann, K.A. (1996) *Arabidopsis AUX1* gene: A permease-like regulator of root gravitropism, *Science* **273**, 948-950.

Bennett, M.J., Marchant, A., May, S.T. and Swarup, R. (1998) Going the distance with auxin: unravelling the molecular basis of auxin transport, *Philos. Trans. R. Soc. Lond. Ser.* B **353**, 1511-1515

Berleth, T. and Sachs, T. (2001) Plant morphogenesis: long-distance coordination and local patterning. Curr. Opin. *Plant Biol.* **4**, 57-62.

Berleth, T., Mattsson, J. and Hardtke, C.S. (2000) Vascular continuity, cell axialization and auxin, *Plant Growth Regul.* **32**, 173-185.

Bernasconi, P., Patel, B.C., Reagan, J.D., and Subramanian, M.V. (1996) The N-1-naphthylphthalamic acid-binding protein is an integral membrane protein, *Plant Physiol.* **111,** 427-432.

Butler, J.H., Hu, S.Q., Brady, S.R., Dixon, M.W. and Muday, G.K. (1998) In vitro and in vivo evidence for actin association of the naphthylphthalamic acid-binding protein from zucchini hypocotyls, *Plant J.* **13**, 291-301.

Cambridge, A.P. and Morris, D.A. (1996) Transfer of exogenous auxin from the phloem to the polar auxin transport pathway in pea (*Pisum sativum* L), *Planta* **199**, 583-588.

Cande, W.Z., Goldsmith, M.H.M. and Ray, P.M. (1973) Polar auxin transport and auxin-induced elongation in the absence of cytoplasmic streaming, *Planta* **111**, 279-296.

Chen, L., Ortiz-Lopez, A., Jung, A. and Bush, D.R. (2001) *ANT1*, an aromatic and neutral amino acid transporter in Arabidopsis, *Plant Physiol.* **125**, 1813-1820.

Chen, R., Hilson, P., Sedbrook, J., Rosen, E., Caspar, T. and Masson, P.H.(1998) The Arabidopsis thaliana AGRAVITROPIC 1 gene encodes a component of the polar-auxin-transport efflux carrier, *Proc. Natl. Acad. Sci.,* USA **95**, 15112-15117.

Cox, D.N. and Muday, G.K.(1994) NPA binding activity is peripheral to the plasma membrane and is associated with the cytoskeleton, *Plant Cell* **6**, 1941-1953.

Davies, P.J. (1995) The plant hormone concept: Concentration, sensitivity and transport, in P.J. Davies, (ed.), *Plant Hormones: Physiology, Biochemistry and Molecular Biology*, 2nd edn. Kluwer Academic Publishers, Dordrecht, Boston, London , pp. 13-38.

Delbarre, A., Muller, P., Imhoff, V. and Guern, J.(1996) Comparison of mechanisms controlling uptake and accumulation of 2,4-dichlorophenoxy acetic acid, naphthalene-1-acetic acid, and indole 3-acetic acid in suspension-cultured tobacco cells, *Planta* **198**, 532-541.

Delbarre, A., Muller, P. and Guern, J.(1998) Short-lived and phosphorylated proteins contribute to carrier-mediated efflux, but not to influx, of auxin in suspension-cultured tobacco cells, *Plant Physiol* **116,** 833-844.

Depta, H., Eisele, K.-H. and Hertel, R.(1983) Specific inhibitors of auxin transport: action on tissue segments and in vitro binding to membranes from maize coleoptiles, *Plant Sci. Lett.* **31**, 181-192.

Depta, H. and Rubery, P.H.(1984) A comparative study of carrier participation in the transport of 2,3,5-trriodobenzoic acid, indole-3-acetic acid, and 2,4-dichlorophenoxyacetic acid by *Cucurbita pepo* L. hypocotyl segments, *J. Plant Physiol.* **115**, 371-387.

Dixon, M.W., Jacobson, J.A., Cady, C.T. and Muday, G.K.(1996) Cytoplasmic orientation of the naphthylphthalamic acid-binding protein in zucchini plasma membrane vesicles, *Plant Physiol.* **112**, 421-432.

Druère, J., Jackson, K., Garbers, C., Söll, D. and DeLong, A (1999) The RCN1-encoded A subunit of protein phosphatase 2A increases phosphatase activity in vivo, *Plant J.* **20**, 389-399.

Friml, J. (2003) Auxin transport – shaping the plant, Curr. Opin. Plant. Biol. **6**, 1-6.

Friml, J. and Palme, K.(2002) Polar auxin transport - old questions and new concepts?, *Plant Mol. Biol.* **49**, 273-284.

Friml, J., Wisniewska, J., Benková, E., Mendgen, K. and Palme, K.(2002) Lateral relocation of auxin efflux regulator PIN3 mediates tropism in *Arabidopsis*, *Nature* **415**, 806-809.

Gälweiler, L., Guan, C., Müller, A., Wisman, E., Mendgen, K., Yephremov, A. and Palme, K.(1998) Regulation of polar auxin transport by AtPIN1 in *Arabidopsis* vascular tissue, *Science* **282,** 2226-2230.

Gaedeke, N., Klein, M., Kolukisaoglu, U., Forestier, C., Müller, A., Ansorge, M., Becker, D., Mamnun, Y., Kuchler, K., Schulz, B., Mueller-Roeber, B. and Martinoia, E.(2001) The *Arabidopsis thaliana* ABC transporter *At*MRP5 controls root development and stomata movement, *EMBO J.* **20**, 1875-1887.

Garbers,C., DeLong, A., Deruère, J., Bernasconi, P. and Söll, D.(1996) A mutation in protein phosphatase 2A regulatory subunit affects auxin transport in A*rabidopsis*, *EMBO J.* **15**, 2115-2124.

Geldner, N., Hamann, T. and Jürgens, G.(2000) Is there a role for auxin in early embryogensis?, *Plant Growth Regul.* **32**, 187-191.

Geldner, N., Friml, J., Stierhof, Y.-D., Jürgens, G. & Palme, K.(2001) Auxin transport inhibitors block PIN 1 cycling and vesicle traficking, *Nature* **413**, 425-428.

Gersani, M. and Sachs, T.(1985) Polarity reorientation in beans expressed by vascular differentiation and polar transport, *Differentiation* **25,** 205-208.

Gil, P., Dewey, E., Friml, J., Zhao, Y., Snowden, K.C., Putterill, J., Palme, K., Estelle, M. and Chory, J.(2001) BIG: a calossin-like protein required for polar auxin transport in *Arabidopsis, Genes Dev.* **15,** 1985-1997.

Goldsmith, M.H.M.(1977) The polar transport of auxin, Annu. Rev. *Plant Physiol.* **28,** 439-478.

Grebe, M., Friml, J., Swarup, R., Ljung, K., Sandberg, G., Terlou, M., Palme, K., Bennett, M.J. and Scheres, B.(2002) Cell polarity signaling in Arabidopsis involves a BFA-sensitive influx pathway, *Current Biol* .**12**, 329-334.

Hager, A., Debus, G., Edel, H.-G., Stransky, H. and Serrano, R.(1991) Auxin induces exocytosis and the rapid synthesis of a high-turnover pool of plasma-membrane H+-ATPase, *Planta* **185**, 527-537.

Hawes, C. and Satiat-Jeunemaitre, B.(1996) Stacks of questions: how does the plant Golgi work?, *Trends Plant Sci.* **1**, 395-401.

Hu, S.Q., Brady, S.R., Kovar, D.R., Staiger, C.J., Clark, G.B., Roux, S.J. and Muday, G.K.(2000) Identification of plant actin binding proteins by F-actin affinity chromatography, *Plant J.* **24**, 127-137.

Katekar, G.F. and Geissler, A.E.(1980) Auxin transport inhibitors: IV. Evidence of a common mode of action for a proposed class of auxin transport inhibitors: The phytotropins, *Plant Physiol.* **66**, 1190-1195.

Lomax, T.L., Muday, G.K. and Rubery, P.H.(1995) Auxin transport, in: P.J. Davies, (ed.), *Plant Hormones: Physiology, Biochemistry and Molecular Biology,* 2nd edn. Kluwer Academic Publishers, Dordrecht, Boston, London, pp. 509-530.

Luschnig, C., Gaxiola, R.A., Grisafi, P. and Fink, G.R. (1998) EIR1, a root-specific protein involved in auxin transport, is required for gravitropism in *Arabidopsis thaliana, Genes Dev.* **12**, 2175-2187.

Luschnig, C.(2001) Auxin transport: why plants like to think BIG, *Curr. Biol.* **11**, 831-833.

Luschnig, C.(2002) Auxin transport: ABC proteins join the club, *Trends in Plant Science* **7**, 329-332.

Martinoia, E., Klein, M., Geisler, M., Bovet, L., Forestier, C., Kolukisaoglu, Ü., Műller-Röber, B. and Schulz, B. (2002) Multifunctionality of plant ABC transporters – more than just detoxifiers, *Planta* **214**, 345-355.

Mayer, U., Buttner, G. and Jurgens, G. (1993) Apical-basal pattern formation in the *Arabidopsis* embryo: studies of the role of the *gnom* gene, *Development* **117,**149-162.

Morris, D.A.(1996) Hormonal regulation of source-sink relationships: an overview of potential control mechanisms, in: Zamski, E. and Schaffer, A.A. (eds.), *Photoassimilate Distribution in Plants and Crops,* Marcel Dekker Inc., New York, Basel, Hong Kong, pp.441-465.

Morris DA (2000) Transmembrane auxin carrier systems - dynamic regulators of polar auxin transport, *Plant Growth Regul.* **32**, 161-172.

Morris, D.A. and Arthur, E.D. (1984) Invertase and auxin-induced elongation in internodal segments in *Phaseolus vulgaris* L., *Phytochemistry* **23**, 2163-2167.
Morris, D.A. and Arthur, E.D.(1985) Invertase activity, carbohydrate metabolism and cell expansion in the stem of *Phaseolus vulgaris* L., *J. Exp. Bot.* **36,** 623-633.
Morris, D.A. and Johnson, C.F.: The role of auxin efflux carriers in the reversible loss of polar auxin transport in the pea (*Pisum sativum* L.) stem. - *Planta* 181: 117-124, 1990.
Morris, D.A., Rubery, P.H., Jarman, J. and Sabater, M. (1991) Effects of inhibitors of protein synthesis on transmembrane auxin transport in *Cucurbita pepo* L. hypocotyl segments, *J. Exp. Bot.* **42**, 773-783.
Morris, D.A. and Robinson, J.S. (1998) Targeting of auxin carriers to the plasma membrane: differential effects of brefeldin A on the traffic of auxin uptake and efflux carriers, *Planta* **205**, 606-612,
Morris, D.A. and Thomas, A.G. (1978) A microautoradiographic study of auxin transport in the stem of intact pea seedlings (*Pisum sativum* L.), *J. Exp. Bot.* **29**, 147—157.
Muday, G.K. (2000) Maintenance of asymmetric cellular localization of an auxin transport protein through interaction with the actin cytoskeleton, *J. Plant Growth Regul.* **19,** 385-396.
Muday, G.K. and DeLong, A. (2001) Polar auxin transport: controlling where and how much, *Trends Plant Sci.* **6**, 535-542.
Muday, G.K. and Murphy, A.S. (2002) An emerging model of auxin transport regulation, *Plant Cell* **14**, 293-299.
Müller, A., Guan, C.H., Gälweiler, L., Tänzler, P., Huijser, P., Marchant, A., Parry, G., Bennett, M., Wisman, E. and Palme, K. (1998) *AtPIN2* defines a locus of *Arabidopsis* for root gravitropism control, *EMBO J.* **17**, 6903-6911.
Noh, B., Murphy, A.S. and Spalding, E.P. (2001) *Multidrug resistance*-like genes of Arabidopsis required for auxin transport and auxin-mediated development, *Plant Cell* **13**, 2441-2454.
Petrášek, J., Elčkner, M., Morris, D.A. and Zažímalová, E. (2002) Auxin efflux carrier activity and auxin accumulation regulate cell division and polarity in tobacco cells, *Planta* **216**, 302-308.
Petrášek, J., Černá, A., Schwarzerová, K., Elčkner, M., Morris, D.A. and Zažímalová, E. (2003) Do phytotropins inhibit auxin efflux by impairing vesicle traffic?, *Plant Physiol.* **131**, *in press.*
Petris, M.J., Mercer, J.F.B., Culvenor, J.G., Lockhart, P., Gleeson, P.A. and Camakaris, J. (1996) Ligand-regulated transport of the Menkes copper P-type ATPase efflux pump from the Golgi apparatus to the plasma membrane: a novel mechanism of regulated trafficking, *EMBO J.* **15**, 6084-6095.
Raven, J.A. (1975) Transport of indoleacetic acid in plant cells in relation to pH and electrical potential gradients, and its significance for polar IAA transport, *New Phytol.* **74**, 163-174.
Richards, S., Hillman, T. and Stern, M. (1996) Mutations in the *Drosophila* pushover gene confer increased neuronal excitability and spontaneous synaptic vesicle fusion, *Genetics* **142,** 1215-1223.
Robinson, J.S., Albert, A.C. and Morris, D.A. (1999) Differential effects of brefeldin A and cycloheximide on the activity of auxin efflux carriers in *Cucurbita pepo* L., *J. Plant Physiol.* **155**, 678-684..
Rubery, P.H. (1990) Phytotropins: receptors and endogenous ligands, Symp. Soc. Exp. Biol. **44,** 119-146.
Rubery, P.H. and Sheldrake, A.R. (1974) Carrier-mediated auxin transport, *Planta* **188**, 101-121.
Ruegger, M., Dewey, E., Hobbie, L., Brown, D., Bernasconi, P., Turner, J., Muday, G. and Estelle, M. (1997) Reduced naphthylphthalamic acid binding in the *tir3* mutant of Arabidopsis is associated with a reduction in polar auxin transport and diverse morphological defects, *Plant Cell* **9**, 745-757.
Sachs, T. (1981) The control of the patterned differentiation of vascular tissues, *Adv. Bot. Res.* **9**, 151-262.
Simpson, F., Whitehead, J.P. and James, D.E. (2001) GLUT4: at the crossroads between membrane trafficking and signal transduction, *Traffic* **2,** 2-11.
Steinman, T., Geldner, N., Grebe, M., Mangold, S., Jackson, C.L., Paris, S., Gälweiler, L., Palme, K. and Jürgens, G. (1999) Coordinated polar localization of auxin efflux carrier PIN1 by GNOM ARF GEF, *Science* **286**, 316-318.
Sussman, M.R. and Gardner, G. (1980) Solubilization of the receptor for N-1-naphthylphthalamic acid, *Plant Physiol.* **66**, 1074-1078.
Swarup, R., Friml, J., Marchant, A., Ljung, K., Sandberg, G., Palme, K. and Bennett, M. (2001) Localization of the auxin permease AUX1 suggests two functionally distinct hormone transport pathways operate in the *Arabidopsis* root apex, *Genes Dev.* **15**, 2648-2653.
Wilkinson, S. and Morris, D.A. (1994) Targeting of auxin carriers to the plasma membrane: effects of monensin on transmembrane auxin transport in *Cucurbita pepo* L. tissue, *Planta* **193**, 194-202.
Zažímalová, E. and Napier, R.M. (2003) Points of regulation for auxin action, *Plant Cell Rep.* , *in press.*

ROLE OF CALCIUM IONS IN PLANT GROWTH AND MECHANISM OF IAA ACTION

S.S. MEDVEDEV AND I.V. MARKOVA
Department of Plant Physiology and Biochemistry, St. Petersburg State University, Oranienbaumskoe 2, 198504 St. Petersburg, Russia

1. INTRODUCTION

Calcium ions have unique properties and a universal ability to conduct different signals which influence responses of cells to stimuli such as hormones, pathogens, light, gravitation etc. (Trewavas and Malho, 1998; Sanders *et al.*, 1999; Roos, 2000; Pandey *et al.*, 2000).

The process of signal transduction usually includes the following steps: perception of a stimulus by a membrane-bound receptor, activation of membrane components of signal transduction: opening of ion channels and synthesis of secondary messengers, increase in the level of second messengers in cell compartments, protein activation by second messengers, protein phosphorylation and signal amplification, and protein dephosphorylation. Secondary messengers are synthesized as a result of the hydrolysis of phosphatidylinositols and/or activation of some enzymes like adenylatecyclase, and most commonly activate proteinkinases. Ca^{2+} plays an important role in every step; it is essential for transducing signals from many stimuli.

The concentration of free Ca^{2+} in the cytoplasm of plant cells under constant conditions is very low varying from 100 to 200 nM. The free Ca^{2+} concentration is much higher inside the organelles. For instance, in the endoplasmic reticulum and mitochondria there is 1 mM of Ca^{2+}. In the cell wall and vacuoles its concentration ranges from 1 to 10 mM (Trewavas and Malho, 1998). Thus, there is a very sharp gradient of free Ca^{2+} from the plasma membrane, tonoplast, membranes of mitochondria and endoplasmic reticulum, which is much higher than any other ion gradient. That is why the increase of calcium level in a precise location in the cytoplasm is a powerful and informative signal. Ca^{2+} ions are effective regulators of metabolic processes in any cell; the regulated systems react to changes from as low as 0.1 nM up to 10 nM in its concentration. At present, more than 100 Ca-binding proteins are known (Zielinski, 1998; Harmon *et al.*, 2000).

Almost every external stimulus leads to an abrupt increase in cytosolic Ca^{2+} concentration by one to two orders of magnitude, to Ca^{2+} interaction with different Ca-binding proteins and to the initiation of Ca-dependent physiological processes. The paradox of Ca-signalling is the necessity of high cytoplasmic calcium, which

I. Macháčková and G.A. Romanov (eds.), Phytohormones in Plant Biotechnology and Agriculture, 157-169

can be lethal if it lasts for a prolonged time. Cells normally use low amplitude Ca-signals or generate very short calcium signals known as calcium waves (Malho *et al.*, 1998; Jaffe, 1999; Allen *et al.*, 2001). A calcium wave initiated in a local site in the cell is the basis of Ca^{2+} signaling because wave transmission is much faster than diffusion. Wave amplitude, frequency and location can be used for signal coding. Therefore the characteristic feature of a calcium signal is its transfer through the cytoplasm or the tissue in the form of calcium waves. Such a mode to transmit information is not only much faster than diffusion, but also provides spatio-temporal regulation of cell functions.

The ability to rapidly regulate Ca^{2+} level in the cell compartments is based on the coordinated functioning of a membrane transport system which includes different types of Ca-channels, Ca-ATPases, and Ca/H-antiporters (Krol and Trebacz, 2000; White, 2000; Geisler *et al.*, 2000; Sze *et al.*, 2000). In the plant plasma membrane there are several classes of Ca-transporting channels: voltage-sensitive, mechano-sensitive and receptor-operated channels (Krol and Trebacz, 2000; White, 2000). Low-selective cation channels, usually designated as K^+/Ca^{2+}, can also act as Ca-channels (White *et al.*, 2000). Four types of Ca-channels have been identified in the vacuolar membrane (Krol and Trebacz, 2000; White, 2000). Two of them are ligand-operated: ITP-dependent and cADP-ribose-dependent. In the tonoplast there are two types of voltage-sensitive Ca-channels. One of them is hyperpolarization-activated, another one is depolarization-activated. Depolarization-activated channels are also called slowly activating vacuolar (SV) channels.

Ca-pumps belong to the superfamily of P-type ATPases that use ATP directly for ion transfer (Geisler *et al.*, 2000; Sze *et al.*, 2000). Two different families of Ca-pumps are classified according to their amino acid sequences. They are either type IIA (endoplasmic type) or type IIB (plasma membrane type). They differ by two criteria: their location in the ER or plasma membrane and their sensitivity to inhibitors. The ER-type is inhibited by cyclopiazonic acid, thapsigargin, and calmodulin whereas the PM-type is activated by calmodulin. Ca^{2+}/H^+ antiporters can potentially provide Ca^{2+} transport against a concentration gradient by using a proton-motive force. CAX1 was the first cloned gene of a Ca^{2+}/H^+ antiporter (Geisler *et al.*, 2000; Sze *et al.*, 2000).

Almost every external stimulus leads to an increase in cytosolic Ca^{2+} level and to Ca^{2+} interaction with different Ca-binding proteins. Some of them provide Ca^{2+} transport inside cell or function as a kind of buffer maintaining a low Ca level in the cytosol. The structure of such proteins are not considerably changed after Ca^{2+} binding. If Ca^{2+} interacts with proteins that are able to perform regulatory functions, then there are considerable changes in protein structure following the formation of a Ca-protein complex. The distinguishing feature of regulatory Ca-binding proteins is the presence of a special Ca-binding site called EF-hand (Zielinski, 1998). The central part of this hand is a Ca-binding loop consisting of 12 amino acid residues. After the formation of a Ca-protein complex the protein is able to regulate activity of many intra- and extracellular processes. Plant Ca-modulated proteins of this type are represented by the well-studied calmodulin (CaM) and Ca-dependent CaM-independent protein kinase.

Calmodulin, containing four Ca-binding centers has been found almost in every cryptogamous and higher plant studied (Zielinski, 1998; Chin and Means, 2000). After Ca^{2+} concentration drops to an initial level, Ca^{2+} ions dissociate from Ca-binding proteins, which leads to inactivation of these proteins. Ca-dependent (CaM-independent) protein kinase (CDPK) is the most investigated plant protein among those regulated by Ca^{2+} ions (Harmon *et al.*, 2000), and it is also found in protists. Until now they were not found in yeast and mammalian cells. CDPK have a protein kinase domain and also the region analogous to calmodulin, with 4 Ca-binding EF-arms.

The generation of a Ca^{2+} signal in a cell is achieved as follows. Ca-channels are normally closed in the resting cell. Under the action of a specific stimulus some populations of Ca-channels are activated which have definite locations, durations of open state, conductivities and other parameters of functioning. Trewavas and Malho (1998) compared Ca^{2+} response to a concrete signal with a finger print because it has unique spatial structure. Ca^{2+} enters a cell from the external medium through Ca-channels in the plasma membrane. Ca^{2+} can also be transported from intracellular Ca-stores through Ca-channels in endomembranes. The most important features of a Ca^{2+} signal are its physical distribution, period, and amplitude. When a Ca-channel is open, then a cloud of high Ca^{2+} concentration is formed near the inner opening of the channel. After the channel closes, this cloud of Ca^{2+} rapidly spreads in different directions. Spatio-temporal characteristics of these elementary events (Ca^{2+}-spikes) depend on the type of Ca-channel and its location, on the architecture of the cell and on cytoplasmic streaming (cyclosis) within the cell.

The transmission of a Ca-signal includes at least two steps. First, the transient locally increases cytoplasmic Ca^{2+} concentration. Second, the extinction of a Ca signal by pumping Ca^{2+}out of the cytoplasm to the extracellular space and intracellular stores. The first step requires the action of signal-sensitive Ca-channels in the plasma membrane and other cell membranes. The second step consumes energy and requires activated Ca-ATPase. Ca^{2+} signals perform at least two functions. They activate local cell processes in close proximity to Ca-channels, or they generate Ca-waves, which act on other channels and thus indirectly activate further intracellular processes. If several cells are interacting, then such intracellular waves can be transmitted to neighboring cells. In this manner intercellular waves are formed that coordinate cellular responses inside tissues and organs. The theoretical basis of our investigation is that interaction of active basipetal auxin transport with polar calcium ion fluxes is the physiological basis of polarity.

The aim of our investigation was to analyse the possibility that intracellular and intercellular IAA signals are transduced through calcium ions and to elucidate the interaction of polar auxin and calcium fluxes during longitudinal growth and gravitropic responses.

2. LONG-DISTANCE POLAR CALCIUM TRANSPORT

Calcium ion movement towards the shoot apex in the absence of transpiration was reported for the first time by de Guzman and de la Fuente (1984) in experiments performed with Ca^{2+}-sensitive electrodes on vertically-oriented sunflower hypocotyl segments (Guzman and de la Fuente, 1984, 1986). Polar calcium transport was also detected in lettuce leaf petioles and young tomato fruits with ^{45}Ca (Banuelos *et al.*, 1987, 1988].

We investigated polar Ca^{2+} transport in sections of maize coleoptiles and pea epicotyls. Polar calcium transport was measured using Ca^{2+}- selective electrodes (prepared at St. Petersburg State University) and ^{45}Ca in the condition of 100% humidity. Growth and gravitropic reaction was estimated using an electromechanical recorder and a microphote. ^{45}Ca was applied in agar blocks at the basal end of coleoptile sections and its transport in the acropetal direction was measured (Medvedev *et al.*, 1989; Markova *et al.*, 1997). Fig. 1 shows the distribution of ^{45}Ca in maize coleoptile and pea epicotyls sections added in the agar block at the apical or basal end. We can see that in the acropetal treatment calcium did not move along the section. When we labelled the basal part with radioactive calcium, the label moved in the acropetal direction at average rate of 16 mm/h.

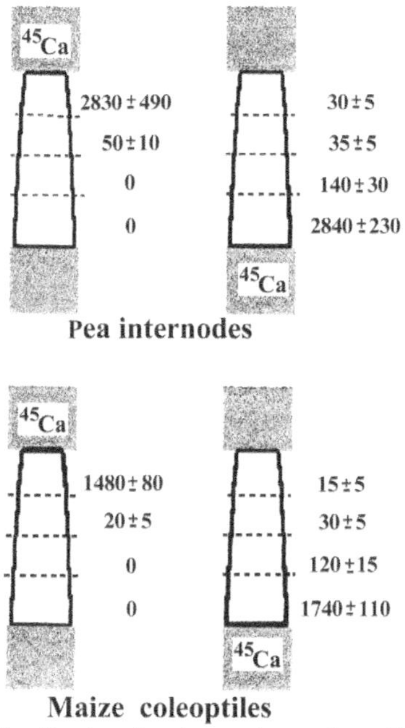

Figure 1. Polar ^{45}Ca distribution in the maize coleoptiles and pea internodes sections (cpm, 60 min).

We should note that the rate of polar transport of calcium is comparable with the rate of IAA polar transport. Polar Ca^{2+} fluxes were also detected with Ca^{2+}-selective electrodes in the acropetal direction in vertically oriented maize coleoptiles and

Arabidopsis inflorescences. It confirmed the polar movement of endogenous Ca^{2+} ions. Ca^{2+} secretion was more intensive in the apical parts of vertically oriented sections than in the basal ones (Fig. 2, 3). Moreover we can see that polar calcium transport is a non-linear process, with an oscillation period of about 60 minutes. Polar calcium transport was suppressed by verapamil, which blocks voltage-gated Ca-channels and orthovanadate, erythrosin B, La^{3+}, which inhibit plasmalemma ATPases. Thus, we believe that polar Ca^{2+} fluxes result from the cooperation of Ca-channels and Ca-ATPases (Fig. 3). Fig. 2 shows secretion of calcium ions from the apical and basal ends of maize coleoptile and *Arabidopsis* inflorescence segments, and demonstrates the possibility of polar acropetal calcium transport in tissues of *Arabidopsis* inflorescence and maize coleoptiles.

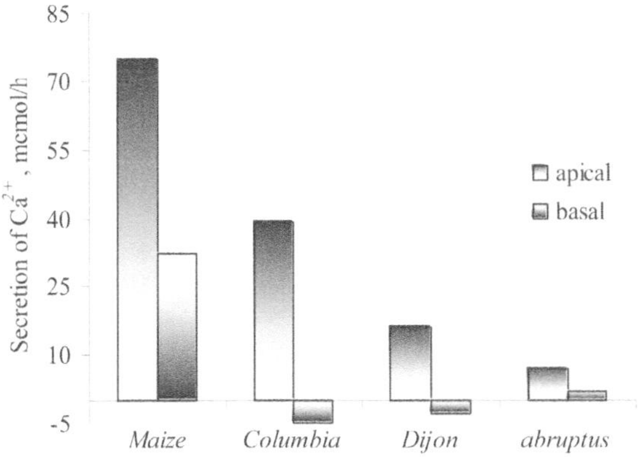

Figure 2. Ca^{2+} secretion from apical and basal ends of maize coleoptile and Arabidopsis inflorescence sections. Columbia and Dijon – races of wild type Arabidopsis; abruptus – PINOID mutant of Dijon. Segments (vertically oriented) were bathed for 1.5 h in medium A containing 15 μM $CaCl_2$ and 1 mM KCl and transferred into the chamber. Apical and basal chamber compartments were filled with medium A so that only the apical and basal end of the segments (2 mm) were in contact with the solution. Ca^{2+} secretion was measured with ion-selective electrodes in darkness at 23° C (Arabidopsis) or 25°C (maize) and 100% relative air humidity. Ca^{2+} secretion was calculated according to (Markova et al., 1997).

The comparative analysis of growth dynamics, hormonal and calcium status was carried out in *Arabidopsis* plants of race Dijon and pinoid type of mutants - *abruptus* (*abr*, which have been kindly supplied by Dr. T.A.Ezhova from Dept. Genetics, Moscow State University). The mutation is located in the 2nd chromosome and results in the formation of an inflorescence without flowers at 27-30 °C, and in the development of abnormal flowers at lower temperatures. It was shown (Bennet *et al.*, 1998) that this mutation is related to auxin transport. Our experiments showed that at 26-28°C the gravitropic reaction of *abr* was 3-4 times lower than that of wild

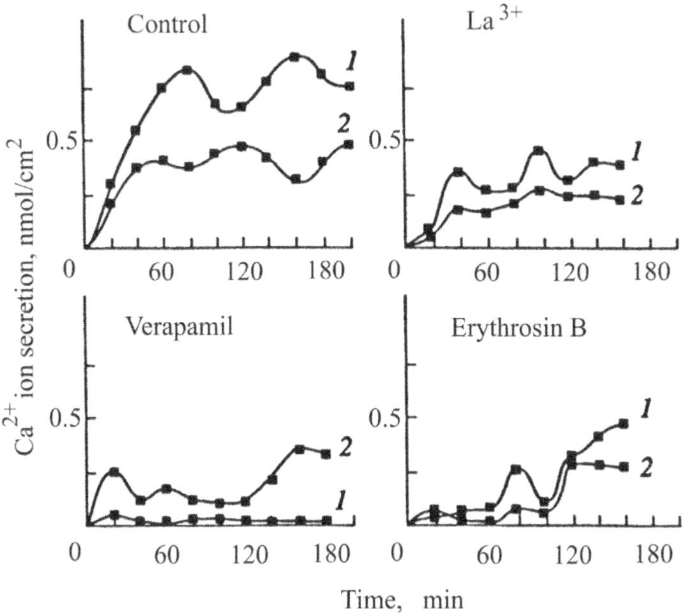

Figure 3. Influence of inhibitors on polar calcium transport in maize coleoptiles. Control – time course of Ca^{2+} ion secretion from the apical and basal ends of maize coleoptile segments. Coleoptile segments were bathed for 1.5 h in medium A (see legend to Fig. 2) containing 15 μM $CaCl_2$, 1 mM KCl and respective inhibitors.

type plants, race Dijon (Medvedev *et al.*, 2002). In mutants *abr* polar calcium transport in the acropetal direction was 2-4 times slower than in wild type plants. Therefore disturbances of polar growth and morphogenesis in mutants *abr* may not only be related to impaired auxin transport but to the disturbances of the polar calcium transport as well. So, we conclude that both a certain level of calcium and its active polar transport are necessary to support IAA polar transport, polar growth and morphogenesis. Naphthylphtalamic acid and triiodbenzoic acid, which are auxin transport inhibitors, suppressed Ca^{2+} polar movement in the acropetal direction, demonstrating the coupling of auxin and calcium polar flows in plant tissues (Tab. 1).

Table 1. The influence of calcium and auxin transport inhibitors on longitudinal growth (micrometer/h) and ^{45}Ca transport (mm/h) in pea internodes.

	^{45}Ca transport	Growth
Control	19 ± 1	69 ± 17
Verapamil, 0.13 mM	11 ± 2	12 ± 3
Na_3VO_4, 0.1 mM	12 ± 2	40 ± 4
TIBA, 0.1mM	7 ± 2	16 ± 5
NPA, 0.1 mM	3 ± 1	20 ± 6
EGTA, 0.1 mM	-	9 ± 6

On the other hand, the Ca-channel blocker verapamil, inhibited both polar calcium transport and the basipetal auxin flux (Tab. 2). Polar acropetal fluxes of ^{45}Ca are measured in an atmosphere of 100 % humidity, which does not allow Ca^{2+}

Table 2. The influence of the verapamil (0.13 mM) on ^{14}C-IAA and ^{45}Ca transport in maize coleoptile sections.

	^{14}C-IAA		**^{45}Ca**	
Segments	*Control*	*Verapamil*	*Control*	*Verapamil*
1 - apical	2100 ± 140	2130 ± 130	20 ± 5	0
2	700 ± 100	260 ± 15	45 ± 5	0
3	100 ± 20	20 ± 5	110 ± 30	40 ± 5
4	20 ± 5	0	800 ± 260	1680 ± 160

movement with transpiration flow. The suppression of the polar transport of reduces the rate of coleoptile section growth and gravitropic reaction (Medvedev *et al.*, 1991). So, we conclude that not only certain level of calcium ions, but also their active polar transport are necessary for polar growth processes.

Thus, in plant tissues polar fluxes of Ca^{2+} ions exist together with active basipetal transport of IAA. Calcium moves acropetally in vertically oriented axial organs of plants and laterally during shoot gravistimulation. We think that oscillating polar Ca-fluxes are the main element of polar growth regulation and perform both "spatial" and "temporal" transduction of hormonal and non-hormonal signals. Polar oscillating calcium fluxes are probably formed in the tissues even earlier than auxin fluxes and determine the direction of IAA movement. Calcium ions evidently transfer the primary information about the polarization vector and create the axis of polarity not only during the alteration of plant spatial orientation (e.g. mutations), but also during changes in vector factors from the environment. Ca^{2+} ions regulate IAA polar transport at three points: (1) IAA influx carrier, (2) IAA efflux carrier, (3) IAA carrier synthesis and/or vesicle transport, which requires association and dissociation of actin filaments. It is known now that vesicle-dependent transport of the PIN1 protein to the basal membrane appears to be along actin tracks (Murphy, 2002). It is necessary to keep in mind that calcium ions are able to control not only the process of active polar auxin transport, but also to take part in IAA intracellular signal transduction.

3. MECHANISM OF IAA ACTION ON CA^{2+} ION LEVEL IN THE CYTOPLASM

The mechanism of any plant hormone involves at least three stages. First, the hormone "perceives" the signal. Second, the signal is transduced through secondary messengers, such as Ca^{2+} ions, to the executive cell mechanisms. Third, the physiological response achieved through the action of «executive» molecular targets (ion channels and pumps, cytoskeletal proteins, genes, etc.) which in their turn switch on a concrete physiological process. The goal of our work was to investigate the role of plasma membrane ion channels in transduction of the IAA signal in plant cells. Vesicles of maize coleoptile plasma membranes were purified in a sucrose density gradient as described earlier (Markova et al., 1995). Calcium transport was investigated with the use of fluorescent probe indo-1 loaded into vesicles. Membrane potential differences in vesicles were created by a of membrane potassium gradient and potassium-valinomycin voltage-clamp (Medvedev et al., 1989). Membrane potential was calculated as the potassium diffusion potential difference measured by diS-C_3-(5)fluorescence. For calculation of free $[Ca^{2+}]$ concentration inside vesicles the BAD program (bound and determined), kindly provided by S. Brooks, was used (Brooks and Storey, 1992). Figure 4 shows that addition of IAA (10 μM) to the medium containing vesicles and Ca^{2+} ions increased fluorescence of indo-1 inside vesicles. Tryptophan in similar conditions didn't change indo-1 fluorescence. Ni^{2+} (an inhibitor of receptor-operated ion channels) reduced the stimulating effect of IAA. Therefore, the presence of IAA-sensitive receptor-operated Ca-channels in plasma membrane of maize coleoptile cells could be measured.

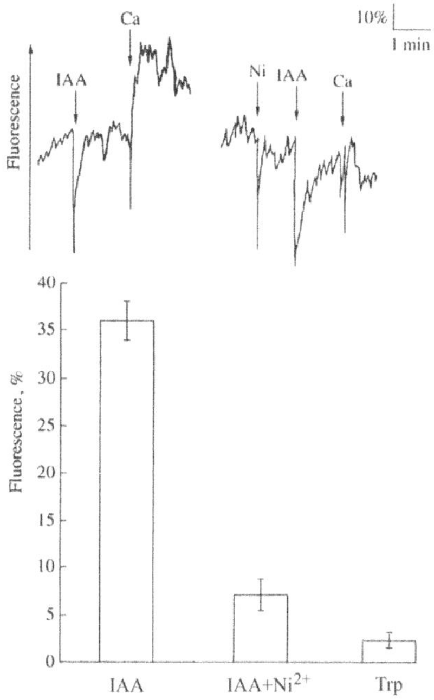

Figure 4. Effect of IAA on calcium ions permeability of the plasma membrane vesicles from maize coleoptile cells.
(a) The characteristic curves of Ca^{2+} permeability estimated by Indo-1. The medium inside the vesicles: 10 mM Tris-Mes, 150 mM sucrose, 0.2 mM EGTA, 150 mM K_2SO_4, 17 mcM Indo-1. The incubation medium: 10 mM Tris-Mes, 150 mM sucrose, 0.2 mM EGTA, 150 mM Na_2SO_4, pH 7.2. The addition of 10 μM IAA, 1 mM Ca^{2+} and 0.5 μM Ni^{2+} indicated by the arrows. The fluorescence of Indo-1 inside vesicles after treatment with ionophore A-23187 in calcium medium (1mM $CaCl_2$) is taken as 100%.
(b) Influence of IAA (10 μM) and tryptophane (trp, 10 μM) on calcium permeability of plasma membrane vesicles.

One of the most important targets of auxin are plasma membrane anion channels (Barbier-Brygoo *et al.*, 1991; Zimmermann *et al.*, 1994). In tobacco protoplasts the auxin-modulated channels were discovered by patch-clamping. These anion channels were activated not only by auxin, but also by antibodies against auxin binding proteins (Barbier-Brygoo *et al.*, 1996). So, IAA action on Ca-fluxes through vesicle membranes as a consequence of membrane potential depolarization, connected with activation of anion channels by IAA, could not be excluded. Earlier we have shown the existence of voltage-gated Ca-channels in the plasma membrane of maize coleoptile cells (Medvedev et al., 1989). The dependence of voltage-gated Ca-channel activity on potential difference is shown in Fig. 5, demonstrating that these channels are activated by membrane depolarization. Direct measurements of cytosolic Ca^{2+} with the fluorescent Ca^{2+} probe fluo3 and confocal microscopy showed that treatment of maize coleoptiles with 2,4-D within 4 minutes caused an increase in cytoplasmic Ca^{2+} from 280 to 380 nM (Gehring et al., 1990). In Fig. 6 our scheme of the mechanism of IAA action in the membrane is shown. It has been thought that the earliest membrane effects of IAA are connected with the activation of anion and Ca-channels (Medvedev et al., 1998). Calcium increase inside the cell activates Ca-dependent K-channels, efflux K^{+}_{cyt} and hyperpolarization of plasma membrane. Increasing of K^+ outside cells activates H^+-ATPase, which pumps out protons and leads to the influx of K^+ up to its initial level in cytoplasm. This scheme is based on the hypothesis that the earliest effects of IAA are connected with the activation of anion and Ca-channels. Ca-channels may be of two different groups: voltage-gated and receptor-operated (Medvedev *et al.*, 1989; Sanders *et al.*, 1999; Crol and Treacz, 2000; White, 2000). Voltage-gated Ca-channels are significant during the transduction of different (i.e. hormonal) signals and activated by membrane depolarization. In plant cell a very effective mechanism of plasma membrane depolarization, and consequently of Ca-channel activation, can be provided by activation of anion channels (Barbier-Brygoo *et al.*, 1999; Crol and Treacz, 2000). In turn the activation of anion channels depend on Ca^{2+} influx in the cell through the plasma membrane. The activity of receptor-operated Ca-channels can be adjusted by hormones. There is also a type of Ca-channels, which is regulated by IP_3 (Sanders *et al.,* 1999). Voltage-gated Ca-channels appeared to be activated after depolarization of the plasma membrane as a result of IAA activation of anion channels and efflux of anions from cells. Thus, an early event in auxin action is activation of Ca-channels and an increase of Ca^{2+} concentration in cytoplasm. The increase of Ca^{2+} level in cytoplasm initiates the whole cascade of processes, that lead to physiological effects of IAA on growth and morphogenesis.

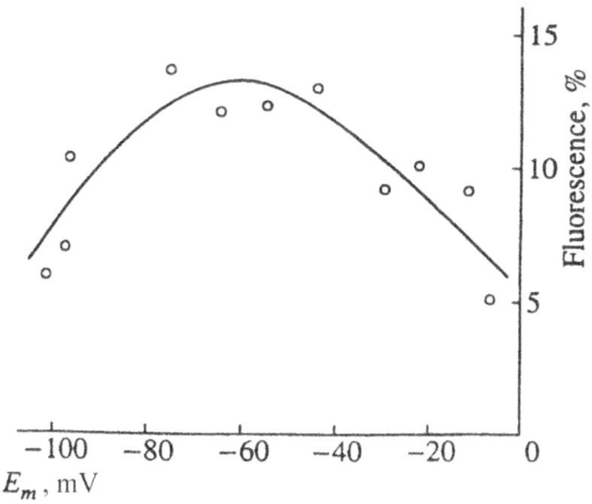

Figure 5. Dependence of Ca^{2+} transport across the vesicle membrane on K^+-diffuse membrane potential (E_m) in maize coleoptile plasma membrane vesicles. The medium inside the vesicles: 1 mM Tris-Mes (pH 7.2), 150 mM sucrose, 150 mM K_2SO_4. The incubation medium: 1 mM Tris-Mes (pH 7.2), 150 mM sucrose, 0.1 mM $CaCl_2$, 150 mM Na_2SO_4, 70 μM chlorotetracycline.

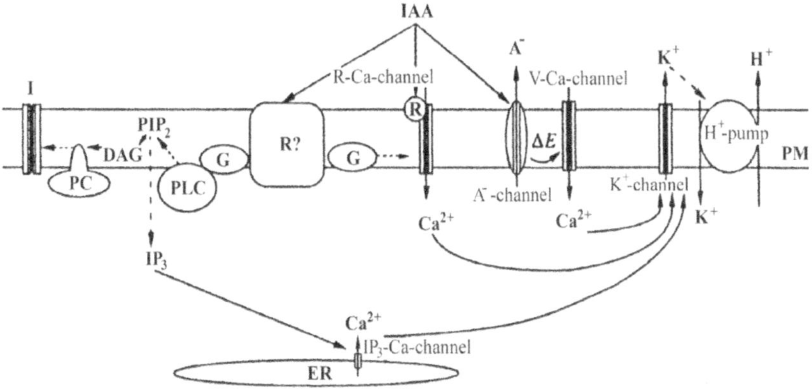

Fig. 6. IAA action on plasma membrane ion fluxes. PM – plasma membrane, ER – endoplasmic reticulum, R – putative IAA receptor, I - ion transporter, PC - protein kinase C, G – G-protein, PLC – phospholipase C, PIP_2 – phosphatidylinositol-4,5,-bisphosphate, DAG – diacylglycerol, IP_3 - phosphatidylinositol-1,4,5,-trisphosphate, R-Ca-channel – receptor-regulated Ca^{2+}-channel, V-Ca-channel – voltage-dependent Ca^{2+}-channel, A^--channel – anion channel, IP_3-Ca-channel - IP_3-regulated Ca^{2+}-channel, K^+-channel – Ca^{2+}-activated K^+-channel, H^+-pump – H^+-ATPase,

So, there are at least three ways that IAA can influence $[Ca^{2+}]_{cyt}$ level: I. Hormone interacts with a receptor. It activates PLC and other enzymes that catalyze the formation of second messengers (IP_3 and cADP-ribose etc.). IP_3 and cADP-ribose activate corresponding Ca-channels and Ca^{2+} release from vacuole and endoplasmic reticulum. II. Activation of anion channels. This leads to plasma membrane depolarization and activation of voltage-gated Ca-channels. III. Direct hormone activation of Ca-channels. On the whole, the presented data and discussion indicate that calcium can act as second messenger in the process of auxin signal trasnsduction both on an intracellular and an intercellular level.

ACKNOWLEDGEMENTS

Authors are grateful to Dr A.Yu.Batov and Dr A.V.Moshkov for participation in some experiments, and to Dr E.I.Sharova for help in preparation of manuscript. This work was supported by grants for Russian Foundation for Basic Research (96-04-48987, 99-04-48695, 02-04-49171), "Universities of Russia" State Scientific and Technical Program (GU-042), INTAS-01-0602.

REFERENCES

Allen, G.J., Chu, S.P., Harrington, C.L., Schumacher, K., Hoffman, T., Tang, T.Y., Grill, E. and Schroeder, J.L. (2002) A defined range of guard cell calcium oscillation parameters encodes stomatal movements, *Nature* **411,** 1053-1057.

Banuelos, G.S., Bangerth, F. and Marschner, H. (1987) Relationship between polar basipetal auxin transport and acropetal Ca 2+ transport into tomato fruits, *Physiol. Plant.* **71**, 321-327.

Banuelos, G.S., Bangerth, F. and Marschner, H.(1988) Basipetal auxin transport in lettuce and its possible involvement in acropetal calcium transport and incidence of tipburn, *J. Plant Nutr.* **11**, 525-533.

Barbier-Brygoo, H., Ephritikhine, G., Klämbt, D., Maurel, C., Palme, K., Schell, J. and Guern, J. (1991) Perception of the auxin signal at the plasma membranes of tobacco mesophyll protoplasts, *Plant J.* **1**, 83-93.

Barbier-Brygoo, H., Zimmermann, S., Thomine, S., White, I.R., Millner, P. and Guern, J. (1996) Elementary auxin response chains at the plasma membrane involve external abp1 and multiple electrogenic ion transport proteins, in A. R. Smith, A.W. Berry, N.V.J. Harpham, I.E. Moshkov, G.V. Novikova, O.N. Kulaeva, M.A. Hall (eds.) *Plant Hormone Signal Perception and Transduction*, Kluwer Acad.Publ. Dordrecht-Boston-London, pp.31-36.

Barbier-Brygoo,H., Franchisse, J.-M., Colcombet, J. and Thomine, S. (1999) Anion channels and hormone signalling in plant cell, *Plant Physiol. Biochem.* **37**, 381-392.

Bennet, M.J., Marchant, A., May, S.T.and Swarup, R. (1998) Going to distance with auxin: unravelling the molecular basis of auxin transport, *Phil.Trans.R.Soc.Lond.B.* **353**, 1511-1515.

Brooks, S.P. and Storey, K.B. (1992) *Analyt. Biochem.* **201**, 119-126.

Chin, D. and Means, A.R. (2000) Calmodulin: a phototypical calcium censor, *Trends in Cell Biol.* **10**, 322-328.

Geisler, M., Axelsen, K.B., Harper, J.F. and Palmgren, M.G. (2000) Molecular aspects of higher plant P-type Ca^{2+} -ATPases, *Biochim. Biophys. Acta.* **1465**, 52-78.

Gehring, C.A., Williams, D.A., Cody, S.H. and Parish, R.W. (1990) Phototropism and geotropism in maize coleoptiles are spatially correlated with increases in cytosolic free calcium, *Nature* **345**, 528-530.

Guzman de, C.C. and Fuente dela, R.K. (1984) Polar calcium flux in sunflower hypocotyl segments. I. The effect of auxin, *Plant Physiol.* **76**, 347-352.

Guzman de, C.C. and Fuente dela, R.K. (1986) Polar calcium flux in sunflower hypocotyl segments. II.The effect of segment orientation, growth and respiration, *Plant Physiol.* **81**, 408-412.

Harmon, A.C., Gribskov, M. and Harper, J.F. (2000) CDPKs – a kinase for every Ca^{2+} signal? *Trends in Plant Sci.* **5**, 154-159.
Jaffe, L.F. (1999) Organization of early development by calcium patterns, *BioEssays*, **21**, 657-667.
Krol E. and Trebacz K. (2000) Ways of ion channel gating in plant cells, *Ann. Bot.* **86**, 449-469.
Malho, R., Moutinho, A., Van der Luit, A. and Trewavas, A.J. (1998) Spatial characteristics of calcium signalling: the calcium wave as a basic unit in plant cell calcium signalling, *Phil. Trans. R. Soc. Lond.* **353**, 1463-1473.
Markova, I.V., Batov, A.Yu., Moshkov, A.V. and Maksimov, G.B. (1995) Calcium-transporting systems in maize coleoptile plasma membrane, *Russian J.Plant Physiol.* **42**, 262-267.
Markova, I.V., Katrichenko, M.I., Shevtsov, Yu.I., Batov, A.Yu. and Medvedev, S.S. (1997) Polar Ca^{2+} fluxes in plant tissues. 2. A mechanism of their formation, *Russian J. Plant Physiol.* **44**, 777-784.
Medvedev, S.S., Markova, I.V., Batov, A.Yu. and Maksimov, G.B. (1989) Polar calcuim ion fluxes and growth of plant tissues, *Soviet Plant Physiol.* **36**, 990-997.
Medvedev, S.S., Maksimov, G.B. and Markova, I.V. (1991) The role of calcuim ions in the regulation of gravitropism, *Experimental Biology*, 71-92.
Medvedev, S.S., Markova, I.V., Batov, A.Y. and Moshkov, A.V. (1998) Membrane mechanism of IAA action, *Biologia* **3**, 31-34.
Medvedev, S.S., Markova, I.V. and Batov, A.Yu. (2002) Calcium in the mechanism of phytohormone action, NATO-Russia Advanced Research Workshop, Moscow, Russia.
Murphy A.S. (2002) An emerging model of auxin transport regulation, *Plant Cell* **14**, 293-299.
Pandey,S., Tiwati, S.B., Upadhuaya, K.C. and Sopory, S.K. (2000) Calcium signalling: linking environmental signals in cellular functions, *Crit. Rev. Plant Sci.* **19**, 291-318.
Roos, W. (2000) Ion mapping in plant cells - methods and applications in signal transduction research, *Planta* **210**, 347-370.
Sanders, D., Brownlee, C. and Harper, J.F. (1999) Communication with calcium, *Plant Cell* **11**, 691-706.
Sze, H., Liang, F., Hwang, I., Curran, A.C. and Harper, J.F. (2000) Diversity and regulation of plant Ca^{2+} pumps: insights from expression in yeast, *Annu. Rev. Plant Physiol. Plant Mol. Biol.* **51**, 433-462.
Trewavas, A.J. and Malho, R. (1998) Ca^{2+} signalling in plant cells: the big network!, *Curr. Opin. Plant Biol.* **1**, 428-433.
White, P.J. (2000) Calcium channels in higher plants, *Biochim. Biophys. Acta.* **1465**, 171- 189.
White, P.J., Pineros, M., Tester, M.and Ridout, M.S. (2000) Cation permeability and selectivity of a root plasma membrane calcium channel, *J. Membrane Biol.* 174, 71-83.
Zielinski, R.E. (1998) Calmodulin and calmodulin-binding proteins in plants, *Annu. Rev. Plant Physiol. Plant Mol. Biol.* **49**, 697-725.
Zimmerman, S., Thomine, S., Guern, J. and Barbier-Brygoo, H. (1994) An anion currents at the plasma membrane of tobacco protoplasts shows ATP-dependent voltage regulation and is modulated by auxin, *Plant J.* **6**, 707-716.

TRANSGENIC PLANTS OF MODIFIED AUXIN STATUS AND ENHANCED PRODUCTIVITY

R.K. SALYAEV[1], N.I. REKOSLAVSKAYA[1], S. MAPELLI[2], V.M. SUMTSOVA[1], N.V. PAKOVA[1] AND A.A. TRUCHIN[1]

[1]*Siberian Institute of Plant Physiology and Biochemistry, Siberian Branch of RAS, PO box 1243, Irkutsk, Russia, e-mail: phytolab@sifibr.irk.ru*

[2]*Istituto Biologia Biotecnologia Agraria, C.N.R., via Bassini 15, Milan, Italy, e-mail: mapo@ibv.mi.cnr.it*

1. INTRODUCTION

Indole-3-acetic acid is the principal phytohormone affecting and regulating growth and developmental events and productivity of plants. However, the exogenous application of IAA does not result in desired changes in the real level and distribution of endogenous auxin between compartments within the cell. The optimization of IAA level for growth stimulation is a result of processes of biochemical controls on rates of biosynthesis, oxidation and conjugation and of transport into and out of cells and organs.

The physiological role of conjugation during development is still unclear and not fully understood. Among several potential donors for the glucosyl moiety UDP-glucose (UDPG) serves exclusively as a reversible conjugating moiety and UDPG-transferase is the main enzyme of this conjugation. UDPG-transferase or by trivial name IAA-glucose synthase is encoded by the gene *ugt/iaglu* from *Zea mays* L., *Lycopersicon esculentum* Mill., *Arabidopsis thaliana* L. and other plants.

Generally, it is well known that UDPG-transferase is able to use other plant growth regulators such as 2,4-D (2,4-dichlorophenoxyacetic acid) as a substrate:

UDPG-transferase β–glucosidase
IAA------------------------→IAA-glucose-------------------→IAA-------------→growth
2,4-D---------------------→2,4-D-glucose---------------- →2,4-D------------→growth

The release of aglycon contributes to the optimal free auxin concentrations for growth stimulation and as a consequence for changes in development and evokes higher productivity. By introducing the gene *ugt(iaglu)* encoding UDPG-transferase, transgenic plants obtained additional amount of IAA stored in the bound form, which might be used in stimulation of growth. In this case, 2,4-D lost its herbicidal effect but the growth stimulating effects of this potent auxin analogue were still preserved.

I. Macháčková and G.A. Romanov (eds.), Phytohormones in Plant Biotechnology and Agriculture, 171-183

For isolation of the gene *ugt/iaglu* from corn the cDNA library was originated in λZapII vector (Stratagene). Immunopositive phage plaques were screened with antibodies raised in rabbit serum against IAA-glucose synthase isolated from corn endosperm. This enzyme is producing 95-98 % of newly synthesised IAA in kernels bound to glucose during development of earcorn. The gene was isolated in the laboratory of Prof. R.S.Bandurski at the Michigan State University (East Lansing, USA).

The main objective was to obtain transgenic plants with enhanced productivity by introducing the gene *ugt /iaglu* encoding the synthesis of UDPG-transferase from corn. The second objective was to make an assessment of physiological and biochemical events leading to the improvement in the productivity of transgenic plants.

2. MATERIALS AND METHODS

2.1. Methods for introducing of foreign DNA into plants

Direct delivery of gene constructs by particle gun
In order to deliver genetic cassettes to recalcitrant species such as wheat, a gene gun was used with tungsten microprojectiles 1.6 μm in size placed on teflon macroprojectile with diameter 4.5 mm. Freshly isolated immature embryos were used for shooting as described earlier (Salyaev *et al.*, 2001). Transformed and nontransformed embryos were placed on P-AGAR with addition of 0.125 M mannitol and 0.125 M sorbitol. On the next day P-AGARs with samples were transferred on MS medium with 100 mg/l of kanamycin for selection.

Agrobacterium-mediated gene transfer

Agrobacterium tumefaciens 699 (chromosomal base EHA105, *nptII, 35S-gus-int*) and *Escherichia coli* DH5α with pBluescript harboring cDNA of the gene *ugt* were used. Triparental mating was performed by standard method (Lichtenstein and Draper, 1988) and transconjugants were selected on the YPD media with appropriate antibiotics. Transconjugant cultures were introduced into axillary buds of plants by pricking the sterile needle with loaded bacteria. The appearing sprouts were placed on agar medium with antibiotics. In other experiments the transgenosis was proved by the exposition of leaf discs from developed potato or tomato leaves to kanamycin solution. Decline in chlorophyll content was measured in acetone extracts after 3-5 days of exposition (Grodzinski and Grodzinski, 1973).

PCR, RT-PCR and Southern blot hybridisation were performed by means of standard procedures (Sambrook *et al.*, 1989) and as it was described earlier (Rekoslavskaya *et al.*, 2001; Salyaev *et al.*, 2001) by using kits and protocols "Ready To Go PCR Beads", "Ready To Go RT-PCR Beads" and ECL System "Gene Image Random Prime Labelling Module" and "Gene Images CDP - Star Detection Module" (*Amersham/Biosiences, England*).

The activity of UDPG-transferase was determined in supernatant fraction after extraction of the plant tissue and purification of the enzyme preparation by methods of Salyaev *et al.*(2001) and of Rekoslavskaya *et al.*(2001).

The amounts of IAA in plant material were determined with HPLC by method (Rekoslavskaya *et al.*, 2001).

The activity of marker enzyme β–glucuronidase (GUS) encoded by the reporter gene *gus* were determined by standard methods (Jefferson *et al.*, 1987).

3. RESULTS AND DISCUSSION

3.1. Direct delivery of gene constructs by gene gun

To study expression of the selective gene *nptII*, nontransformed and transformed regenerants obtained from spring wheat embryos (*Triticum aestivum* L. var. Tulunskaya 12) were placed on MS agar medium with 100 mg/l of kanamycin. Transgenic wheat plantlets grew faster and were higher in comparison to nontransfromed ones which were chlorotic, smaller and later died in the presence of kanamycin (100 mg/l) in medium.

PCR with primers to the fragment of the gene *ugt* (234 bp) from corn (Rekoslavskaya *et al.*, 2001; Salyaev *et al.*, 2001) revealed the homology in genomic DNA of transgenic wheat transformed with the gene *ugt* from corn to this fragment. There was no band of appropriate size of 234 bp in genomic DNA of nontransformed wheat (Fig.1). Southern blot hybridisation with amplificates synthesized on wheat DNA with the same primers was performed earlier (Salyaev *et al.*, 2001) and confirmed the homology of the gene *ugt* used as a probe to the amplified fragment synthesized with these primers.

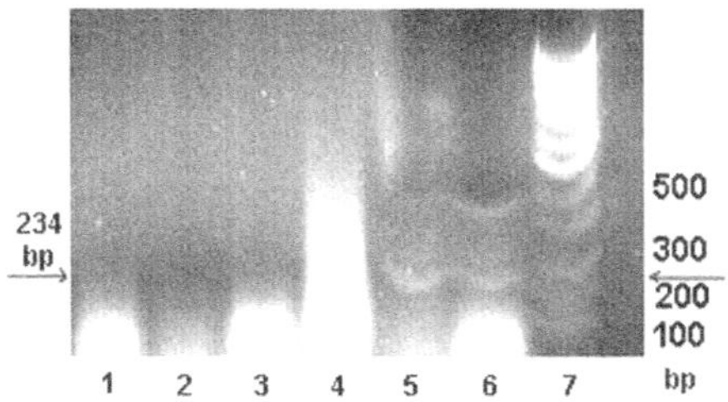

Figure 1. PCR with primers to 234 bp fragment of the gene ugt on genomic DNA from nontransformed (1-3) and transgenic (4-6) wheat.
Lines 1, 4 – stem; lines 2, 5 – roots; lines 3,6 – spike; line 7 - standard DNA "ladder 100 bp".

Nontransformed and transgenic plantlets placed in keramsite support were grown until spikes developed. The growth activity of transgenic wheat plants was higher in comparison to nontransformed ones (Fig.2). Therefore it was decided to determine the activity of UDPG-transferase and the IAA content in wheat plants. In transformed wheat plantlets there was found higher specific activity of the target enzyme UDPG-transferase (38.7±4.4 nmol IAA-glucose.mg^{-1}protein h^{-1}] in transgenic plants and 18.8±5.8 in control plants) and also an increased IAA content almost in all parts of transgenic wheat plants (Tab. 1).

Table 1. IAA content in nontransformed and transgenic wheat (nmol.organ^{-1})

Organ	Nontransformed	Transgenic
Leaf	1.0±0.1	3.0±0.3
Stem	6.7±0.1	19.9±2.6
Shoot	0.5±0.2	2.0±0.1
Root	0.9±0.1	2.9±0.2
Spike	0.1±0	0.3±0.0

Figure 2. Nontransformed control and transgenic spring wheat plants after growing on keramsite support.

CONCLUSION

Thus, it was achieved the integration and expression of the gene *ugt* in plants of spring wheat by means of bioballistic methods.

3.2. The Agrobacterium mediated transformation

The scheme of triparental mating is presented in Fig 3. Upon electrophoresis onlyone type of plasmid of about 30 kb was found in plasmid DNA preparation. It was concluded that transconjugant plasmid was a cointegrate forming after recombination on homologous site of *bla* gene (β-lactamase) which imparted the resistance to ampicillin both in plasmid of *E.coli* and *A.tumefaciens*. Therefore Ti-plasmid with transferred target genes occured as a cointegrative vector after triparental mating.

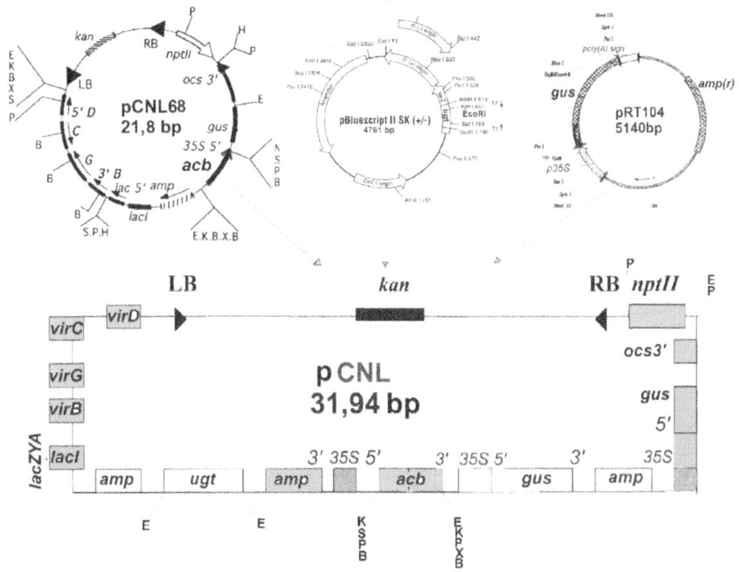

Figure 3. Scheme of obtaining of transconjugant of triparental mating.

Solanum tuberosum L., var. Borodyanski and *Lycopersicon esculentum* Mill. var.Ventura were used. In some experiments with 2,4-D *Solanum demissum* L. was used as well.

3.2.1. Transformation of potato

Transgenic potato plantlets with introduced gene *ugt* from corn revealed fast growth during *in vitro* cloning (Tab. 2).

Table 2. The dynamics of growth of potato plantlets in vitro, var Borodyanski

Variant	Days	after	cutting
	0	9	14
Length of stem (mm)			
Nontransformed	0	11±2	42±13
Transgenic	0	17±3	68±8
Total root length (mm)			
Nontransformed	0	5±2	176±123
Transgenic	0	103±33	340±69

Evidence for the presence of genetic cassettes in potato seedlings introduced via agrobacterial infection was achieved during the study of integration and expression of target *ugt* and marker genes *nptII* and *gus*. The integration and expression of *nptII* gene in DNA isolated from triparental transconjugants and from transformed potato were proved in several ways. The presence of selective gene *nptII* in plasmid DNA of transconjugants of triparental mating and in genomic DNA from transformed potato were established with PCR (Fig.4 A,B).

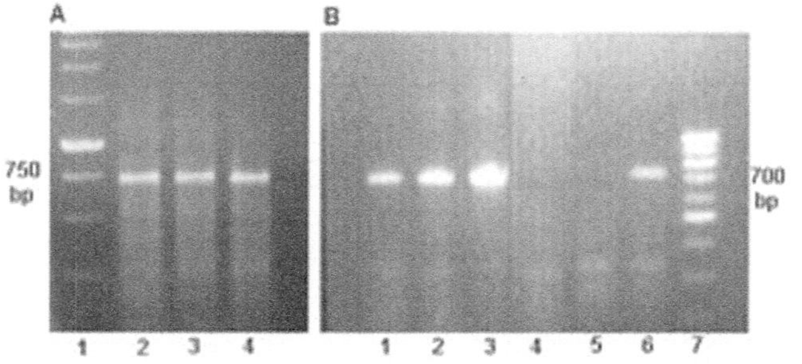

Figure 4. PCR of the selective gene nptII on DNA vector plasmid pBin19 (A) and on DNA of transconjugants and transgenic potato plant (B).
A – line 1 – standard DNA "ladder 1 kb"; lines 2-4 – individual clones of E.coli with pBin19 (nptII);
B – lines 1- 3 - individual clones of transconjugants on the base A.tumefaciens 699 (nptII); line 4 – A.tumefaciens EHA105 (-nptII); line 5 – nontransformed potato, line 6 – transgenic potato, line 7 – standard DNA "ladder 100 bp".

In addition, potato plantlets were grown in the presence of 25 –50 mg/l of kanamycin in MS medium. Transgenic potato plants revealed resistance to the presence of kanamycin in agar medium; nontransformed plantlets grew slower and were more sensitive to kanamycin than the transgenic ones. Green leaves detached from transgenic plants grown in pots in greenhouse had higher chlorophyll content (2 times) and longer maintained green colour during exposition to kanamycin solutions in comparison to control ones.

Specific activity of GUS was 0.127±0.039 imp·sec^{-1}·mg-equivalent^{-1} of enzyme in nontransformed potato sprouts and 0.430±0.002 imp·sec^{-1}·mg-equivalent^{-1} of enzyme in transgenic potato after 10 min of incubation with substrate 1mM 4-methyl-β-D-glucuronide (MUG).

PCR (A) and Southern blot hybridization (B) of DNA isolated from potato var Borodyanski are presented in Fig.5. There was a homology in genomic DNA from both nontransformed and transgenic potato plants, but the intensity of bands were higher in the case of transgenic potato plants. RT-PCR technique showed the expression of the gene *ugt* in transgenic potato plants (data not shown).

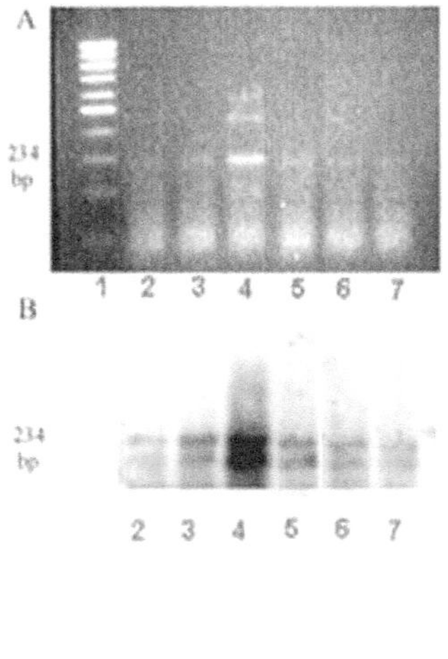

Figure 5. PCR (A) on DNA isolated from nontransformed and transgenic shoots and tubers of potato var. Borodyanski and Southern blot hybridisation of amplificates (B) with the probe synthesised with primers to the fragment 234 bp of the gene ugt cloned in pBluescript E.coli DH5α;
Line 1 – standard DNA "ladder 100 bp"; line 2 – nontransformed tuber; line 3 – nontransformed shoot; line 4 – transgenic shoot; line 5 – transgenic tuber; line 6 – plasmid of transconjugant; line 7 – pBluescript with cloned gene ugt. For Southern blot hybridisation (B) the indications of DNA samples were the same with exception line 1 which was not used for blot transfer.

Contents of free and bound IAA are presented in Tab. 3. Bound alkali labeled IAA was not found in methanolic extracts from nontransformed potato plants. But the amounts of free and bound auxin were higher as much as 6 times in transgenic potato plants. This bound auxin was considered to be IAA-glucose. The amount of IAA was also higher in transgenic *S.demissum* plants in comparison with nontransformed plants.

Table 3. Contents of free and bound IAA and the activity of UDPG-transferase in Solanum sp.

Variant	Free IAA (pmol.g^{-1} fw)	Bound IAA (pmol.g^{-1} fw)	Activity of UDPG-transferase (nmol·mg^{1}protein h^{1})
Solanum tuberosum L., var.Borodyanski			
Nontransformed	5.1±0.6	0	85
Transgenic	8.2±5.1	22.8±0	1283
Solanum demissum L.			
Nontransformed	2.6±0.6	<0.1	73
Transgenic	6.7±0	<0.1	119

During growth in soil after spraying with 2.4-D transgenic potato plants (var.Borodyanski) developed larger leaf areas and stems especially in basal parts. Nontransformed potato plants showed the increase of internode growth in the upper part. Leaves of nontransformed plants became damaged after 2,4-D treatment. Perhaps, there was a difference in direction of assimilates flow after 2,4-D application to nontransformed and to transgenic potato plants.

Enzyme preparations from control and transgenic *S.demissum* L. were used to study the substrate specificity to IAA and 2,4-D. It was found that both substrates were used in binding with glucose (Rekoslavskaya *et al.*, 2001). Tuber formation was higher in transgenic plants (Tab. 4).

Table 4. Tuber formation after 2,4D (100 mg/l) treatments of S.tuberosum L. var.Borodyanski transferred from in vitro to field beds

Variant	Mass of tubers (g. plant^{-1})
Nontransformed	218±98
Nontransformed + 2.4D	203±22
Transgenic	432±97
Transgenic + 2.4D	747±97

To study photosynthesis, gas exchange and transpiration, potato tubers were planted in soil at beginning of May in the open air. At the end of May the plants were of a good size, with leaves wide enough for measurement. Using portable equipment LCA3 (Leaf Chamber Analysis system) water transpiration, stomatal conductance and net carbon dioxide assimilation were measured. On each plant, 2 leaves were taken for 12-15 measurements. The measurements are not destructive so the plants had no damage or stress. The measurements were performed in the morning, between 9 and 11 hour. At this time photosynthetic active radiation (PAR) values were constant enough.

The data of Tab. 5 show that transgenic plants have higher net photosynthesis and also high transpiration, in comparison to the control ones. In consequence, transgenic plants have high accumulation of photosynthates and high growth rate. However, high transpiration means high lost of water.

Table 5. Comparison of transpiration, gas conductivity and photosynthesis in nontransformed and transgenic potato var.Borodyanski

Variant	Plants	Transpiration (mmol · $m^{-2}s^{-1}$)	Gas conductivity (mol · $m^{-2}s^{-1}$)	Photosynthesis (μmol · $m^{-2}s^{-1}$)
MAY 29	PAR 1000-1200 μmol · $m^{-2}s^{-1}$; T(°C) 22-23			
Nontrans-formed	1	7.30±0.53	0.64±0.12	6.45±0.49
	2	8.21±0.74	0.94±0.28	15.74±2.02
Transgenic	1	9.02±0.29	1.13±0.13	16.35±0.83
	2	8.45±0.38	0.85±0.11	16.98±1.92
JULY 21	PAR 1000-1200 μmol · $m^{-2}s^{-1}$; T(°C) 25-26			
Nontrans-formed	1	6.38±0.05	0.42±0.01	7.12±0.29
	2	6.32±0.06	0.37±0.01	6.11±0.13
Transgenic	1	9.01±0.19	0.17±0.01	15.13±0.28
	2	9.10±0.03	0.98±0.02	17.88±0.59

The higher activity of photosynthesis in transgenic plants correlated with the larger leaf area: control plants had leaf area of about 5850 cm^2, while the transgenic ones about 9000 cm^2. The yield of potato tubers var.Borodyanski collected from field beds was higher in transgenic plants: it represented on average 5 kg/plant, while in the control only 3.7 kg/plant.

PCR (A,B) and Southern blot hybridization (C,D) with DNA isolated from tubers were performed (Fig.6). It was found that bands of amplified fragments of 234 bp of the transgene *ugt* were about 2 times more intensive in transgenic tubers (B) as compared to nontransformed ones (A). Homology of these amplified fragments from DNA of transgenic potato tubers was much higher (D) with fragment of 234 bp of the gene *ugt* from corn than in nontransformed potato (C).

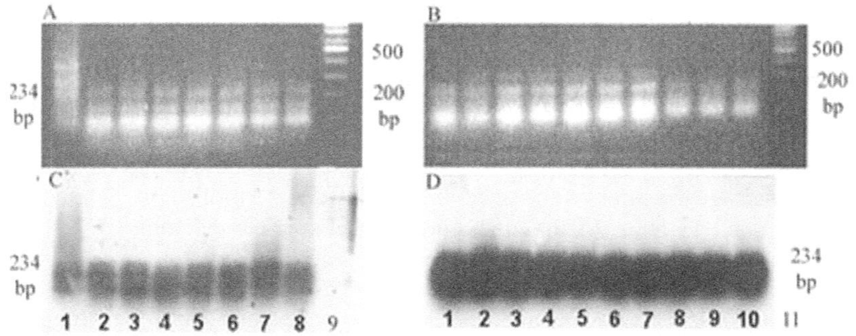

Figure 6. PCR (A,B) and Southern blot hybridization (C,D) performed with DNA isolated from nontransformed (A,C) and transgenic (B,D) tubers harvested in 2001 year.
A,C – lines 1-7 – DNA of tubers from individual nontransformed plants; line 8 – pBluescript with the gene ugt; line 9 - standard DNA "ladder 100 bp"
B,D - lines 1-7 – DNA of tubers from individual transgenic plants; lines 8-10 – pBluescript with the gene ugt, line 11 - standard DNA "ladder 100 bp".

3.2.2. Transformation of tomato

Tomato seedlings were infected in apical part between cotyledons with needle coated with the transconjugant cells with plasmid DNA harboring genes *nptII, gus* and *ugt.*

The integration and expression of the marker gene *nptII* were established with PCR and RT-PCR (data not shown) and after incubation of detached leaves on solution with 50-300 mg/l of kanamycin in which the transgenic leaves maintained the green colour about 2 times longer in comparison to controls.

Specific activity of GUS was 1.31 and 198.18 imp·sec^{-1}·mg-$equivalent^{-1}$ of enzyme in cytosol and plastid fractions, correspondingly, from transgenic tomato leaves. There was no fluorescence in the same fractions isolated from nontransformed tomato leaves after 1 hour of incubation with 1 mM MUG.

PCR and Southern blot analyses of transgene *ugt* were performed earlier (Rekoslavskaya *et al.*, 1998; Rekoslavskaya *et al.*, 2001) confirming the integration of the gene *ugt* into tomato plants upon transformation with triparental transconjugant. As an evidence for expression of the gene *ugt* we found higher IAA content and the increase in specific activity of UDPG-transferase in transformed tomato (Tab. 6).

Table 6. IAA content and specific activity of UDPG-transferase in nontransformed and transgenic tomato var. Ventura

Variant	IAA ($nmol.g^{-1}$ dm)	Specific activity of UDPG-transferase ($nmol.mg^{-1}$ protein. h^{-1})
Nontransformed	569±6	139.7
Transgenic	1263±329	286.3

Transgenic plants grew faster and developed larger leaf area which was as much as 247.3±37.1 cm^2 per 1 leaf in transgenic tomato and 161.3±12.6 cm^2 per 1 leaf from non-transformed plant. Thus, there was a correlation between photosynthesis and leaf area both in potato and tomato plants.

Measurement of leaf gas exchange (Tab. 7) gave the indication that Ventura transgenic tomato plants have higher net carbon dioxide assimilation, lower stomatal conductance and water transpiration showing a possible higher efficiency of water use and some increase in photosynthesis.

Table 7. Comparison of leaf gas exchange, transpiration and photosynthesis between non-transformed and transgenic tomato plants

Variant	Transpiration ($mmol \cdot m^{-2}s^{-1}$)	Gas conductivity ($mol \cdot m^{-2}s^{-1}$)	Photosynthesis ($\mu mol \cdot m^{-2}s^{-1}$)
	30th June		
Nontransformed	11.07±0.01	1.79± 0.73	11.20±0.22
Transgenic	9.07±0.01	0.83±0.20	13.35±0.26
	31st July		
Nontransformed	12.40±0.13	1.65±0.68	12.56±0.67
Transgenic	9.89±0.09	0.83±0.24	15.55±0.41

The faster growth, large net photosynthesis were coincided with higher productivity of transgenic tomato (Tab. 8).

Table 8. Productivity of tomato plants var. Ventura

Variant	Green fruits (g)	Red fruits (g)	Number of fruits	Total mass (g)
Nontransformed				
1st plant	1748	2680	118	4428
2nd plant	1349	1589	64	2948
3rd plant	2794	1005	91	3799
4th plant	2479	593	74	3072
Transgenic				
1st plant	4998	950	123	5948
2nd plant	6260	2068	177	8328
3rd plant	2795	2449	118	5244
4th plant	4868	2085	164	6953

Besides wheat, potato and tomato during 1998 - 2001 the genetic transformation of cucumber, pepper, egg-plants and lettuce have been performed. In order not to overload the article, only one table with summarizing data of the productivity of nontransformed and transgenic plants is included (Tab. 9).

Table 9. Harvest of nontransformed and transgenic plants of cucumber, pepper, egg plants and lettuce (g.plant^{-1})

Variant	Cucumber	Pepper	Egg-plants	Lettuce
Non-transformed	7338	1352	2079	1000
	7736	1317	-	1455
	6605	1200	-	-
Transgenic	20609	1715	3771	1478
	46559	2831	-	1750
	13119	1520	-	-

CONCLUSION

The gene *ugt* was found to be of interest for changing of auxin metabolism by increasing the level of bound and free IAA due to introducing of the gene *ugt* encoding the synthesis of UDPG - transferase from corn. It has beneficial effect on growth and productivity of transgenic plants.

REFERENCES

Grodzinski, A.M. and Grodzinski, D.M. (1973) *Short Newsletter on Plant Physiology*, [In Russian], Naukova dumka, Kiev, 591 pp.

Jefferson, R.A., Kavanagh T.A. and Bevan, M.V. (1987) GUS fusion: β–glucuronidase as a sensitive and versatile gene fusion marker in higher plants, *EMBO J.* **6**, 3901-3907.

Lichtenstein, K.and Draper, D. (1988) Genetic transformation of plants, in D. Glover (ed.), *Cloning of DNA. Methods*, Mir, Moskva, pp. 315-380.

Rekoslavskaya, N.I., Gamanetz, L.V., Bryksina, I.V., Mapelli, S. and Salyaev, R.K. (1998) Obtaining of transgenic tomato (*Lycopersicon esculentum* Mill.) and potato (*Solanum tuberosum* L.) by transfer of the gene *ugt* from corn, *Rep. Tom. Genet. Coop.* **48,** 40-42.

Rekoslavskaya, N.I., Salyaev, R.K. and Vysotskaya, E.F. (2001) Application of target genes and their complexes to plants genetic transformation, in R.K. Salyaev (ed.), *Genome Study and Plant Genetic Transformation.* Nauka, Novosibirsk, pp. 83-105.

Salyaev, R.K., Rekoslavskaya N.I., Sumtsova V.M., Pirog O.V. and Kusnetsova E.V.(2001) Acquisition of transgenic winter wheat via introduction of *ugt* gene by bioballistic method, in R.K. Salyaev (ed.), *Genome Study and Plant Genetic Transformation,* Nauka, Novosibirsk, pp. 123-131.

Sambrook, J., Fritsch, E.F. and Maniatis,T. (1989) *Molecular Cloning*, 2nd ed., Vol.1-3, Cold Spring Harbor, New York, 545 pp.

CELL WALL EXTENSIBILITY CHANGES BY HORMONAL AND NON-HORMONAL FACTORS

ELENA I. SHAROVA
Department of Plant Physiology and Biochemistry, Faculty of Biology and Soil Sciences, Saint-Petersburg State University, 7-9, Universitetskaya emb., 199034, St-Petersburg, Russia

1. INTRODUCTION

Growth reactions are often accompanied by changes in cell wall extensibility, for example, the activation of coleoptile elongation by IAA (Cleland, 1967) and its inhibition by ABA (Kutschera and Schopfer, 1986), activation of leaf expansion by light (Van Volkenburgh and Davies, 1983) and its inhibition under high soil salinity (Tomos and Pritchard, 1994). Nevertheless, changes in wall extensibility are often found to be very small in comparison with growth effects. The disproportion could be explained by poor correlation between physical and physiological extensibility, in other words, the ability of the cell wall to extend irreversibly during cell growth (Cosgrove, 1993). Rapid and reversible growth effects of hormones and some environmental factors are likely to be connected with unstable changes in the conditions in the apoplast that are lost during tissue preparation for extensibility measurement. For example, IAA induced wall acidification is negligible if "acid" creep at pH 5 is used as a criterion of wall extensibility. To evaluate the effect of such changes on wall extensibility it is useful to test it *in vitro*. It is known that pH change in the physiological range can rapidly and reversibly change wall extensibility (Rayle and Cleland, 1992). Apoplastic H_2O_2 concentration was shown to be another natural factor that influenced wall extensibility *in vitro* (Schopfer 1996), although H_2O_2 (1 mM) during 1 h of treatment decreased the extension of frozen-thawed maize coleoptiles by only 13 %. The objective of our work was to find a measure of wall extensibility that correctly reflected cell growth ability and to use it to investigate the regulation of wall extensibility.

2. MATERIALS AND METHODS

Four-day-old etiolated maize seedlings (Zea mays L. Cv. ZPTK-196) grown at 27°C were used in most experiments. Coleoptile segments were cut 4 mm below the tip, peeled to remove cutinized epidermis and recut from basal end to 13 mm (cell wall extension measurement) or to 10 mm length. Segments were preincubated for 4 h in the basal medium (0.5 mM $CaCl_2$ in 1.5 mM K-phosphate-citrate buffer, pH 6) before any experimental treatment. Segment elongation was determined over a 1-2h

I. Macháčková and G.A. Romanov (eds.), Phytohormones in Plant Biotechnology and Agriculture, 185-191

magnification. Each sample consisted of 20 segments in 4 ml of basal medium. To register apoplastic peroxidase release into the medium, 40 segments were incubated in 4 ml of basal medium. Every 20 min the solution was replaced with a fresh one. Peroxidase activity in the removed solution was assayed in a reaction mixture containing 35 mM guaiacol and 3.5 mM H_2O_2 in 20 mM Na-acetate buffer, pH 5. Absorption was measured at 470 nm. To register hormone-induced pH change in the incubation medium, 40 segments were incubated in 4 ml of 10 mM KCl adjusted to pH 6 at the beginning of incubation.

To assay cell wall extension *in vitro*, 13-mm-long peeled coleoptile segments (13 mm x 4 mm) from leaves of *Vicia faba* L. and *Setcreasea purpurea* Boom. plants grown in a green house under natural light were used. Leaf segments were abraded with emery cloth before use. Segments were frozen at -60°C after the standard preincubation procedure, thawed at room temperature and fixed between two clamps on an adjustable table (Fig.1A). The lower immobile clamp was U-shaped and formed a cup filled with 10 mM Na-acetate buffer, pH 6.5 with tested reagents. The upper mobile clamp was suspended on a lever arm. The latter was put in touch with an angular displacement voltage transducer (diode lamp Mechanotron 6MX1C (Turkin *et al.*, 1980)) using macro- and microscrews of the table. A constant load of 10 g (leaves) or 20 g (coleoptiles) was applied to the opposite arm, and the resulting segment extension was continuously registered (Fig.1B).

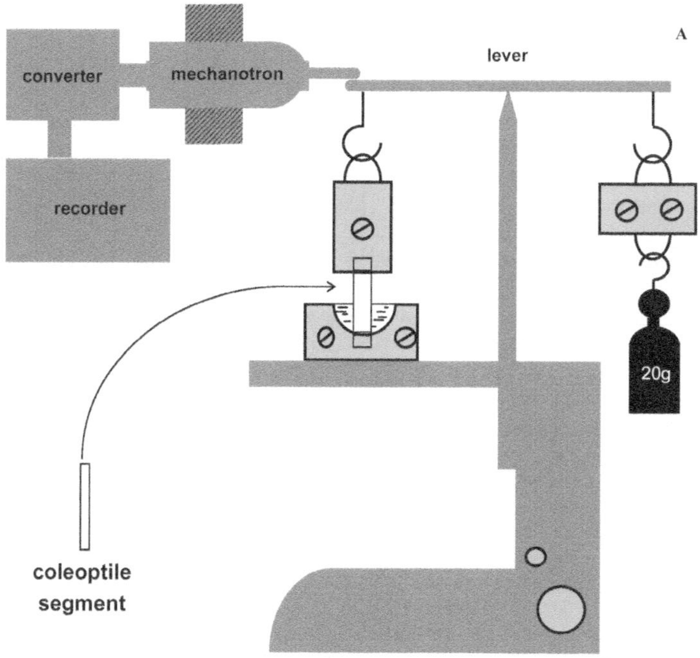

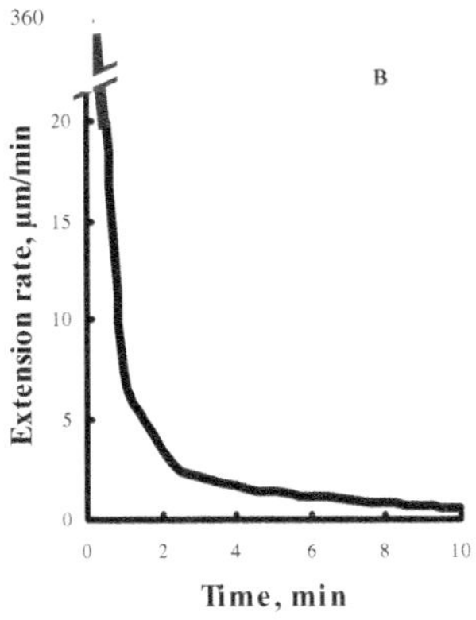

Figure 1. Scheme of the constant load extensometer (A) and kinetics of cell wall extension after the load application (B). See Materials and methods for details.

In figures the results of typical experiments are shown. In tables mean values (± SE) are given. The experiments were repeated 3-4 times except for extension measurements, which were replicated at least 10 times for each kind of plant material and treatment.

2. RESULTS AND DISCUSSION

Among different methods for *in vitro* cell wall extensibility estimation, Cosgrove (1993) considered wall creep under a constant load to be the best one. To test it as an indicator of physiological wall extensibility, we measured wall extension for 10 min at different pH values just after (rapid extension) and 10 min after (slow extension; creep, in the strict sense of the word) the load application. It was shown that acid creep measured 10 min after the load application was the value that decreased in parallel with growth cessation, especially if it was expressed in % of extension at pH 6 (Tab. 1, 2). Similar observations were made earlier with bean leaves (Van Volkenburgh *et al.*, 1985) and oat coleoptile segments (Cleland, 1983), although the changes found were not so sharp, possibly because the extension was measured from the time of load application.

Table 1. Decrease of "acid" creep under 10 g load of frozen-thawed segments from leaves of Vicia faba and Setcreasea purpurea during leaf development. pH was changed 10 min after the load application and creep was measured for the next 10 min.

Leaf number, beginning from the shoot tip	Acid creep (pH 4) for 10 min, % of control (pH 6)	
	Vicia faba	*Setcreasea purpurea*
1	731 ± 27	523 ± 22
2	610 ± 24	481 ± 19
3	395 ± 11	309 ± 17
4	299 ± 16	228 ± 13
5	122 ± 7	163 ± 11

Table 2. Decrease of growth response of maize coleoptile segments to IAA and parallel decrease of acid creep under 20 g load of the frozen-thawed segments during aging of excised coleoptiles. pH was changed 10 min after the load application and acid creep was measured for the next 10 min.

Time after excision, h	IAA (10^{-5}M)-induced elongation for 1 h, % of control	Acid creep (pH 5) for 10 min, % of control (pH 6)
4	278 ± 14	562 ± 32
14	183 ± 9	280 ± 19
24	121 ± 8	175 ± 11

IAA and ABA action on the elongation of maize coleoptile segments was accompanied by changes in acid creep (Tab. 3), but the latter were small in comparison with growth effects. It is evident that IAA-induced wall acidification did not occur in conditions of creep measurement. Therefore IAA effect on the extensibility should be estimated by the degree of creep acceleration under acidification. ABA growth inhibitory effect was as strongly dependent on protein synthesis and vesicular secretion as IAA growth effect but differed by its independence on pH (Tab. 3). ABA caused no change in pH of the incubation medium and inhibited growth to the same extent at pH6 and pH 3.2 (Tab. 3).

Table 3. Comparison of some IAA and ABA effects on maize coleoptile segments.

Hormonal treatment	Elongation for 2 h, % of control (-hormone)			
	Basic medium	Cyclo-heximide, 10^{-5}M	Brefeldin A, 20 mg/l	1.5mM buffer, pH3.2
IAA, 10^{-5} M	256±12	110±3	162±6	125±6
ABA, 10^{-4} M	71±3	105±4	95±3	71±3

Hormonal treatment	pH of the medium after 2 h	Effect of 2 h hormonal treatment (% of control) on:	
		Peroxidase release to the medium	Acid creep (pH5) under 20 g load
IAA, 10^{-5} M	5.21±0.07	100±6	122±5
ABA, 10^{-4} M	5.85±0.08	75±4	83±4

We showed that ABA decreased apoplastic peroxidase activity (Tab. 3), using the previously described method to evaluate the dynamics of its activity (Sharova and Souslov, 2000). Peroxidases are usually considered as factors of wall rigidification due to their role in lignification and oxidative link formation (Fry, 1986). Peroxidases utilize H_2O_2 and also produce it, oxidizing NADH in the apoplast (Penel, 1997). Therefore the correlation between peroxidase and H_2O_2 change in apoplast can be both positive (Lin and Kao, 2001) and negative (Gomez *et al.*, 1995). H_2O_2 (0.5-50 mM) rapidly inhibited acid creep. The extent of the effect depended on H_2O_2 concentration. The inhibitory effect of 2 mM H_2O_2 was mostly irreversible and decreased by peroxidase inhibitors azide and ascorbate (Fig.2A, Tab. 4). So, it was peroxidase-dependent. H_2O_2 at 20 mM inhibited creep reversibly (Fig.2B). The effect was almost independent of peroxidase activity but was diminished by the chaotropic agent urea, inactive with 2 mM H_2O_2 (Tab. 4). We suggest that the effect of 20 mM H_2O_2 was physico-chemical and connected with the ability of H_2O_2 to form hydrogen bonds. Thus, wall extensibility is strongly dependent on H_2O_2 concentration in apoplast. Further investigation is necessary to determine the role of this negative wall extensibility factor in different growth reactions.

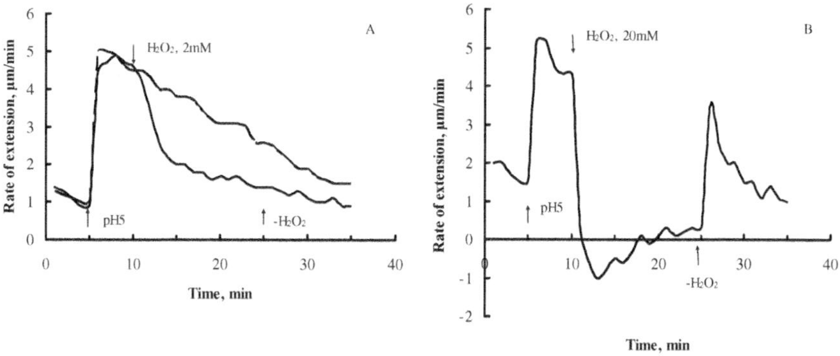

Figure 2. Influence of H_2O_2 (2 mM – A, 20 mM – B) on extension of peeled and frozen-thawed maize coleoptile segments under a 20 g load. Recording was started 5 min after the application of the load (zero time). The segment was incubated in 10 mM Na-acetate buffer at pH 6 (first 5 min of recording) and pH 5 (the remainder of time). H_2O_2 was present in the medium between the 10-25 min interval (solid line).

Table. 4. The influence of azide, ascorbate and urea on the inhibitory effect of H_2O_2 on cell wall extension.

Treatment	Extension (pH 5) for 15 min in H_2O_2, % of control ($-H_2O_2$)	
	H_2O_2, 2 mM	H_2O_2, 20 mM
H_2O_2 only	64 ± 2	0 ± 4
+ azide, 1 mM	75 ± 3	2 ± 6
+ ascorbate, 0.3 mM	72 ± 3	4 ± 5
+ urea, 1 M	68 ± 3	57 ± 5

ACKNOWLEDGEMENT

This work was supported by grant from Ministry of Education, Russia (E00-6-166).

REFERENCES

Cleland, R.E. (1967) Auxin and the mechanical properties of the cell wall, *Proc. New York Acad. Sci.* **144**, 3-8.

Cleland, R.E. (1983) The capacity for acid-induced wall loosening as a factor in the control of *Avena* coleoptile cell elongation, *J. Exp. Bot.* **34**, 676-680.

Cosgrove, D.J. (1993) Wall extensibility: its nature, measurement and relationship to plant cell growth, *New Phytol.* 124, 1-23.

Fry, S.C. (1986) Cross-linking of matrix polymers in the growing cells of angiosperms, *Annu. Rev. Plant Physiol.* **37**, 165-186.

Gomez, L.D., Casano, L.M. and Trippi, V.S. (1995) Effect of hydrogen peroxide on degradation of cell wall associated proteins in growing bean hypocotyls, *Plant Cell Physiol.* **36**, 1259-1264.

Kutschera, U. and Schopfer, P. (1986) Effect of auxin and abscisic acid on cell wall extensibility in maize coleoptiles, *Planta.* **167**, 527-535.

Lin, C.C. and Kao, C.H.: Abscisic acid induced changes in cell wall peroxidase activity and hydrogen peroxide level in roots of rice seedlings, *Plant Sci.* **160**, 323-329.

Penel, C. (1997) Production and roles of hydrogen peroxide, in H. Greppin, C. Penel and P. Simon (eds.) *Traveling Shot on Plant Development*, Univ. Geneva, pp. 219-238.

Rayle, D.L. and Cleland, R.E. (1992) The acid growth theory of auxin-induced cell elongation is alive and well, *Plant Physiol.* **99**, 1271-1274.

Schopfer, P. (1996) Hydrogen peroxide – mediated cell-wall stiffening *in vitro* in maize coleoptiles, *Planta* **199**, 43-49.

Sharova, E.I. and Souslov, D.V. (2000) Regulation of peroxidase secretion during plant cell elongation, in *Proc.18th Symp. Plant Biol.*, Keimyung Univ., Taegu, Korea, pp. 21-32.

Tomos, D. and Pritchard, J. (1994) Biophysical and biochemical control of cell expansion in roots and leaves, *J. Exp. Bot.* **45**, 1721-1731.

Turkin, N.I., Priyatkin, S.A., Andreev, A.P. and Polevoi, V.V. (1980) New method to measure short-term growth of plant axial organs, *Soviet Plant Physiol.* **27**, 437-440.

Van Volkenburgh, E. and Davies, W.J. (1983) Inhibition of light-stimulated leaf expansion by abscisic acid, *J. Exp. Bot.* **34**, 835-845.

Van Volkenburgh, E., Schmidt, M.G. and Cleland, R.E. (1985) Loss of capacity for acid-induced wall loosening as the principal cause of the cessation of cell enlargement in light-grown bean leaves, *Planta* **163**, 500-505.

COMPLEX HORMONAL REGULATION OF DEVELOPMENTAL PROCESSES

HORMONAL REGULATION OF CRASSULACEAN ACID METABOLISM (CAM) AND INTER-ORGAN STRESS SIGNAL TRANSDUCTION

VL. V. KUZNETSOV, A.G. KRUGLOVA, O.I. MOLODYUK, V.V. KARYAGIN, A.B. MESHCHERYAKOV, V.V. RAGULIN, V. YU. RAKITIN AND V.P. KHOLODOVA
K.A. Timiryazev Institute of Plant Physiology Russian Academy of Sciences, Botanicheskaya 135, 127 276 Moscow, Russia

Plants respond to damaging effects of unfavourable environmental factors by dramatic changes in cell metabolism. The synthesis of a set of new macromolecules is involved in long-term adaptations to extreme conditions. *De novo* formation of some new enzymes in response to stress allows plants to maintain homeostasis and fulfill their ontogenetic cycle in environments which are deleterious for life.

A unique model to investigate mechanisms of adaptation and stress signal transduction is the common ice plant (*Mesembryanthemum crystallinum* L.). *M. crystallinum* is a facultative halophyte which shifts from C3-type of photosynthesis to *Crassulacean* acid metabolism (CAM) under certain stress conditions. High salinity, osmotic or dehydration stresses induce the expression of genes encoding enzymes of CAM. CAM is known to be a very effective adaptive mechanism, providing an extremely efficient mechanism of photosynthesis. The night-time CO_2 fixation allows plants to close stomata during the day thereby reducing evaporative water loss (Thomas *et al.*, 1993; Kholodova *et al.*, 2002). One of the actual problems in this field is to understand what kind of signals can provide stress-dependent switches of specific metabolic pathways. The most likely candidates for this role are plant hormones. Among them, abscisic acid (ABA) is usually considered to be a stress hormone. The role of cytokinins in plant adaptation to environmental stress is usually connected with their counteraction to ABA. The data available in the literature on hormonal regulation of CAM are very controversial (Schmitt and Piepenbrock, 1992; Thomas et al., 1992, 1993) and do not yet allow the problem to be answered.There is another more intriguing problem. It is known that many stress factors, such as drought or salinity, affect only certain parts or tissues of plants. Nevertheless, cell metabolism may change in the whole plant, including in cells, which are located some distance from the place of stress factor action. However, until now, the identity of inter-organ stress signals remains unknown. ABA is a potential candidate for such a role. The data presented here indicate an involvement of ABA in long-distance stress signal transduction. Nevertheless, a hydraulic signal may play the role of the primary messenger in stress adaptation to salinity.

I. Macháčková and G.A. Romanov (eds.), Phytohormones in Plant Biotechnology and Agriculture, 195-203

In our experiments, the switch on CAM-photosynthesis was determined by measuring the activity of phosphoenolpyruvate carboxylase (PEPC - the key enzyme of CAM), and/or diurnal changes in organic acid content (mainly malate), namely its accumulation during night (dark) period and subsequent decarboxylation in the day time. To induce stress, 400 mM NaCl was used in most experiments described. Three to five days after NaCl addition to the rooting solution 7-8 weeks old common ice plants switched on CAM-type photosynthesis and exhibited typical PEPC activity together with night organic acid accumulation (Fig. 1 A,B).

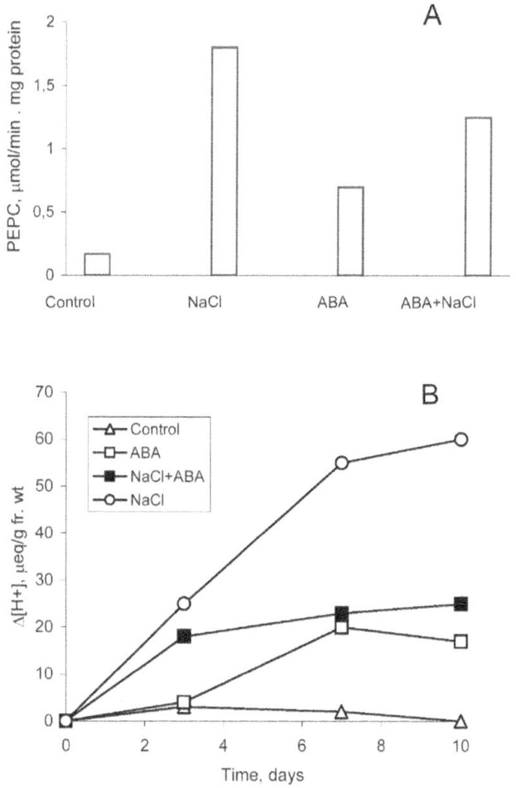

Figure 1. CAM activity during 400 mM NaCl stress and in the presence or absence of ABA
A . activity of PEPC at the 10th day of experiment; B changes of titratable acidity

The role of ABA in adaptation to stress factors was investigated in experiments in which the hormone was supplied exogenously through roots. ABA (10^{-5} M) added to the nutrient medium initiated a shift from C3- to CAM-type of photosynthesis. Though ABA induced CAM activity, the shift started usually later and its level was lower than in plants of the same age under salinity (Fig. 1B). Only mature plants (more than 8 weeks old) treated with ABA showed CAM activity comparable or sometimes even higher than that of plants treated with NaCl (data not shown). ABA participation in CAM induction was shown by measuring ABA contents in leaves after NaCl addition to the rooting medium (Fig. 2 A,B). Its level started to increase after 3-6 hours of salinization (Fig.2B). The first 48 hours (Fig. 2A) ABA concentration rose to a level 20-40-fold higher than in leaves of the control plants and was maintained at this level for 3 days before falling again to rather low values. The data obtained enable us to propose that the mechanism of ABA induction of CAM in the common ice plants was similar to that in response to NaCl. However, when added with NaCl, ABA not only showed an additive effect, but decreased PEPC activity and organic acid accumulation induced by NaCl alone decreased (Fig. 1 A,B). The reasons of this situation became clear when the physiological responses of the common ice plants to NaCl and ABA alone and in combination are considered. ABA induced a rapid and strong decrease in leaf transpiration to 10 % of control after 6 hours of salinization. This seems to be the result of the primary action of ABA in stomatal closure. The inhibition of transpiratory water loss resulted in maintenance of a relatively higher water content in the ABA-treated leaves as compared to leaves of the control plants. ABA had no significant effect on leaf osmotic potential but in combination with NaCl, ABA maintained higher osmotic potential than when NaCl was applied alone (Fig. 3).Thus, the inductive effect of NaCl in changing the type of photosynthesis from C3 to CAM was mainly caused by its influence on water status of the plants, specifically by a decrease of water content in leaf tissues and a sharp decrease in osmotic potential due to the accumulation of inorganic ions and compatible osmolytes such as proline (Kuznetsov et al., 2000). ABA alone or added with NaCl shortened wasteful water evaporation and improved the water status of the plants. This was seen as a smaller decrease of water content and a reduced drop in osmotic potential in treatment in which NaCl and ABA were added together. Moreover, accordingly to the data published by Dörffling *et al.* (1974), ABA also increases water uptake by roots, possibly by aquaporine activation (Hose *et al.*, 2000). Thus, the processes induced by these two agents, NaCl and ABA, were the reverse of each other, resulting in intermediate but not in additive level of CAM activity in response to combined treatment. Hence, in spite of the fact that ABA alone can induce CAM, which was delayed and moderate, given together with NaCl, ABA caused mainly stress-protective effects but not stress-inductive ones. Nevertheless when given alone,exogenous ABA could induce a shift to CAM in the common ice plant. It may be due to shortage of CO_2 in mesophyll tissue of the leaves induced by stomata closure.

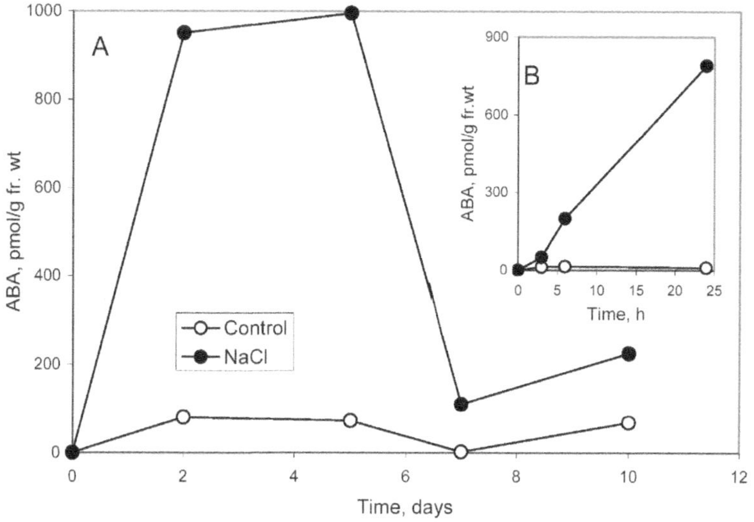

Figure 2. Endogenous ABA levels in leaves of control and 400 mM NaCl treated plants
A .in the long-term experiment; B . the initial hours of treatment

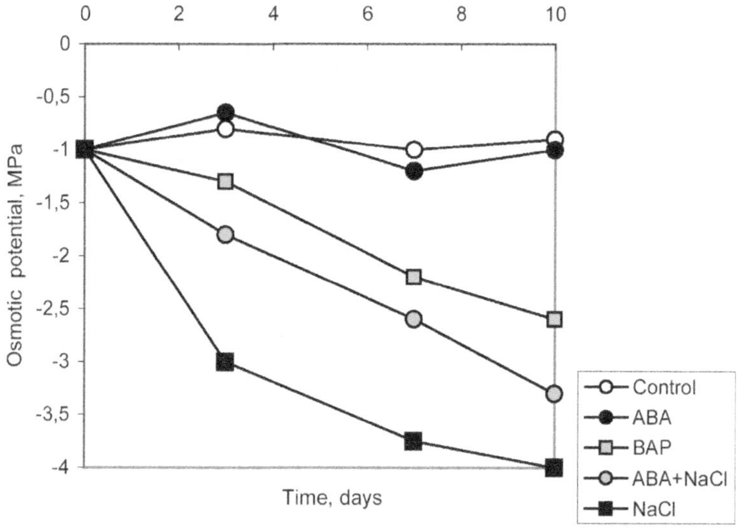

Figure 3. Changes in osmotic potential of leaf sap as affected by NaCl and phytohormones

The ability of the very different hormone, the cytokinin 6-benzylaminopurine (BAP), to switch the common ice plants to CAM, was even more unexpected. When applied exogenously through the roots, BAP (10^{-6} M) was highly effective in inducing CAM activity (Fig. 4 A,B). Used alone or added together with NaCl, BAP induced a rapid switch to CAM and its activity was sometimes the highest one, even in comparison to 400 mM NaCl. This effect was showed on the level of mRNA for PEPC: BAP induced a very rapid expression of the gene for PEPC, comparable to the effect of NaCl. Although the effect of BAP was unexpected, it was easily understandable. The cytokinin treatment of the common ice plants resulted in changes of several basic physiological parameters, principally similar to the effects of NaCl. There was an inhibition of transpiration, a significant decrease in leaf water content and even a decrease of osmotic potential (Fig. 3). In experiments with detached leaves it was shown that BAP inhibited water flow into the plants. There may be several explanations of these effects of cytokinin on CAM induction: 1) it may be a non-specific effect of cytokinin due to acceleration of plant development (Thomas *et al.*, 1993); 2) being added together with NaCl, cytokinin seems to become a participant in a chain of signal transduction; and at last 3) it is impossible to exclude that exogenously applied cytokinin can act like a stressor (see review of Pospisilova, 2000). Hence, the phytohormones, ABA and cytokinin, are not only involved in regulation of separate, individual biochemical reactions or physiological processes, but also the hormones themselves, without any additional stressors, can induce the whole new adaptive metabolic system and cause CAM-type photosynthesis. Pleiotropic action of hormones observed in the investigation, was evidenced by the situation where ABA showed stress-protective properties and cytokinin acted as a stress stimulus.

However, to evaluate the role of phytohormones in the transduction of intracellular or/and inter-organ stress signals it is necessary to investigate early events in plant adaptation. The synthesis of the CAM-type isoform of PEPC was first recorded on the second day of salinization and the mRNA of PEPC appeared even earlier (Thomas *et al.*, 1992; Taybi and Cushman, 1999). In our experiments with whole plants, a significant accumulation of mRNA for PEPC was found 6 hours after 400 mM NaCl addition to the roots. However, the process probablystarted much earlier Thus, probably within 1-2 hours of NaCl addition to roots the expression of genes responsible for key enzymes of CAM-photosynthesis in leaves may be initiated. But in what way information about the sharp decrease in water potential in the rooting medium was perceived by the roots, transduced then into modification of cell metabolism, and transmitted to leaves to initiate the shift from C3- to CAM-type of photosynthesis? And what is the role of phytohormones in these processes? The potential-driven water efflux out of roots resulted in changes in the state of the cell plasma membrane which seemed to be perceived by a transmembrane osmosensory system. The existence of a transmembrane osmosensory system was demonstrated in *E. coli*, *Saccharomyces cerevisiae* and *A.thaliana* by Shinozaki and coworkers (see: Urao *et al.*, 2000). A so-called two-component osmosensory system with hybrid histidine kinase and some additional intermediate molecules is active in the absence

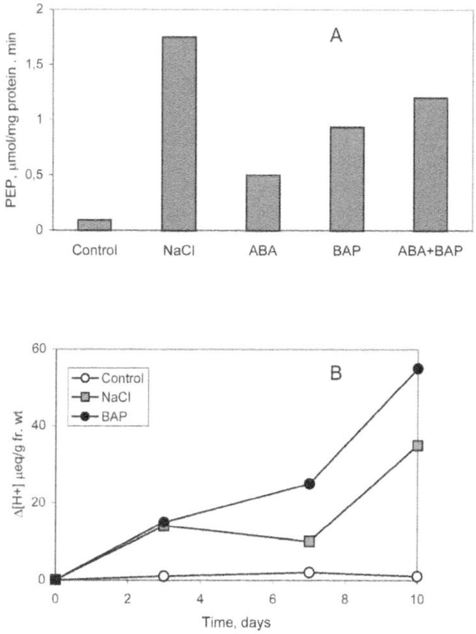

Figure 4. CAM activity during 400 mM NaCl stress and in the presence or absence of BAP
A . activity of PEPC at the 10th day of experiment; B changes of titratable acidity

of an external signal. But under high osmolarity, it (ATHK1) transforms to an inactive state, which in turn activates a MAP kinase cascade. In our experiments, for the first time activation of MAPK was shown for *M. crystallinum.* Under salinity, a protein with characteristic molecular weight of 20 kD was shown in the common ice plant roots. Its transient kinase activity was directed to specific protein MBP (mitogen basic protein). This activity was practically absent initially, before salinization, but it increased very rapidly up to 5-10 min after 400 mM NaCl addition and then returned to its initial level. Activation of MAPK may be a reason for the induction of water deficit gene expression, particularly those responsible for ABA synthesis. An increase of endogenous ABA content in roots started very rapidly (Fig. 5). In 30 min its level exceeded 3.5 times the initial value, and after 24 hours of NaCl treatment it had increased 25-fold. This rapid and sharp accumulation of ABA in roots and especially in xylem exudate, in response to an environmental stress (soil drought or salinity) suggest that ABA may be consideredas a primary inter-organ signal molecule (Tardieu *et al.*, 1996). Nevertheless, some experimental data contradict this view.

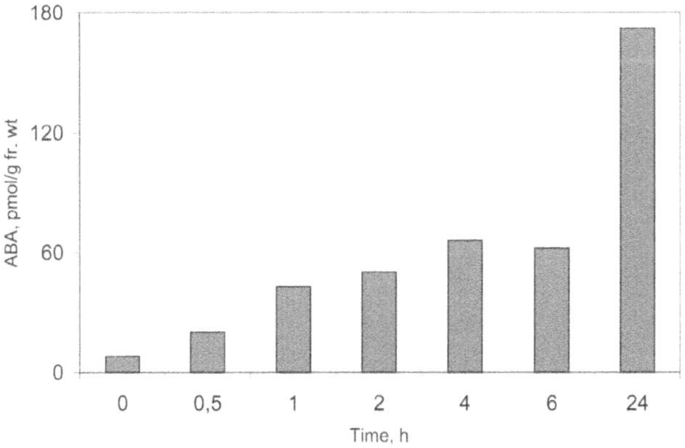

Figure 5. Endogenous ABA levels in roots of 400 mM NaCl treated plants

The sharp decrease of water potential in the rooting medium induced by NaCl (from -0.08 MPa to .1.9 MPa at 400 mM NaCl) decreased the water content in plants. Weight loss could exceed 10% in 3 hours and was even more after 6 hours of NaCl treatment. Thus when transpiration was inhibited (to 30% after 3 h NaCl and to 10% after 6 h) water flow through xylem was strongly limited and for some time even reversed in direction. Water was lost from the plant during initial period of salinization and attempts to collect xylem exudate failed during the first 2-3 hours after NaCl addition. Therefore a period of up to an hour or more there was no water flow from roots to leaves and hence ABA could not be translocated into aerial parts of a plant during initial period of salinization. On the other hand, non-chemical signals could be involved in inter-organ communication in such circumstances, e.g. hydraulic signals. Hydraulic signals are self-propagating changes in water pressure (Malone, 1993), and the front of pressure waves can spread throughout the xylem within seconds. The hydraulic pressure waves in the xylem and in the continuous water phase of a plant will be followed immediately by a physical displacement of water in living cells. In conditions in which there is a rapid decrease of water potential in the rooting medium, caused for example by NaCl, water mass flow will be directed out of cells and result in a drop in cell turgor pressure. In the common ice plant a significant decrease of turgor pressure in cells of leaves was registered with the help of a pressure probe. By 10 minutes after NaCl addition to the rooting medium, turgor pressure had already declined in mesophyll cells to 0.17 MPa and in epidermal bladder cells to 0.32 MPa (Fig. 6). Minimum recorded values as low as 0.13 MPa for mesophyll cells and 0.07 MPa epidermal bladder cells were recorded after 2 hours of NaCl treatment. Later, after a lag-phase of 3-6 hours a significant accumulation of ABA occurred in the leaves (see Fig. 2). Thus we proposed that the same sequence of events took place in the leaves as occurred earlier in the roots in response to NaCl. These were: osmotic and/or turgor pressure signal perception by the plasma membrane; its conversion into MAPK cascade; expression of genes encoding synthesis of ABA and PEPC; and finally induction of CAM activity. In spite of the fact that ABA took part in the adaptive response of the plants to salinity, the role of a primary inter-organ signal belonged not to the hormone but more likely to hydraulic signaling processes.

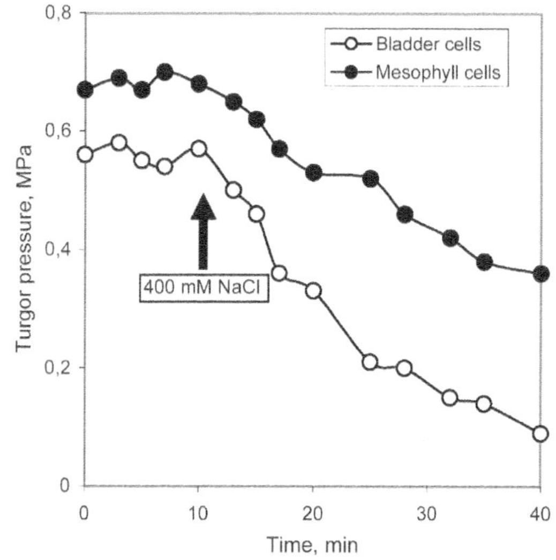

Figure 6. Turgor pressure in leaf cells and its short-term changes after NaCl addition to rooting medium

REFERENCES

Dörffling, K., Sonka, B. and Tietz, D. (1974) Variation and metabolism of abscisic acid in pea seedlings and after water stress. *Planta* **121**, 57-66.

Hose, E., Steudle, E. and Hartung, W. (2000) Abscisic acid and hydraulic conductivity of maize roots: a study using cell- and root-pressure probes, *Planta* **211**, 874-882.

Kholodova, V.P., Neto, D.S., Meshcheryakov, A.B., Borisova, N.N., Alexandrova,S.N. and Kuznetsov, Vl.V. (2002) Can stress-induced CAM provide for performing the developmental program in *Mesembryanthemum crystallinum* plants under long-term salinity?, *Russ. J. Plant Physiol.* **49**, 336-343.

Kuznetsov, Vl.V., Neto, D.S., Borisova, N.N., Dam, Z.B., Rakitin, V.Yu., Alexandrova, S.N. and Kholodova, V.P. (2000) Stress-induced CAM development and the limit of adaptation potential in *Mesembryanthemum crystallinum* plants under extreme conditions, *Russ. J. Plant Physiol.* **47**, 190-198.

Malone, M. (1993) Hydraulic signals, *Phil. Trans. R. Soc. Lond.* **341**, 33-40.

Pospíšilová, J., Synková, H. and Rulcová, J. (2000) Cytokinins and water stress, *Biol. Plant.* **43**, 321-328.

Schmitt, J.M. and Piepenbrock, M. (1992) Regulation of phosphoenolpyruvate carboxylase and Crassulacean acid metabolism induction in *Mesembryanthenum crystallinum* L. by cytokinin: modulation of leaf gene expression by roots?, *Plant Physiol.* **99**, 1664-1669.

Tardieu, F. (1996) Drought perception by plants. Do cells of droughted plants experience water stress?, *Plant Growth Regul.* **20**, 93-104.

Taybi, T. and Cushman J.C. (1999) Signaling events leading to *Crassulacean* acid metabolism induction in the common ice plant, *Plant Physiol.* **121**, 545-555.

Thomas, J.C. and Bohnert, H.J. (1993) Salt stress perception and plant growth regulators in the halophyte *Mesembryanthenum crystallinum* L., *Plant Physiol.* **103**, 1299-1304.

Thomas, J.C., McElwain, E.F. and Bohnert, H.J. (1992) Convergent induction of osmotic stress-responses, *Plant Physiol.* **100**, 416-423.

Urao, T., Yamaguchi-Shinozaki, K. and Shinozaki, K. (2000) Two-component systems in plant signal transduction, *Trends Plant Sci.* **5**, 67-74.

MONOMERIC G-PROTEINS AND MAP KINASES ARE INVOLVED IN ETHYLENE SIGNAL TRANSDUCTION

I.E. MOSHKOV[1,3], G.V. NOVIKOVA[1], L.A.J. MUR[2], A.R. SMITH[2] AND M.A. HALL[2]

[1]*Timiryazev Institute of Plant Physiology, Russian Academy of Sciences, Botanicheskaya, 35, Moscow 127276, Russia*

[2]*Institute of Biological Sciences, University of Wales, Aberystwyth, Ceredigion SY23 3DA, UK*

[3]*Corresponding author, e-mail iim995@hotbox.ru; fax 007 (095) 977 8018*

1. INTRODUCTION

Great progress has been made in the elucidation of components of the ethylene signal transduction pathway, mainly through studies on *Arabidopsis thaliana* mutants. Such components include five partially functionally redundant receptors (Chang *et al.*, 1993; Hua *et al.*, 1995; Sakai *et al.*, 1998), a protein (CTR1) having homology with mitogen-activated protein kinase kinase kinases (MAPKKK) (Kieber *et al.*, 1993), further downstream components such as the EIN series (Johnson and Ecker, 1998; Alonso *et al.*, 1999) and ethylene response element binding proteins (Solano *et al.*, 1998). A significant feature of the system is that the receptors appear to be negative regulators, that is, they are active in the absence of ligand and inactive in its presence (Hua and Meyerowitz, 1998).

Implicit in the presence of a putative MAPKKK in the transduction sequence is that protein phosphorylation via a mitogen activated protein kinase (MAPK) cascade(s) is involved in mediating responses to ethylene. It has been shown in tobacco (Raz and Fluhr, 1993), peas (Berry *et al.*, 1996) and *A. thaliana* (Novikova *et al.*, 1999) that ethylene upregulates protein phosphorylation overall and that in the dominant receptor mutant *etr1-1* the process is downregulated relative to wild type (Novikova *et al.*, 1999). Equally, there is evidence that ethylene affects protein kinase activity. Thus, in the *ctr1-1* mutant of *A. thaliana* where the mutation is in CTR1 (Kieber *et al.*, 1993), an 'ethylene-treated' phenotype is observed, which has been taken to mean – in accord with negative regulation by the receptors – that the activity of the kinase is downregulated by ethylene. In tobacco, ethylene activates PK12, a protein kinase of the LAMMER type and increases levels of its transcript (Sessa *et al.*, 1996).

Here we present the data showing that described above ethylene signal transduction pathway is not the only one.

I. Macháčková and G.A. Romanov (eds.), Phytohormones in Plant Biotechnology and Agriculture, 205-216

2. ETHYLENE UPREGULATES MAP KINASE OF ERK1 TYPE

In vivo ethylene treatment of etiolated pea seedlings or wild-type *Arabidopsis* rosette leaves led to a marked increase in *in vitro* phosphorylation of myelin basic protein (MBP), which is a preferable substrate for MAP kinases *in vitro* (Fig. 1A, B). In-gel MBP phosphorylation revealed a mol. mass for ethylene-activated MBP kinase of 47 kDa (Fig. 1C, D). This MBP kinase could be precipitated with either anti-ERK1 or phosphotyrosine antibodies (Fig. 1) proving that it belongs to MAP kinases of ERK1 type.

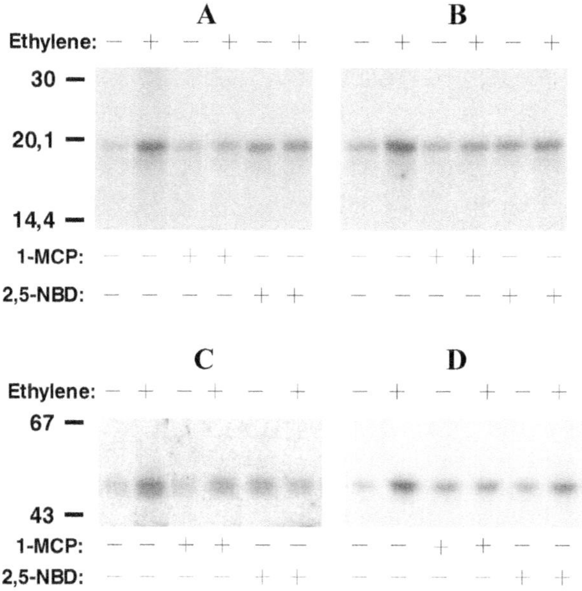

Figure 1. MAP kinase activity in etiolated pea epicotyls as affected by ethylene, 1-MCP and 2,5-NBD. Intact seedlings were treated with ethylene (1 μL L^{-1}, 20 min), 1-MCP (100 nL L^{-1}, 2 h), 2,5-NBD (2,000 μL L^{-1}, 2 h) or ethylene binding inhibitors (as indicated above) followed by ethylene treatment. MAP kinase activity was assayed after immunoprecipitation with either anti-ERK1 (A and C) or anti-phosphoTyr (B and D) antibodies in vitro (A and B) or in situ (C and D).

To demonstrate specificity of ethylene effect on the MAP kinase, receptor-directed inhibitors 2,5-norbornadiene (2,5-NBD) and 1-methylcyclopropene (1-MCP) were used. The application of the inhibitors *in vivo* resulted in some increase in MAP kinase activity but this activation was always much smaller than that caused by ethylene and – what is the most important – the pretreatment of plants with the inhibitors prior ethylene application lowered ethylene-induced alteration of MAP kinase activity (Fig. 1) demonstrating ethylene- and receptor-dependent activation of MAP kinase.

In *Arabidopsis* ethylene-insensitive receptor mutant *etr1-1*, firstly, the constitutive level of MAP kinase activity was much lower as compared to wild-type plants (Fig. 2); secondly, ethylene treatment did not affect MAP kinase activity. In *ctr1-1* mutant with constitutive ethylene phenotype, MAP kinase activity was high and comparable with that in ethylene-treated wild type (Fig. 2). Thus, it appears that ethylene insensitivity correlates with low MAP kinase activity while ethylene phenotype (or ethylene treatment of wild type) – with high enzyme activity. This conclusion was supported when MAP kinase activity was assayed in dominant and recessive ethylene receptor mutants. As Fig. 3 shows, in all cases MAP kinase activity was lower in dominant ethylene-insensitive mutants than in corresponding recessive mutants, which possess wild-type response to ethylene (Hua and Meyerowitz, 1998).

Activation of MAP kinases of ERK type occurs due to phosphorylation on Thr and Tyr residues in the phosphorylation motif Thr-Glu-Tyr. Using antibodies to non-phosphorylated and phosphorylated forms of animal ERK1 MAP kinases (PhosphoPlus Kit, New England Biolabs), we found that ethylene treatment did not alter the enzyme amount but considerably increased portion of phosphorylated form of MAP kinase (Fig. 4).

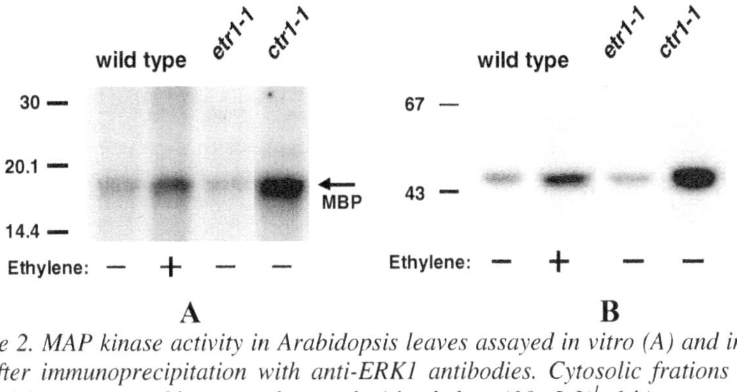

Figure 2. MAP kinase activity in Arabidopsis leaves assayed in vitro (A) and in situ (B) after immunoprecipitation with anti-ERK1 antibodies. Cytosolic frations were derived from untreated leaves and treated with ethylene (10 μL L^{-1}, 1 h).

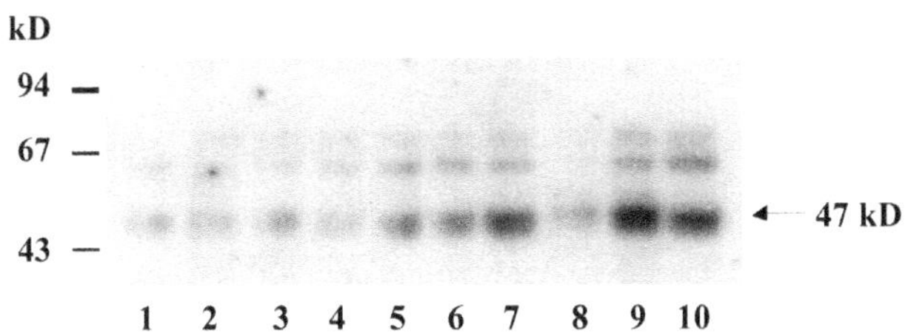

Figure 3. MAP kinase activity assayed in situ in wild-type and Arabidopsis mutants. 1 – wild type, 2 – etr1-1 (dominant receptor mutant), 3 – etr1-6 (recessive receptor

(Figure 3. continued) mutant), 4 – *etr2-1* (dominant receptor mutant), 5 – *etr2-3* (recessive receptor mutant), 6 – *ein4-1* (dominant receptor mutant), 7 – *ein4-4* (recessive receptor mutant), 8 – *ers2-3* (recessive receptor mutant), 9 – *etr1-6; etr2-3; ein4-4* ('triple' recessive receptor mutant), 10 – *ctr1-1*.

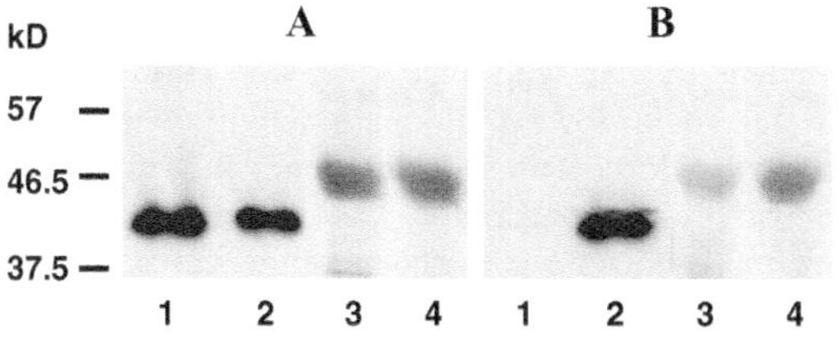

Figure 4. Western blots of wild-type MAP kinase probed with antibodies against non-phosphorylated (A) and phosphorylated (B) of MAP kinase (ERK1) (PhosphoPlus Kit, New England Biolabs). 1 – non-phosphorylated and 2 – phosphorylated control peptides (PhosphoPlus Kit, New England Biolabs), 3 – MAP kinase activity in untreated Arabidopsis wild-type leaves, 4 – MAP kinase activity in ethylene-treated Arabidopsis wild-type leaves.

One of the characteristic features of MAP kinases (at least in animal cells) is their rapid and transient activation. In etiolated pea epicotyls, even 5-min treatment resulted in substantial increase in MAP kinase activity (Fig. 5). The activity peaked at 20 min, then dropped by 30 min and rose again by 60 min of exposure to the phytohormone. Values of activation in different experiments varied from 2.5 to 5 fold but the pattern was very constant. The second peak was always broader and never exceeded the first one.

Thus, in etiolated pea epicotyls and *Arabidopsis* wild-type leaves, ethylene treatment led to activation of a protein kinase, which fits to a number of criteria for an ERK-type MAP kinases, namely: MBP serves as a preferable exogenous substrate; the enzyme interacts with antibodies to ERK1 and phosphotyrosine; it demonstrates rapid and transient activation in response to ethylene; activation is due to phosphorylation on Thr and Tyr residues in phosphorylation motif Thr-Glu-Tyr.

The results obtained raise the question about the MAP kinase position in ethylene signal transduction chain. It is rather unlikely that this MAP kinase belongs to CTR1-dependent pathway since CTR1 is considered as a negative regulator, i. e. it is active in the absence of ethylene and ethylene inactivates CTR1 (and therefore the downstreame cascade) (Kieber *et al.*, 1993). The only reasonable explanation for these data is that ethylene-activated MAP kinase operates in other MAP kinase cascade(s). It is unlikely that ethylene-activated cascade directly links to the receptors since CTR1 has been reported to interact directly to ETR1 (Clark *et al.*, 1998). In animal cells, most membrane receptors signal to downstream MAP kinase cascades via monomeric G-proteins (Denhardt, 1996). On the other hand, we demonstrated that ethylene treatment of etiolated pea epicotyls or *Arabidopsis* wild-type leaves resulted in an increase in phosphorylation and activity of nucleoside diphosphate kinase (Novikova *et al.*, 1999, 2003), which may provide GTP for G-proteins.

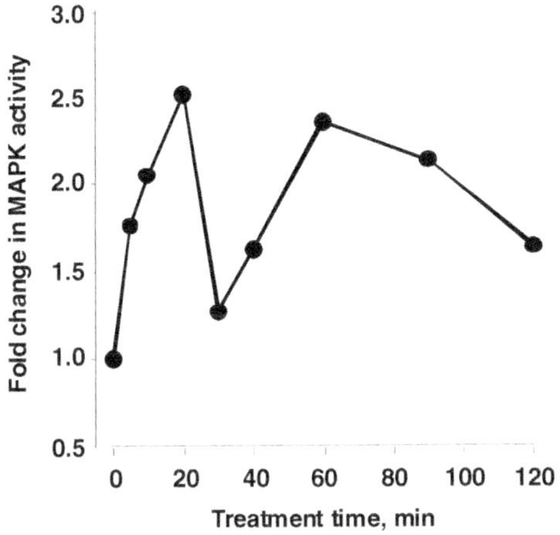

Figure 5. Time course of ethylene-modulated MAP kinase activity. Pea seedlings were treated with 1 μL L^{-1} ethylene for different time periods followed by isolation of cytosolic proteins. MAP kinase activity was assayed in vitro. Experimental points were derived from scans of autoradiographs of phosphorylated MBP SDS-PAGE separations.

3. ETHYLENE REGULATES GTP BINDING TO MONOMERIC G-PROTEINS AND TRANSCRIPTION OF SEVERAL G-PROTEIN GENES

For GTP binding assessment proteins from membrane fractions pelleted at 130,000 *g* were consequently solubilised with 750 mM potassium chloride and 1% Triton X-100 and GTP binding was analysed with [α-^{32}P]GTP (Löw *et al.*, 1992). Specific GTP binding was observed to polypeptides ranged between 20 and 30 kD (Novikova *et al.*, 1997, 1999) that is appropriate for monomeric G-proteins.

Ethylene treatment led to activation of GTP binding. Both 1-MCP (Fig. 6) and 2,5-NBD (not shown) significantly decreased ethylene-promoted binding showing receptor dependence and specificity albeit it should be noted that both compounds caused some activation of GTP binding but in much smaller scale than ethylene. The pattern of GTP binding kinetics (Fig. 6) is very similar to that observed for ethylene-regulated MAP kinase (Fig. 5) suggesting possible involvement of both components into the same signalling pathway. Importantly, binding activation was very rapid indeed: even 2-min treatment doubled the binding indicating close linkage of G-proteins with ethylene receptors. To our knowledge, this is the fastest biochemical response to ethylene treatment of intact tissue.

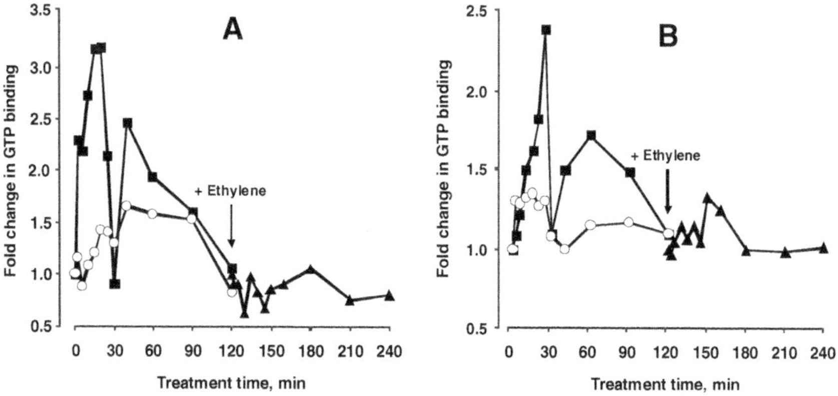

Figure 6. Time course of GTP binding to monomeric G-proteins extracted with 750 mM KCl (A) and 1% Triton X-100 (B) as affected by ethylene (1 μL L^{-1}, ■), MCP (100 nL L^{-1}, ○) and when ethylene was applied after pretreatment with MCP (▼). The arrow indicates time point of ethylene application. Experimental points are derived from scans of 1D autoradiographs of SDS-PAGE separations of labeled proteins.

However, this represents only the overall picture. To resolve GTP binding to individual proteins 2D-electrophoresis was employed. It revealed very complicated picture (Fig. 7). Firstly, there was a large number of components observed that is not surprising since, for example, in *Arabidopsis* data base there are over 30 genes for monomeric G-proteins (The Arabidopsis Genome Initiative, 2000). Secondly, the effect of ethylene on GTP binding was differential. We could quantify GTP binding to several components in KCl-extracted fractions from pea epicotyl membranes. Two components (spots 8 and 9) showed the bimodal response but further four (spots 2, 6, 12 and 14) showed a transient decrease before returning almost to control levels by 40 min. In the Triton fractions, because of the hydrophobic nature of proteins we could not separate all the components rigorously. However, quantification of groupings on the 2D gels, assigned on the basis of molecular mass and pI, indicated a somewhat different pattern. Thus, three groups (groups 5, 6, and 7) showed bimodality comparable to the KCl fractions, with significant effects being observed already within 2 min, peaks at 10 and 60 min and return to baseline at 20. A further bimodal activation peaked at 20 and 60 min (group 3) but four groups (1, 2, 4 8) showed a unimodal response with a peak at 20 min. No components in the Triton fraction were downregulated.

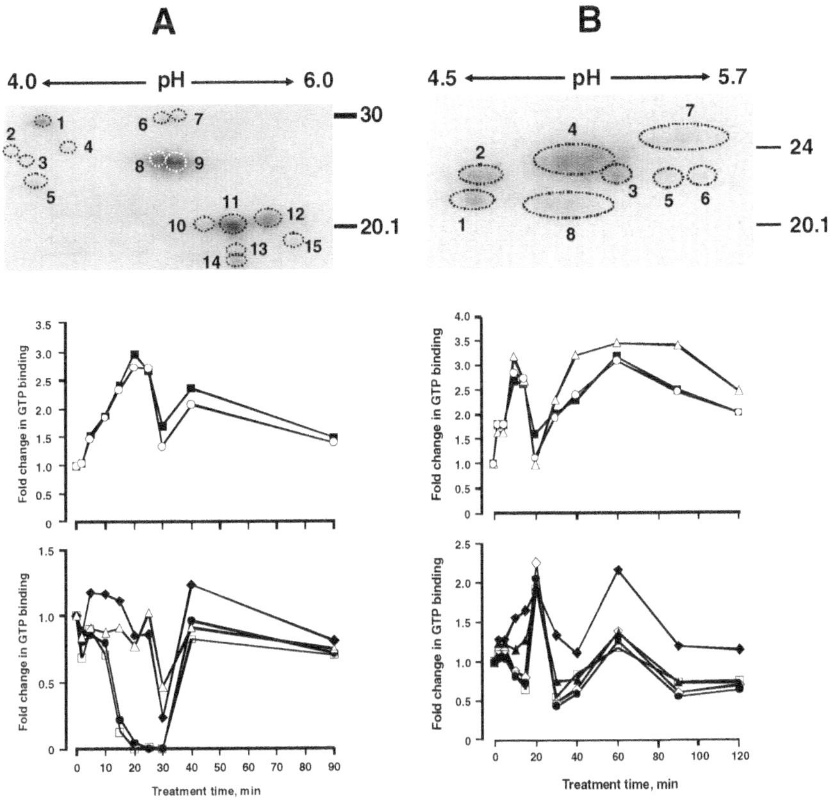

Figure 7. 2D separations of GTP-binding proteins. Proteins were consequently extracted with 750 mM KCl (A) and 1% Triton X-100 (B) from untreated or ethylene-treated (1 μL L^{-1}) epicotyls and separated by 2D gel electrophoresis. Quantification of GTP binding during time course: in panel A – 2 (●), 6 (◆), 8 (■), 9 (○), 12 (△), 14 (□); in panel B – 1 (◇), 2 (●), 3 (◆), 4 (▲), 5 (■), 6 (△), 7 (○), 8 (□).

When we compared in 2D gels the GTP-binding patterns in *Arabidopsis* wild type and ethylene-response mutants (Fig. 8), we found that they were very similar. In *etr1-1*, the binding was much lower than in wild type while in *ctr1-1* it was very high, comparable to that in wild type treated with ethylene. Hence, it can be concluded that the level of GTP binding correlates with the physiological response to ethylene and phenotype. In other words, it is increased in wild type as a result of ethylene treatment, in *etr1-1*, which is insensitive to ethylene, it is very low and not affected by ethylene, and in *ctr1-1*, which shows an ethylene phenotype in air, it is very high and also not affected by ethylene. This mirrors the situation for the MAP kinase.

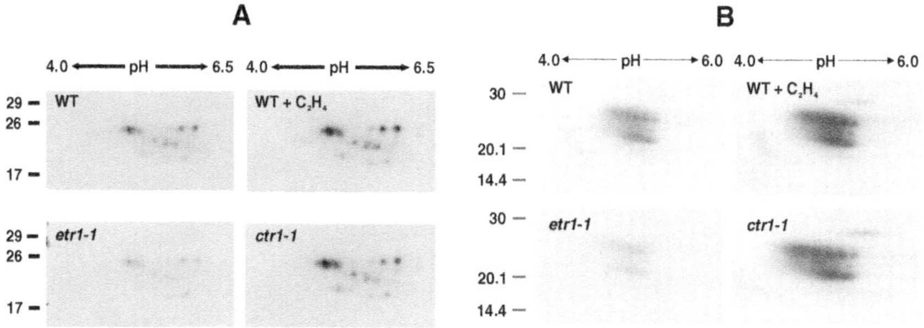

Figure 8. 2D PAGE separation of monomeric G-protein components from Arabidopsis leaf membranes consequently extracted with 750 mM KCl (A) and 1% Triton X-100 (B).

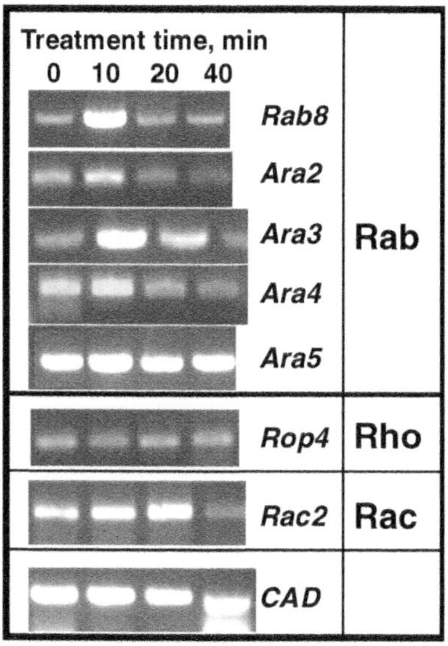

Figure 9. Transcriptional analysis of monomeric G-protein expression following ethylene treatment. Wild-type Arabidopsis plants were treated with ethylene for 0, 10, 20 and 40 min and RNA was isolated from each of three Arabidopsis plants per time point. First-strand cDNA was constructed from each plant. Specific oligonucleotide primers were used to detect transcript levels of Rab8, Ara3, Ara2, Ara4, Ara5, Rop4 and Rac2 in 1 g of first-strand cDNA. Amplifications from each cDNA from each plant were repeated three times (n = 9), representative results of which are shown. Control amplifications were carried out using oligonucleotide primers to cinnamyl alchohol dehydrogenase (CAD).

Using RT-PCR we examined the effect of ethylene on the expression of several genes of Rab class and Rho/Rac classes (Fig. 9). Three patterns of effect can be observed. In the cases of *Rop4* and *Rac2* there was no effect – although this may bemisleading as several genes are involved and changes in one may conceal changes in another. The second case (*Ara2, Ara4* and *Ara5*) is where constitutive levels of transcription were relatively high but was then downregulated by ethylene. And lastly, there were two cases (*Rab8* and *Ara3*) where transcription was markedly but transitionally upregulated by ethylene. Additionally (Fig. 10), in *etr1-1* transcription of *Rab8* was not affected by ethylene and in *ctr1-1* the constitutive level of expression was much higher than in wild type and also was not regulated by ethylene. Hence, we observed differential effects of ethylene at two different levels: at the level of GTP binding to proteins and at the level of regulation of gene expression.

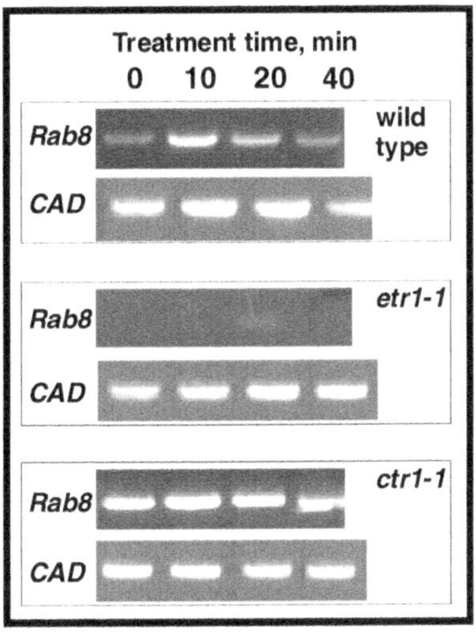

Figure 10. Rab8 expression in Arabidopsis wild-type and in ethylene signaling mutants. For details see the legend to Fig. 9.

We have observed rapid activation of transcription of the *Rab8* gene in *Arabidopsis* and therefore undertook immunoprecipitation studies with antibodies to this protein; the results are shown in Figure 11.

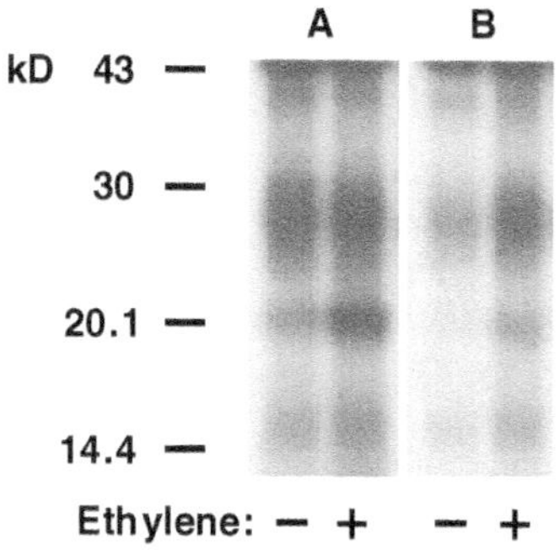

Figure 11. Immunoprecipitation of [α-^{32}P]GTP-labeled monomeric G-proteins with anti-Rab8 antibodies. Membrane proteins were consequently solubilised with 750 mM KCl (A) and 1% Triton X-100 (B). Fractions were derived from untreated etiolated pea epicotyls and treated with ethylene (1 μL L^{-1}, 20 min).

Two diffuse bands were revealed, but with much higher abundance of antigens in the KCl-solubilised fraction. However, the degree of activation in the two fractions was quite different. Thus, the components with higher molecular masses in the KCl-solubilised fraction showed no response to ethylene but the lower molecular mass components showed a significant promotion. Conversely, in the Triton-solubilised fraction the high molecular mass components showed some activation but this increases significantly in the lower molecular mass components. This latter is much greater than the overall activation in the Triton-solubilised fraction at this point in the time course (Fig. 6).

4. CONCLUSION

The data presented here clearly demonstrate that ethylene affects the activation of both monomeric G-proteins and protein kinase(s) in etiolated pea epicotyls and *Arabidopsis* wild-type leaves. Assuming CTR1 as a negative regulator, it is very unlikely that these components belong to the CTR1-dependent pathway of ethylene signal transduction. Therefore we suggest an existence of another ethylene signal transduction pathway, which initiates by the same receptors as for CTR1-dependent pathway. Hua and Meyerowitz (1998) showed that ethylene receptors are negative regulators and there is no contradiction between negative role of the receptors for CTR1-dependent pathway and for that which we found since there are examples when the same receptors can be negative regulators for one pathway but at the same time positive for another. In the absence of ethylene the CTR1-dependent pathway is active while the other one is inactive. Binding of ethylene to the receptors leads to inactivation of the first pathway but activation of another one through monomeric G-proteins and ethylene-activated MAP kinase cascade. We also suggest that the ratio of the activities of both cascades are responsible for the final response to ethylene rather than a linear on/off switch. How and at what level these pathways may interact remains to be elucidated.

ACKNOWLEDGEMENTS

This work was in part supported by INTAS grant no. 99-01200 and Russian Foundation for Basic Research grant no. 02-04-48414.

REFERENCES

Alonso, J.M., Hirayama, T., Roman, G., Nourizadeh, S. and Ecker, J.R. (1999) EIN2, a bifunctional transducer of ethylene and stress responses in *Arabidopsis*, *Science* **284**, 2148-2152.

Berry, A.W., Cowan, D.S.C., Harpham, N.V.J., Hemsley, R.J., Novikova, G.V., Smith, A.R. and Hall, M.A. (1996) Studies on the possible role of protein phosphorylation in the transduction of the ethylene signal, *Plant Growth Regul.* **18**, 135-141.

Chang, C., Kwok, S.F., Bleecker, A.B. and Meyerowitz, E.M. (1993) *Arabidopsis* ethylene-response gene *ETR1*: similarity of product to two-component regulators, *Science* **262**, 539-544.

Clark, K.L., Larsen, P.B., Wang, X. and Chang, C. (1998) Association of the *Arabidopsis* CTR1 Raf-like kinase with the ETR1 and ERS ethylene receptors, *Proc. Natl. Acad. Sci. USA* **95**, 5401-5406.

Denhardt, D.T. (1996) Signal transduction protein phosphorylation cascades mediated by Ras/Rho proteins in the mammalian cell: the potential for multiplex signaling, *Biochem. J.* **318**, 729-747.

Hua, J., Chang, C., Sun, Q. and Meyerowitz, E.M. (1995) Ethylene insensitivity conferred by *Arabidopsis ERS* gene, *Science* **269**, 1712-1714.

Hua, J. and Meyerowitz, E.M. (1998) Ethylene responses are negatively regulated by a receptor gene family in *Arabidopsis thaliana*, *Cell* **94**, 261-271.

Johnson, P.R. and Ecker, J.R. (1998) The ethylene gas signal transduction pathway: a molecular perspective, *Ann. Rev. Genet.* **32**, 227-254.

Kieber, J.J., Rothenberg, M., Roman, G., Feldmann, K.A. and Ecker, J.R. (1993) CTR1, a negative regulator of the ethylene response pathway in *Arabidopsis*, encodes a member of the Raf family of protein kinases, *Cell* **72**, 427-441.

Löw, A., Faulhammer. H.G. and Sprinzl. M. (1992) Affinity labeling of GTP-binding proteins in cellular extracts, *FEBS Lett.* **303**, 64-68.

Novikova, G.V., Moshkov, I.E., Smith, A.R., Kulaeva, O.N. and Hall, M.A. (1999) The effect of ethylene and cytokinin on guanosine 5′-triphosphate binding and protein phosphorylation in leaves of *Arabidopsis thaliana*, *Planta* **208**, 239-246.

Novikova, G.V., Moshkov, I.E., Smith, A.R. and Hall, M.A. (2003) Is nucleoside diphosphate kinase a possible component of ethylene signal transduction pathway?, *Biochemistry* (Moscow), in press.

Novikova, G.V., Moshkov, I.E., Smith, A.R. and Hall, M.A. (1997) The effect of ethylene on GTP binding in extracts from pea seedlings, *Planta* **201**, 1-8.

Raz, V. and Fluhr, R. (1993) Ethylene signal is transduced via protein phosphorylation events in plants, *Plant Cell* **5**, 523-530.

Sakai, H., Hua, J., Chen, Q.G., Chang, C., Medrano, L.J., Bleecker, A.B. and Meyerowitz, E.M. (1998) *ETR2* is an *ETR1*-like gene involved in ethylene signaling in *Arabidopsis*, *Proc. Natl. Acad. Sci. USA* **95**, 5812-5817.

Sessa, G., Raz, V., Savaldi, S. and Fluhr, R. (1996) PK12, a plant dual-specificity protein kinase of the LAMMER family, is regulated by the hormone ethylene, *Plant Cell* **8**, 2223-2234.

Solano, R., Stepanova, A., Chao, Q. and Ecker, J.R. (1998) Nuclear events in ethylene signaling: a transcriptional cascade mediated by ETHYLENE INSENSITIVE3 and ETHYLENE RESPONSE-FACTOR1, *Genes Dev.* **12**, 3703-3714.

The Arabidopsis Genome Initiative (2000) Analysis of the genome sequence of the flowering plant *Arabidopsis thaliana, Nature* **408**, 796-815.

ROSETTE LEAF SENESCENCE IN WILD TYPE AND AN ETHYLENE-INSENSITIVE MUTANT OF *ARABIDOPSIS THALIANA* DURING INFLORESCENCE AND FRUIT DEVELOPMENT

I. SERGIEV[1], D. TODOROVA[1], V. ALEXIEVA[1], E. KARANOV[1], A. SMITH[2] AND M. HALL[2]

[1] *Acad. M. Popov Institute of Plant Physiology, Bulgarian Academy of Sciences, Acad. G. Bonchev Str., Bl. 21, 1113 Sofia, Bulgaria;*

[2] *Institute of Biological Sciences, University of Wales, Edward Liwyd Bluilding, Aberystwyth, Ceredigion, SY23 3DA, UK*

1. INTRODUCTION

Senescence is an important developmental process in plants, which eventually leads to whole plant, organ, tissue and cell death through highly regulated, endogenously controlled degenerative processes (Chandlee, 2001). Leaf senescence is a key developmental step in the life of an annual plant, as it is the time during which material built up by the plant during its growth phase is mobilized into the developing seed to prepare for the next generation (Smart, 1994; Smart *et al.*, 1995). Obvious visual symptoms for leaf senescence are the loss of chlorophyll pigments, desiccation and eventual abscission. Cellular and molecular events contributing to these visual symptoms include chloroplast disintegration, a decline in photosynthesis and damages of proteins and nucleic acids. Other internal symptoms of senescence are loss in the ability to accumulate protein and nucleic acids because of enhanced degradation and/or diminution of synthesis, and loss of plasma membrane and endomembrane structure with associated increases in permeability to inorganic and organic solutes, as well as an increase in the levels of active oxygen species, the later forming organic free radicals (Smart, 1994; Buchanan-Wollaston, 1997). The free radicals impose a significant impact for the deterioration of cell constituents during plant senescence. They are highly reactive, self-propagating, potentially damaging, and are also formed during the normal metabolic processes. Their effective removal is of importance for the well-being of the plant organisms (Fridovich, 1976). One of the mechanisms by which plants defend against free radical mediated damage is the

induction of the superoxide dismutase (SOD). Peroxidase and catalase then breakdown the formed hydrogen peroxide, and thus prevent the further formation of potent free radicals (Dhindsa *et al.*, 1981).

I. Macháčková and G.A. Romanov (eds.), Phytohormones in Plant Biotechnology and Agriculture, 217-228

The initiation and progression of leaf senescence can be influenced by number of external factors, such as various environmental stresses (Abarca *et al.,* 2001). The internal senescence inducing factors appear to be hormonal in nature. Studies have clearly implicated cytokinins and ethylene as significant regulators of the senescence program in leaves and other plant tissues including fruits and flowers (Chandlee, 2001).

In the ethylene insensitive genotypes the senescence processes are normally taking place, but the visual symptoms appear later or are missing, i.e. the leaves are yellowing in a lesser degree and the degradation of the chlorophyll-protein complexes is going on in a slower rate than in the wild type (Abarca *et al.,* 2001). Such mutants, defective in their normal senescence program express the so called "stay green" phenotypes (Chandlee, 2001).

The current article presents a comparative study on the senescence of rosette leaves of a wild type of *Arabidopsis thaliana* Heinh. Plants, and an ethylene insensitive mutant (*eti5*). The changes in chlorophyll and protein content, lipid peroxidation and protein oxidation, as well as the activities of superoxide dismutase, catalase, guaiacol peroxidase and glutathione-S-transferase were followed in order to evaluate the senescence related metabolic alterations in both genotypes during flowering and silique maturation.

2. MATERIALS AND METHODS

Plant material, growth conditions and mesurements. Wild type and ethylene insensitive mutant (*eti5*) plants of *Arabidopsis thaliana* (L.) Heinh, were grown in a growth chamber on soil/perlite mixture (3:1) under the following conditions: 16/8 h day/night photoperiod; 70 $\mu mol.m^{-2}.s^{-1}$ photon flux density; 26/22°C day/night temperature; 60% air humidity. Plants were daily irrigated. In the experiments, rosette leaves (leaf nodes 3 to 7) from plants at stages of development "flowering" and "silique maturation" were used as a source material for all analyses.

Plastid pigments were evaluated according to Arnon (1949). Malondialdehyde content was measured as a parameter reflecting the biomembrane integrity deterioration (Kramer *et al.*, 1991). The amount of carbonyl groups as products of protein breakdown was determined according to Levine (1994).

The hydrogen peroxide content was determined by a reaction with 1M potassium iodide (Jessup *et al.*, 1994). Soluble protein content was determined according to Bradford (1976). The activities of the enzymes measured were determined as follows: catalase (Brennan and Frenkel, 1977), guaiacol peroxidase (Dias and Kosta, 1983), superoxide dismutase (Bauchamp and Fridovich, 1971), glutathione-S-transferase (Gronwald *et al.,* 1987).

Statistics. All experiments were repeated three times with two or three replicates each. The data reported are mean values statistically evaluated according to Fisher's procedure.

3. RESULTS AND DISCUSSION

The changes in the plastid pigment content in rosette leaves of wild type and mutant *Arabidopsis* plants during flowering and silique maturation are presented on Fig. 1.

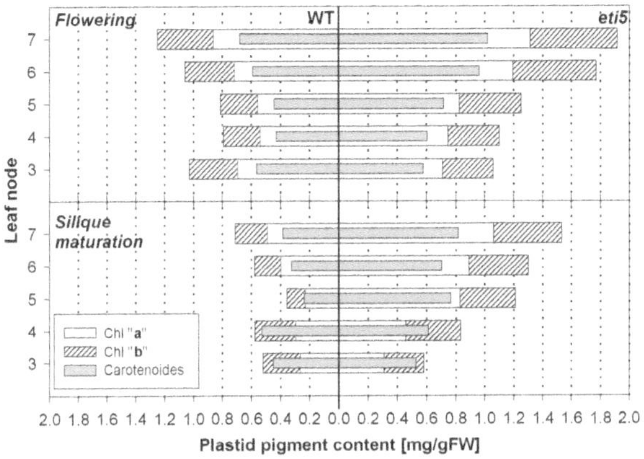

Figure 1. Pigment content in rosette leaves of wild type and eti5 mutant of Arabidopsis thaliana L. (Heynh.) plants during flowering and silique maturation. LSD 5% = 0.146

It was established that during flowering, the mutant *eti5* possessed higher amounts of chlorophyll "a" than the wild type. The same applies also for chlorophyll "b" and carotenoids. As expected, in both genotypes the amounts of plastid pigments in the younger leaves (higher nodes) were found to be greater than those from the older leaves (lower nodes), and obviously this pattern of distribution is age-related. During silique maturation there was an overall decrease of the chlorophyll and carotenoids content in both genotypes, but while in the wild type there was a reduction in the content of both chlorophyll "a" and "b" (carotenoids were less affected), in *eti5* the major decrease of the photosynthetic pigments was mainly due to a decline of chlorophyll "a". Thus, the wild type appears to loose its photosynthetic pigments in an advanced rate than the mutant.

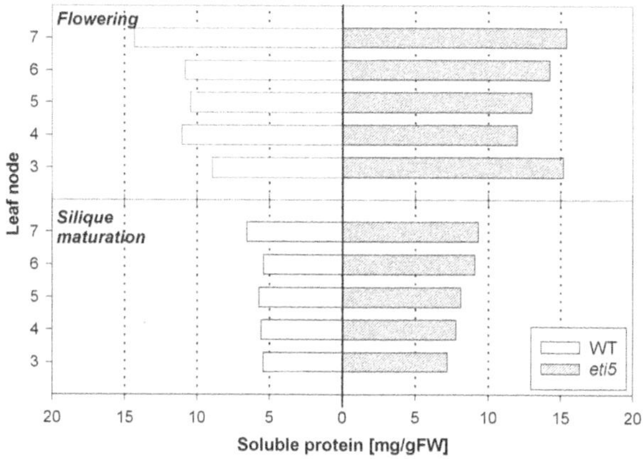

Figure 2. Protein content in rosette leaves of wild type and eti5 mutant of Arabidopsis thaliana L. (Heynh.) plants during flowering and silique maturation. LSD 5% = 3.75

During the normal processes of senescence along with the breakdown of biomembranes and plastid pigments there is a significant decline in the proteins and nucleic acids. The changes in the soluble protein content (Fig. 2) and the products of protein oxidation, expressed as carbonyl group equivalents (Fig. 3) were also followed in order to relate the senescence associated loss of macromolecules with the changes in the photosynthetic pigments and the biomembrane integrity. It was found that during flowering, as well as during silique maturation stage of development, the *eti5* mutant contained superior amounts of soluble proteins than the wild type. On the other hand, the leaves of *eti5* consisted in less quantities of carbonyl groups than the wild type during flowering, but there was no significant difference between both genotypes during silique maturation. Within leaf nodes in both genotypes, the older leaves accumulated more carbonyl groups than the younger ones, showing an inverse relation between the leaf node and protein damage, as well as indicating that greater oxidative injuries of proteins are taking place with advancing of age. As with the plastid pigments, the decline in the soluble protein content during silique maturation is well correlated with a build-up in the products of protein oxidation.

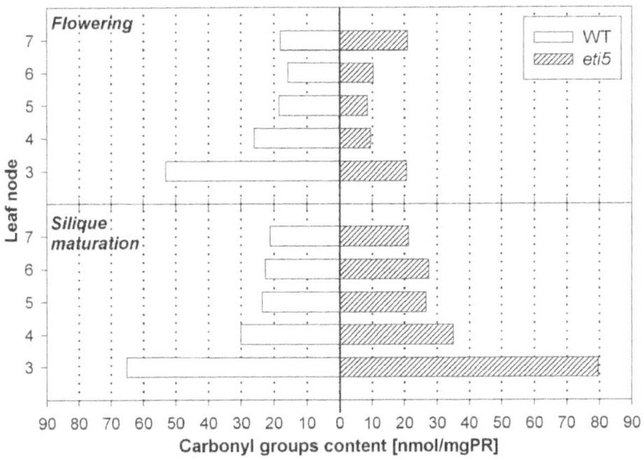

Figure 3. Carbonyl content in rosette leaves of wild type and eti5 mutant of Arabidopsis thaliana L. (Heynh.) plants during flowering and silique maturation. LSD 5% =5.12

Leaf senescence is an oxidative process and the breakdown of chlorophyll and biomembranes is associated by a significant increase in the free radical production (Nooden *et al.*, 1997). The generation of free radicals in the presence of physiological concentrations H_2O_2 leads to the production of the highly oxidising and damaging hydroxyl radical. The formation of this radical *in vivo* causes a lipid peroxidation (Bray *et al.*, 2000). In senescing plant tissues, the lipid peroxidation plays a key role in the loss of membrane integrity (Gidrol *et al.*, 1989). The basal level of lipid peroxidation in intact plant tissues is thought to reflect the activities of membrane electron transport chains, the enzymatic production of oxygen radicals as well as the efficiency of the systems eliminating active oxygen species and the intermediate products of lipid peroxidation (Merzlyak, 1991).

It was found that during flowering, the wild type of *Arabidopsis* plants (with exception of the oldest leaf nodes) contained higher concentrations of thiobarbituric acid reaction products (MDA) the mutant plants (Fig. 4). Throughout the next stage of development the amounts of MDA were considerably increased in both genotypes as compared to the flowering stage, but similarly, the wild type of *Arabidopsis* contained more MDA than the mutant. Likewise the protein oxidation, the lipid peroxidation showed analogous tendency of distribution within the leaf nodes and is in inverse relation to the protein and chlorophyll content.

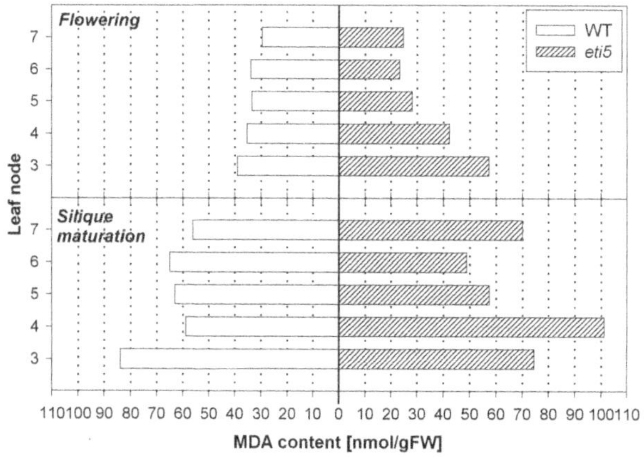

Figure 4. Level of lipid peroxidation (measured as malondialdehyde equivalents) in different rosette leaves of wild type and eti5 mutant of Arabidopsis thaliana L. (Heynh.) plants during flowering and silique maturation. LSD 5% = 3.75

Levels of free radicals and active oxygen species such as hydrogen peroxide increase during senescence as well as in response to environmental stresses (Merzlyak and Hendry, 1994). It is well known that the hydrogen peroxide possesses a dual function in the plant organisms. Its amounts can be altered during the processes of senescence and as a consequence of the influence of number of stress factors, but also it plays a considerable role as a signal molecule in the normal physiological processes. A significantly higher amount of hydrogen peroxide was detected in the rosette leaves of the flowering wild type as compared to the mutant plants (Fig. 5). Surprisingly, during the next stage studied, this difference was compensated by a decrease of the hydrogen peroxide in the wild type and an augment in *eti5*. The observed discrepancy could probably be due to a reduced H_2O_2 production in the wild type because of cessation of the normal function of the photosynthetic apparatus and the other electron transporting chains where the hydrogen peroxide appears as a side product.

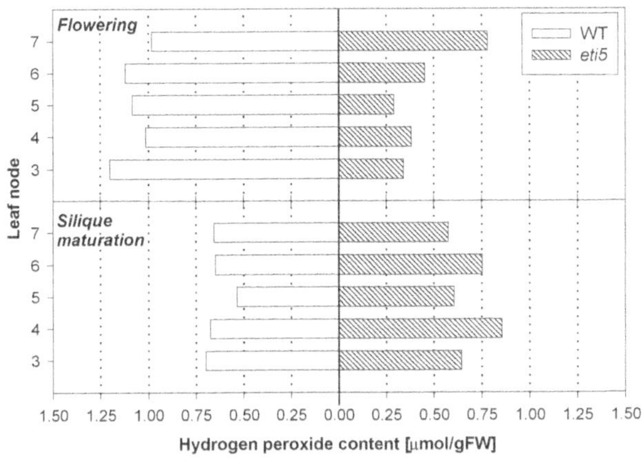

Figure 5. Hydrogen peroxide content in rosette leaves of wild type and eti5 mutant of Arabidopsis thaliana L. (Heynh.) plants during flowering and silique maturation. LSD 5% = 0.32

Protection against free radicals *in vivo* is provided by mechanisms that prevent the formation of reactive oxygen species or their detoxification by antioxidant enzymes. It is well known that senescing leaves retain to a great extent their defence potential both against activated oxygen species and toxicity of lipid peroxidation products. Senescing leaves contain high concentrations of carotenoids and, according to some reports, possess relatively high activities of SOD and peroxidase (Merzlyak *et al.*, 1993) to provide protection against photodestruction. The superoxide radicals, appearing as by-products of the oxidative metabolism, and interacting with the hydrogen peroxide lead to the formation of the hydroxyl radicals, which are thought to be primarily responsible for the oxygen toxicity in the cell. Thus, the dismutation of superoxide radicals is an important step in the cell protection, and in this process SOD is considered a key enzyme (Prasad *et al.*, 1994). During flowering, the superoxide dismutase activity in the rosette leaves of the wild type of *Arabidopsis* was found to be noticeably higher than in *eti5* mutant plants (Fig. 6). Throughout silique maturation, there was an overall increase in the superoxide dismutase activity in both genotypes, but nevertheless, this was better expressed in the wild type. Taking in account the fact that SOD is an inducible enzyme system (Kumar and Knowels, 1993), the observed difference between the wild type and *eti5* could be related to a greater need of the wild type for scavenging of reactive oxygen species. In opposite to SOD activity, we found that the mutant *Arabidopsis* plants possessed a higher catalase activity than wild type during the flowering stage of development (Fig. 7). Although there was a general stimulation of catalase activity in both genotypes during silique maturation, this increment was better expressed in the mutant.

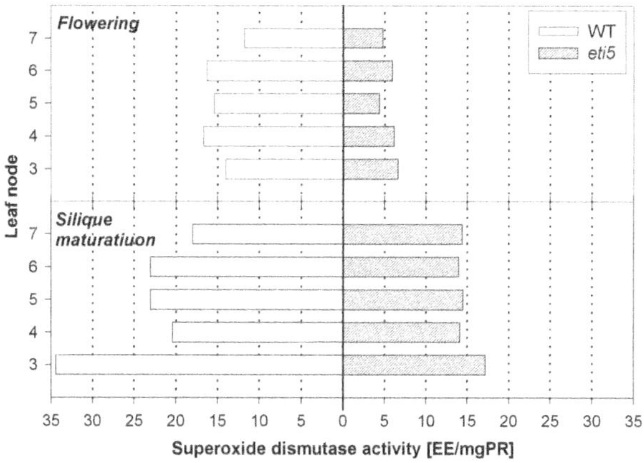

Figure 6. Superoxide dismutase activity in rosette leaves of wild type and eti5 mutant of Arabidopsis thaliana L. (Heynh.) plants during flowering and silique maturation. LSD 5% =1.86

Among the other enzymes involved in the metabolism of reactive oxygen species, an increase in the peroxidase activity takes place in many aging plant systems. However, the role of this enzyme system is at least dual, it can serve both as a source of free radicals (e.g. during oxidation of polyphenols and pyridine nucleotides) and as a factor responsible for elimination of the hydrogen peroxide (Halliwel and Gutteridge, 1989). The data for peroxidase activity are presented in Fig. 8. Generally, with few exceptions the changes in the guaiacol peroxidase activity in both genotypes followed a similar pattern as the catalase. Nevertheless, some authors report a decreased activity of catalase, peroxidase and SOD during senescence, our data showed that in rosette leaves of *Arabidopsis* there is an increase in the activity of the antioxidant enzyme systems studied, and this is consistent with the observed by Kumar and Knowels (1993) senescence related stimulation of these enzyme systems in potato.

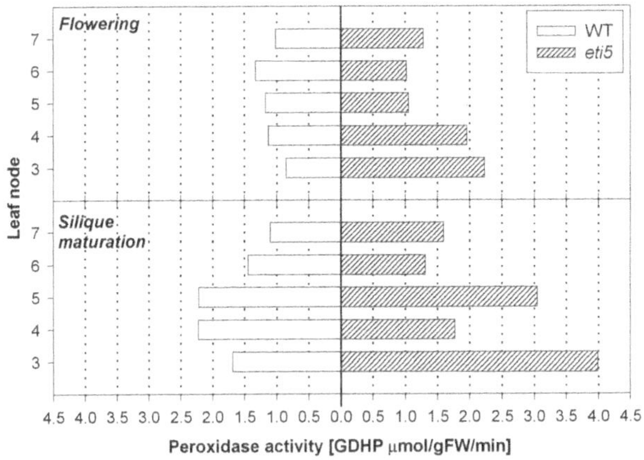

Figure 7. Catalase activity in rosette leaves of wild type and eti5 mutant of Arabidopsis thaliana L. (Heynh.) plants during flowering and silique maturation. LSD 5% = 0.38

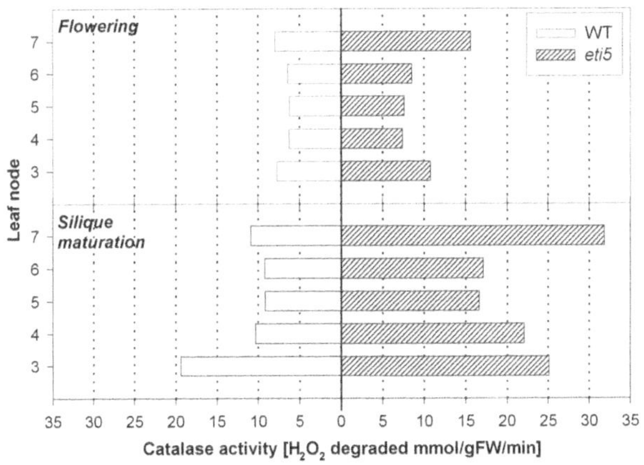

Figure 8. Guaiacol peroxidase activity in rosette leaves of wild type and eti5 mutant of Arabidopsis thaliana L. (Heynh.) plants during flowering and silique maturation. LSD 5% = 2.06

Coming back to the hydrogen peroxide, the higher amounts of H_2O_2 in the wild type during flowering could be related to the augmented activity of superoxide dismutase. During silique maturation, nevertheless the SOD activity was even higher, there was decrease of the hydrogen peroxide quantities - this correlates with the amplified activity of catalase and peroxidase. On the other hand, the flowering leaves of *eti5* were characterised by lower activity of SOD, lower amounts of H_2O_2 respectively, and higher activities of catalase and peroxidase as compared to the wild type. During silique maturation the mutant possessed an increased SOD, catalase and peroxidase activity, accompanied by increased H_2O_2 content.

Glutathione-S-transferase (GST) is known to be involved mainly in the detoxification of xenobiotics by conjugation with glutathione. Alternatively, this enzyme could have some role in sulphur metabolism during senescence - gene encoding GST shows an increased expression in senescing leaves (Smart *et al.*, 1995). Recently the glutathione-S-transferase is discussed as an enzyme system which takes part in the detoxification of organic peroxides (Ivanov, 2003). Although there were no considerable changes in GST activity in both genotypes during flowering and silique maturation (Fig. 9), it should be noted that the rosette leaves of the mutant possessed higher GST activity than the wild type.

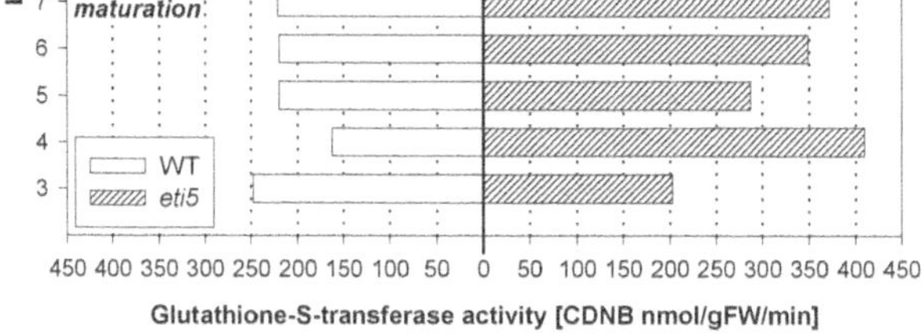

Figure 9. Glutathione-S-transferase activity in rosette leaves of wild type and eti5 mutant of Arabidopsis thaliana L. (Heynh.) plants during flowering and silique maturation. LSD 5% = 27.6

To conclude, the mutant *eti5* contained higher amounts of plastid pigments and soluble protein, higher activity of antioxidant enzyme defense systems, and the senescence induced oxidative damages are taking place in lower rates than in the wild type. The current data are in agreement with the observed by Harpham *et al.* (1991) distinctive delay in the senescence program of the ethylene insensitive mutant *eti5*, and introduce some new senescence related physiological characteristics.

ACKNOWLEDGEMENTS

This work was supported by the INCO-COPERNICUS Programme, Project ERBIC15CT96091.

REFERENCES

Abarca, D., Martin, M. and Sabater, B. (2001) Differential leaf stress responses in young and senescent plants, *Physiol Plant*, **113**, 409-415.

Arnon, D. (1949) Copper enzymes in isolated chloroplasts. Polyphenol oxidase in *Beta vulgaris, Plant Physiol*, **24**, 1-15.

Bauchamp, C. and Fridovich, I. (1971) Superoxide dismutase: Improved assay and an assay applicable to acrylamide gels, *Anal. Biochem*. **44**, 276-287.

Bradford, M. (1976) A rapid and sensitive method for the quantification of microgram quantities of protein utilizing the principle of protein-dye binding, *Anal Biochem*, **72**, 248-254.

Bray, E., Bailey-Serres, J. and Weretilnyk, E. (2000) in: *Biochemistry and Molecular Biology of Plants*, B.Buchanan, W. Gruissen, R. Jones (eds.), John Wiey&Sons, pp. 1158-1203.

Brennan, T. and Frenkel, C. (1977) Involvement of hydrogen peroxide in regulation of senescence in pear, *Plant Physiol*, **59**, 411-416.

Buchanan-Wollaston, V. (1997) The molecular biology of leaf senescence, *J Exp Bot*, **48,** 181-199.

Chandlee, J. (2001) Current molecular understanding of the genetically programmed process of leaf senescence. *Phisiol Plant*, **113**, 1-8.

Dhindsa, R., Plumb-Dhindsa, P. and Thrope, T. (1981) Leaf senescence: correlated with increased levels of membrane permeability and lipid peroxidation, and decreased levels of superoxide dismutase and catalase, *J Exp Bot*, **32**, 93-101.

Dias, A. and Kosta, M. (1983) Effect of low salt concentrations on nitrate reductase and peroxidase of sugar beet leaves. *J Exp Bot*, **34**, 537-543.

Fridovich, I. (1976) Oxygen radicals, hydrogen peroxide and oxygen toxicity. in: *Free-radicals in biology*, W. Pryor (ed.), Academic Press, New York, pp 239-277.

Gidrol, X., Serghini, H., Noubhani, A., Mocquot, B. and Mazilak, P. (1989) Biochemical changes induced by accelerated aging in sunflower seeds. I. Lipid peroxidation and membrane damage. *Phisiol Plant*, **76**, 591-597.

Green, P. (1994) Ribonucleases in plants. *Ann Rev Plant Physiol Plant Mol Biol*, **45**, 421-445.

Gronwald, J., Fuerst, P., Eberlein, C. and Egli, M. (1987) Effect of herbicide antidotes on gluthatione content and glutathione S-transferase activity of sorghum shoots. *Pestic Biochem Physiol*, **29**, 66-76.

Halliwel, B. and Gutteridge, J. (1989) Free Radicals in Biology and Medicine. (2nd ed), Oxford: Claredon Press.

Harpham, N., Berry, A., Knee, E., Roveda-Hoyos, G., Raskin, I., Sanders, I., Smith, A., Wood, C. and Hall, M. (1991) The effect of ethylene on the growth and development of wild type and mutant *Arabidopsis thaliana* (L.) Heynh, *Ann Bot*. **68**, 55-61.

Ivanov, S. (2003) *Effect of Sub-Herbicide Concentrations of Antrazine and 2,4-D and their Interactions with High Temperatures on some Physiologobiochemical Processess in Pea Plants, PhD Thesis*, Sofia (in Bulgarian).

Hensel, L., Grbic, V., Baumgarten, D. and Bleecker, A. (1993) Developmental and age-related processes that influence the longevity and senescence of photosynthetic tissues in *Arabidopsis. The Plant Cell*, **5,** 553-564.

Jessup, W., Dean, R. and Gebicki, J. (1994) Iodometric determination of hydroperoxides in lipids and proteins. *Meth Enzymol*, **233,** 289-303.

Kramer, G., Norman, H., Krizek, D. and Mirecki, R. (1991) Influence of UV-B radiation on polyamines, lipid peroxidation and membrane lipids in cucumber, *Phytochem*, **30**, 2101-2108.

Kumar, G. and Knowles, N. (1993) Changes in lipid peroxidation and lipolytic and free-radical scavenging enzymes during the aging and sprouting of potato (*Solanum tuberosum*) seed-tubers. *Plant Physiol*, **102**, 115-124.

Levine, R. (1994) Carbomyl assays for determination of oxidatively modified proteins. *Meth Enzymol*, **233**, 346-363.

Merzlyak, M. (1991) Syndrome of lipid peroxidation in plants. in: *Plant lipid biochemistry, structure and utilization,* P. Quinn and J. Harwood (eds.)., London: Portland Press Limited, pp. 281-288.

Merzlyak, M., Hendry, G., Atherton, N., Zhigalova, T., Pavlov, V. and Zhiteneva, O. (1993) Pigment destruction, lipid peroxidation and free radical formation in leaves during autumn senescence, *Biokhimya*, **58,** 240-249.

Merzlyak, M. and Hendry, G. (1994) Free radical metabolism, pigment degradation and lipid peroxidation in leaves during senescence, in*: Oxygen and environmental stress in plants,* R.Crawford, G. Hendry and B. Goodman (eds.), Proceedings of the Royal Society of Edinburgh, 102B, 459-471.

Nooden, L., Guiamet, J. and John, I. (1997) Senescence mechanisms, *Physiol Plant* **101**, 746-753.

Prasad, T., Anderson, M., Martin, B. and Steward, C. (1994) Evidence for chilling-induced oxidative stress in maize seedlings and a regulatory role for hydrogen peroxide, *Plant Cell* **6,** 65-74.

Smart, C. (1994) Gene expression during leaf senescence, *New Phytol* **126**, 419-448.

Smart, C., Hosken, S., Thomas, H., Greaves, J., Blair, B. and Schuch, W. (1995) The timing of maize leaf senescence and characterization of senescence-related cDNAs, *Physiol Plant* **93,** 673-682.

Wittenbach, V. (1983) Effect of pod removal on leaf photosynthesis and soluble protein composition of field-grown soybeans. *Plant Physiol*, 73, 121-124.

INVOLVEMENT OF HORMONES IN THE INCREASE IN PLANT PRODUCTIVITY INDUCED BY LOCALISED FERTILIZER PLACEMENT

G.R. KUDOYAROVA, I.I. IVANOV, V.K. TRAPEZNIKOV AND S.YU. VESELOV
Institute of Biology, Ufa Research Centre of RAS, Prospekt Oktyabrya 69, Ufa, 450054 Russia

1. INTRODUCTION

Localized placement of fertilizers, whereby they are introduced into the soil as a discreet band, is used in agriculture due to its beneficial effect on plant growth and productivity. However this effect is not always reproducible. An increase in the efficiency of fertilizer usage following this method of application requires a better understanding of the physiological mechanisms possibly involved. Under natural conditions mineral placement in the soil is also patchy and this patchiness is thought to be an important characteristic of root environment (Jackson and Caldwell, 1996). Thus the study of the effect of heterogeneous distribution of mineral nutrients in soil on plant growth is not only important for improvement of the methods of application of fertilizers but also could contribute to a better understanding of the processes of plant adaptation to the natural soil environment. The studies of root morphology showed increased proliferation of roots in nutrient-rich soil patches (Drew and Saker, 1978; Robinson, 1996). This root growth response is believed to be important for increasing plant productivity, since it enables more efficient acquisition of mineral nutrients (Drew and Saker, 1978; Fransen *et al.*, 1999). However, comparison of the effect of localized and even distribution of fertilizers on productivity of plants grown at different levels of soil moisture showed that beneficial effects of localized nutrition were more pronounced when plants experienced drought (Usov *et al.*, 1988). Increased drought resistance induced by localized fertilizer placement enabled us to speculate that changes in root morphology in response to heterogeneous distribution of nutrients might be important not only for the uptake of ions but also of water. It was also of interest to study possible involvement of hormones in the response of plants to localized fertilizer placement. It is well established that the level of mineral nutrition influences the content of hormones in plants (Kudoyarova and Usmanov, 1991; Munns and Cramer, 1996; Walch-Liu *et al.*, 2000; Takei *et al.*, 2001). Both accumulation of ABA and a decline in cytokinin content are believed to be responsible for stimulation of root growth, which is a characteristic growth response tov mineral nutrition deficit (Kuiper *et al.*, 1989; Chapin, 1991; Teplova *et*

I. Macháčková and G.A. Romanov (eds.), Phytohormones in Plant Biotechnology and Agriculture, 229-235

al., 1998). We showed earlier that localized fertilizer placement may also induce changes in hormone content in plants (Ivanov *et al.*, 1998). However, the observed hormonal reaction was mostly related to shoot and not root growth response. Recently Forde (2002) suggested that IAA might be responsible for enhanced proliferation of that part of the root system in the region of soil with increased concentration of nutrients. However, this suggestion was not supported experimentally. The aim of the present work was to study the effect of heterogeneous distribution of nutrients on hormone content in shoots and roots of wheat plants in an attempt to relate them to changes in root growth and drought resistance.

2. MATERIALS AND METHODS

Spring wheat (*Triticum durum* L., cv. Bezenchukskaya 139) was used. In field experiments a granulated NPK compound fertiliser was applied using two methods: 1) homogeneous distribution in a layer 0-18 cm and 2) localised placement in bands 8-10 cm deep with 15 cm between bands. The level of NPK application was (N) 70, (P) 64 and (K) 70 kg ha^{-1} respectively. Seeds were sown at a depth of 5-6 cm. Five 10 m^2 plots were used for each treatment. Grains harvested from each plot were weighed and the average yield was calculated.

In glasshouse experiments pots were filled with 10 kg of a mixture of soil and sand (2/1). Fertiliser (0.3 g kg^{-1}) was either mixed with the whole volume of substrate or introduced as a strip at a depth of 10 cm in one of the pot compartments. Water content was maintained either at 60 % (normal conditions) or 30 % (drought conditions) of full moisture capacity. Plants were grown until booting and leaves were then sampled for ABA estimation.

To estimate ABA and cytokinins in roots of wheat seedlings were planted in split-root pots (volume of each compartment = 0.5 l) with 5 % Hoagland-Arnon nutrient solution. Plants were grown under an irradiance of 90 Wm^{-2} PAR with a 14-h photoperiod. The concentration of solution in the whole pot or one of the compartments (n=5) was increased by up to 300 % when the seedlings were one week old. Control plants were maintained on an unchanged nutrient solution. Samples were collected 30 h and one week after increasing nutrient concentration.

Phytohormones were extracted with 80 % ethanol from frozen and homogenized samples, alcohol removed from the filtered extract in vacuum, purified and immunoassayed with the help of antibodies raised against ABA and zeatin riboside as described by Mustafina *et al.* (1997).

3. RESULTS

Heterogeneous distribution of nutrients caused similar effects on root growth both in field and solution culture experiments (Trapeznikov *et al.*, 1999). Roots in contact with the higher concentration of nutrients (high-salt roots) were shorter and had more laterals than those supplied with a low concentration (law-salt root), while roots of plants supplied uniformly with an average concentration of nutrients had intermediate length (Fig. 1). Solution culture experiments showed t hat ABA was lower in low-salt roots (Table 1).

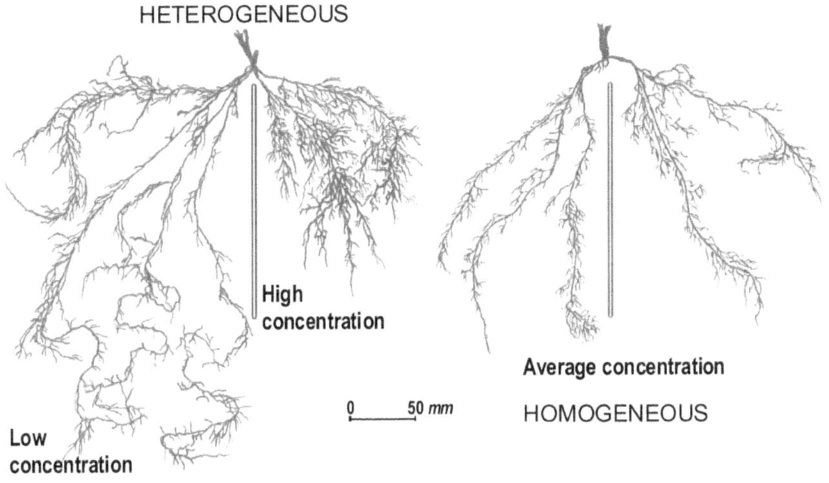

Figure 1. Wheat root system in split-root solution culture.

Table 1. ABA and cytokinin content in roots of wheat seedlings 30 h after the addition of nutrients to one of the compartments, ng g^{-1} f.w.

Nutrition and status of roots	ABA	Cytokinins (zeatin + zeatin riboside)
Homogeneous: low-salt (5% H-A)	52±8	0.5±0.1
Heterogeneous: low-salt (5% H-A)	31±5	3.2±0.6
Heterogeneous: high-salt (300% H-A)	120±22	25±4

At 60 % of full soil moisture capacity shoot dry matter production at the stage of plant booting was increased by localized fertilizer placement by 36 % as compared to homogeneous nutrition (Table 2), which was registered. At the same stage transpiration was slightly higher with localized nutrition than with homogeneous, while both ABA and cytokinin content in leaves were increased by localized nutrition (Table 2). Decrease in soil water content to 30 % of full soil moisture capacity, led to a several-fold increase in ABA content (up to 1100 ng) at uniform nutrition, while plants given localized nutrition accumulated less ABA (up to 680 ng) than those given a homogeneous application.

Table 2. Shoot dry matter production, content of phytohormones in leaves, and evapotranspiration of wheat plants at stage of booting (glasshouse, 60% level of total soil moisture capacity)

Nutrition	Shoot biomass, *mg plant^{-1}*	ABA, *ng g^{-1} f.w.*	Cytokinins (Zeatin + ZR riboside), *ng g^{-1} f.w.*	Evapotranspiration, *g pot^{-1} d^{-1}*
Homogeneous	210±13	36±6	196±39	32.3±2.2
Heterogeneous	286±20	119±26	508±52	42.4±3.1

Z – zeatin, ZR- Zeatin riboside

Increase in crop yield induced by localized nutritient application as compared to homogeneous was more pronounced with decrease in rainfall (Fig. 2).

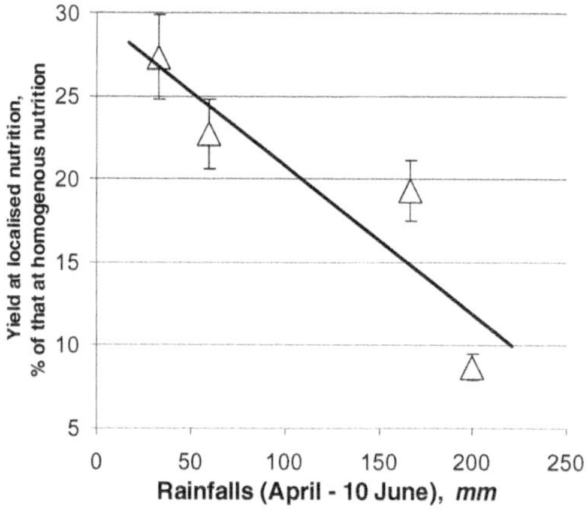

Figure 2. Effect of rainfall on increase in crop yield induced by localization as compared to homogenoeous nutrition.

4. DISCUSSION

In our experiments roots in contact with increased concentrations of nutrients were shorter and had more laterals. The importance of enhanced root proliferation in nutrient-rich soil patches to increase the efficiency of ion uptake as discussed by Drew and Saker (1978) and by Jackson and Caldwell (1996). It would be of interest to establish, how this growth response is achieved. ABA inhibited elongation of maize roots when applied to well-watered plants (Sharp and Le Nobble, 2002), and treatment of plants with cytokinins also made their roots shorter (Cary *et al.*, 1995).

Thus, accumulation of ABA and cytokinins in roots supplied with increased concentrations of nutrients might be responsible for the inhibition of elongation of high-salt roots, since both these hormones are known to influence root growth in this way. It is more difficult to relate changes in the content of cytokinins to root branching. Since these hormones are believed to inhibit lateral root formation (Li *et al.*, 1992), intensive root branching with increased concentration of nutrients cannot be explained by an elevated concentration of cytokinins in high-salt roots (see Table 1). The effect might be due to accumulation of ABA in the roots, since this hormone was reported to increase root branching (Torrey, 1976). Drew and Saker (1978) suggested that increased root branching was due to the inhibition of elongation of the main axes leading to a decrease in correlative inhibition of laterals by the apical meristem.

Increased proliferation of roots in nutrient-rich soil patches has been reported in several previous publications (e.g. Drew and Saker, 1978; Robinson, 1996) and the changes in root morphology were discussed in regard to their importance for enhanced capacity of plants for ion uptake. The results of our experiments draw attention to the part of root system outside the nutrient-rich soil patches. They were longer than both roots in contact with high concentration of nutrients and those totally fed with intermediate concentration (Fig. 1). Increased elongation of these roots, which might be due to reduced concentration of both ABA and cytokinins, enables their penetration deeper into soil. This should result in increased ability of plants to absorb water from lower soil layers when upper layers become dry under drought conditions. This possibility is supported by our measurements of ABA concentration in leaves of water-stressed wheat plants. Drought is well known to induce ABA accumulation in plants (Davies, 1995), and in our experiments decrease in water supply elevated the concentration of this hormone. However, under localized nutrition leaves accumulated less ABA suggesting that the physiological impact of water stress was reduced in these plants.

It is of interest that in well-watered plants ABA content was higher in leaves under localized nutrition than under homogeneous nutrition, that is opposite to results observed under drought conditions. The increase in ABA content in leaves of wheat plants might be due to elevated export of hormone from roots induced by localized nutrition of well-watered plants. The problems of measuring the concentration of hormones in the xylem sap of wheat plants prevented investigation of the effect of localized nutrition on hormone delivery from roots. However, in previous work was found that heterogeneous distribution of fertilizers increased delivery of ABA from the roots of maize plants (Ivanov *et al.*, 2000). The elevated capacity of roots to supply shoots with ABA might be due to the increase in production of hormone by high-salt roots, which was manifested in higher concentration of ABA in these roots (Table 1). ABA is well known to be involved in stomatal closure and to inhibit leaf growth (Davies, 1995). Consequently increased ABA concentration in leaves, which was observed under localized nutrition, should inhibit transpiration and growth of wheat seedlings. However a contrary effect of localized fertilizers placement was observed in our experiments. Above-ground biomass production was greater under localized nutrition, while transpiration was higher than under homogeneous nutrition. This might be due to the effect of cytokinins, which content was also increased by localized nutrition, since these hormones are well-known antagonists of ABA (Evans, 1984). Increased level of mineral nutrition was shown to stimulate production of cytokinins by roots (Michael, Beringer, 1980). More cytokinins were delivered from locally fed roots of maize plants than from homogeneously fed ones (Ivanov et al., 1998). Cytokinins stimulate leaf extension growth, cause stomata to remain open, and activate photosynthesis (Cherniad'ev, 2000). So their accumulation in parallel with ABA might overcome inhibitory effect of this hormone. As a result localized fertilizer placement increased crop yield not only under conditions of drought but also (although to a lesser extent) in better-watered plants.

Thus changes in concentration of hormones induced by heterogeneous nutrition are likely to be responsible for root morphology, enabling both acquisition of nutrients from nutrient-rich soil patches and water uptake from lower soil layers. This contributes to increased plant productivity and drought resistance.

ACKNOWLEDGEMENTS

Authors are grateful to Prof. P.J. Davies for his useful comments, which helped, as we hope, to improve this manuscript.

REFERENCES

Cary, A. J., Liu, W. and Howell, S. H. (1995) Cytokinin action is coupled to ethylene in its effects on the inhibition of root and hypocotyl elongation in Arabidopsis thaliana seedlings, *Plant Physiol.* **107**, 1075-1082.

Chapin, F.S. (1991) Integrated responses of plants to stress. A centralized system of physiologicalresponses, *BioScience* **41**, 29-36.

Cherniad'ev, I.I. (2000) [Ontogenetic changes in the photosynthetic apparatus and effect of cytokinins], in Russ., *Prikl. Biokhim. Mikrobiol.* **36**, 611-625.

Davies, P.J. (1995) The plant hormones. Their nature, occurrence and functions, in P.J. Davies (ed.), *Plant Hormones. Physiology, Biochemistry and Molecular Biology,* Kluwer Acad. Pub., Dordrecht, pp 1-12.

Drew, M.C. and Saker, L.R. (1978) Nutrient supply and the growth of seminal root system in barley. III. Compensatory increases in growth of lateral roots, and in rates of phosphate uptake, in response to a localised supply of phosphate, *J. Exp. Bot.* **29**, 435-451.

Evans, M.L. (1984) Functions of hormones at the cellular level of organisation, in T.K. Scott (ed.), *Encycl. of Plant Physiol. Hormonal Regulation of Development*, Springer Verlag, Berlin, pp.23-79.

Forde, B.G. (2002) The role of long-distance signalling in plant responses to nitrate and other nutrients, *J. Exp. Bot.* **53**, 39-43.

Fransen, B.G., De Kroon, H., De Kovel, C.G.F. and Van Den Bosch, F. (1999) Disentangling the effects of root foraging and inherent growth rate on plant biomass accumulation in heterogeneous environments: a modelling study, *Ann. Bot.* **84**, 305-311.

Ivanov, I.I., Kudoyarova, G.R. and Trapeznikov, V.K. (1998) The effect of local application of fertilizer on the content of cytokinins in xylem sap of maize, *Biol. Plant.* **41**, 587-590.

Ivanov, I.I., Trapeznikov, V.K. & Kudoyarova, G.R. (2000) [Cytokinins and abscisins in xylem exudate of roots differing in salt status], in Russ., *Agrokhim.* **N3**: 25-29.

Jackson, R.B. and Caldwell, M.M. (1996) Integrating resource heterogeneity and plant plasticity: Modelling nitrate and phosphate uptake in a patchy soil environment, *J. Ecology* **84**, 891-903.

Kudoyarova, G.R. and Usmanov, I.Y. (1991) Hormones and Mineral Nutrition, *Fiziol. Biokhim. Kult. Rast.* **23**, 232-244.

Kuiper, D., Kuiper, P.J.C., Lambers, H., Schuit, J. and Staal, M. (1989) Cytokinin concentration in relation to mineral nutrition and benzyladenine treatment in *Plantago major* ssp. Pleiosperma, *Physiol. Plant.* **75**, 511-517.

Li, Y., Hagen, G. and Guilfoyle, T.J. (1992) Altered morphology in transgenic tobacco plants that overproduce cytokinins in specific tissues and organs, *Dev. Biol.* **153**, 386-395.

Michael, J. and Beringer, H. (1980) The role of hormones in yield formation, in *Physiological Aspects Crop Production,* Worblaufen, Bern, pp.85-116.

Munns, R. and Cramer, G.R. (1996) Is coordination of leaf and root growth mediated by abscisic acid? Opinion, *Plant Soil* **185**, 33-49.

Mustafina, A.R., Veselov, S.Yu., Valcke, R. and Kudoyarova, G.R. (1997) ABA and cytokinins content in shoots of wheat seedlings during their dehydration, *Biol. Plant.* **40**, 291-293.

Robinson, D. (1996) Resource capture by localized root proliferation: why do plants bother?, *Ann. Bot.* **77**, 179-185.

Sharp, R.E. and Le Nobble, M.E. (2002) ABA, ethylene and the control of shoot and root growth under water stress, *J. Exp. Bot.* **53**, 33-37.

Takei, K., Sakakibara, H., Taniguchi, M. and Sugiyama, T. (2001) Nitrogen-dependent accumulation of cytokinins in root and the translocation to leaf: implication of cytokinin species that induces gene expression of maize response regulator, *Plant Cell Physiol.* **42**, 85-93.

Teplova, I.R., Veselov, S.Yu. and Kudoyarova, G.R. (1998) Dynamics of ABA and IAA content in roots and shoots of wheat seedlings under nitrogen deficiency, in *Proceedings of the 5-th Symposium of the International Society of Root Research,* Kluwer Acad. Pub., Dordrecht, pp. 625-633.

Torrey, J.G. (1976) Root hormones and plant growth, *Annu. Rev. Plant Physiol.* **27**, 407-413.

Trapeznikov, V.K., Ivanov, I.I. and Talvinskaya, N.G. (1999) *[Local nutrition of plants]*, in Russ., Gilem, Ufa, 240 pp.

Usov, V.P., Ivanov, I.I., Trapeznikov, V.K., Talvinskaya, N.G. and Shendel, G.V. (1988) [Dependence of productivity and grain quality of spring wheat from conditions of cultivation on a background of complex of influences in an ontogenesis], in Russ., *Agrokhim.* **N12**, 46-52.

Walch-Liu, P., Neumann, G., Bangerth, F. and Engels, C. (2000) Rapid effects of nitrogen form on leaf morphogenesis in tobacco, *J. Exp. Bot.* **51**, 227-237.

DISTRIBUTION OF PHYTOHORMONES AMONG DIFFERENT PLANT ORGANS DETERMINES THE DEVELOPMENT OF COMPETITIVE SINKS IN RADISH PLANTS

N. BUKHOV, A. KOTOV, S. MAEVSKAYA, E. EGOROVA AND I. DROZDOVA
Timiryazev Institute of Plant Physiology Russian Academy of Science, Botanicheskaya 35, 127276 Moscow, Russia

1. INTRODUCTION

Phytohormones are known to mediate many responses of plants to changes in environmental cues including light climate. Light quality governs various stages of plant growth and development (Short and Briggs, 1994; Smith, 1995). One of the most important features of plant photoregulation is involvement of light quality in the determination of sink strength (Aksenova *et al.*, 1994). Several sinks, such as newly developing leaves, petioles and stem, and storage organ(s) compete for photosynthetic metabolites in a plant. Some plant organs dominate others in their ability to attract assimilates from donor leaves, depending on their sink strength (Drozdova *et al.*, 1987). Correspondingly, various hormones have been proposed to stimulate metabolite flow to above- and/or underground plant organs (Palmer and Smith, 1969; Metzger, 1988). It is possible therefore, that light quality changes the direction of assimilate flow in a plant through alterations in hormonal status of plant organs. We show here that hormones belonging to different classes (gibberellins, cytokinins, and auxin) act in concert to determine preferential development of either above-ground sink (petioles and stem) under red light (RL) or underground sink (storage organ) under blue light (BL).

2. MATERIALS AND METHODS

Radish (*Raphanus sativus* L., cv Rubin) plants were grown in phytotron chamber on a top-soil under 16/8 h light/dark cycles at 22/20°C day/night temperature and equal photon fluxes of BL or RL (170 μmol m^{-2} s^{-1}). Phytohormones were extracted from plant tissues and then analyzed by ELISA as previously reported (Bukhov *et al.*, 1996).

I. Macháčková and G.A. Romanov (eds.), Phytohormones in Plant Biotechnology and Agriculture, 237-241

3. RESULTS AND DISCUSSION

Fig. 1 shows ontogenetic patterns of dry matter accumulation in leaf blades, petioles plus stem, and hypocotyl plus roots of BL-or RL-grown plants. At all harvest dates, no difference was found between plants grown under BL or RL in leaf blade dry matter. On the contrary, RL favored the accumulation of dry matter in petioles and stem, while BL promoted the development of storage organ (swollen hypocotyl).

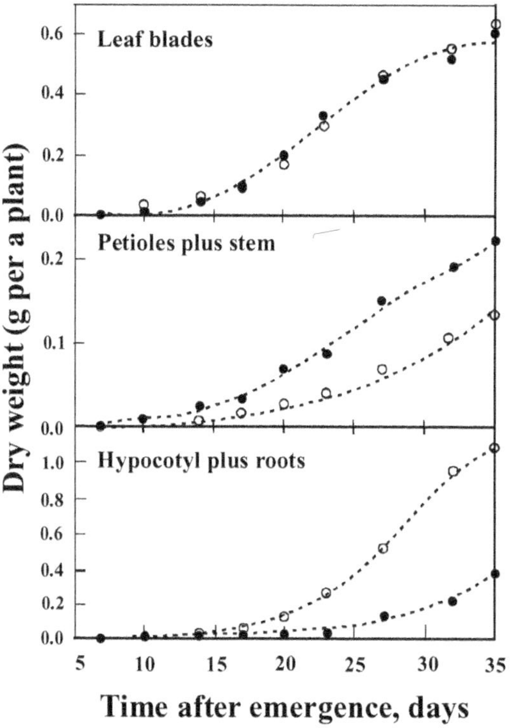

Figure 1. Ontogenetic patterns of dry matter accumulation (g/plant) in leaf blades (upper panel), petioles (plus stem for RL-grown plants, middle panel), and hypocotyl plus roots (bottom panel) of radish plants grown under blue (empty symbols) or red (dark symbols) light. Each point represents a dry weight averaged over 6 individual plants.

The specific distribution of dry matter within BL- or RL-grown plants indicates that assimilate flow from donor leaves was directed under BL mainly to the hypocotyl thus supporting the formation of storage tissues, whereas assimilate flow to petioles and developing stem dominated under RL. The data in Fig. 2 show that contrasting levels of different hormones in shoots and hypocotyl may determine the difference in distribution of dry matter between BL- and RL-grown plants. This figure demonstrates the contents of gibberellins (A), cytokinins (B, C), and indole-3-acetic acid (IAA, D) in 11-, 14-, and 21-day-old radish plants grown under BL or RL.

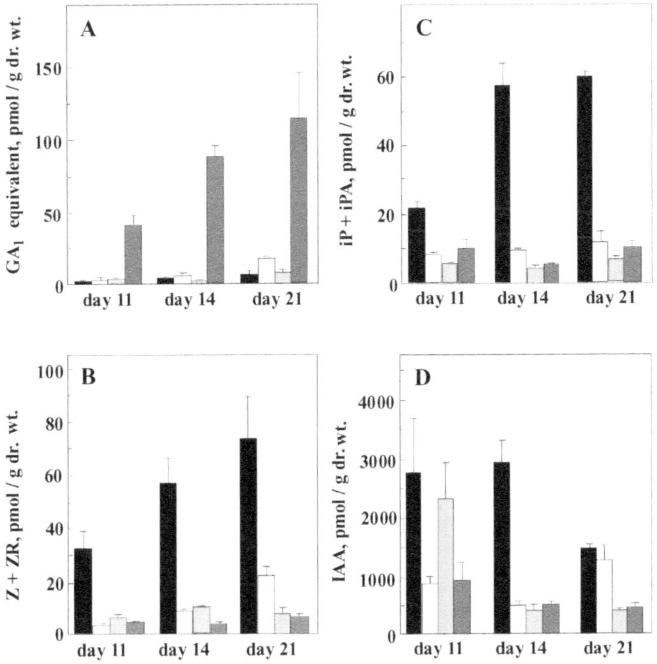

Figure 2. Contents of endogenous gibberellins (A), zeatin plus zeatin riboside (B), isopentenyladenine plus isopentenyladenosine (C), or indole-3-acetic acid (D) in a shoot or in a hypocotyl of 11-, 14-, or 21-day-old radish plants grown under blue or red light. Hormone contents were expressed as pmol g^{-1} dr. wt. Black columns – hypocotyl, BL; white columns – hypocotyl, RL; light grey columns – shoot, BL; dark grey columns – shoot, RL.

In young, 11-day-old plants, an approximately 10-times higher level of gibberellins was found in shoots of RL-grown plants (Fig. 2.A). This dramatic difference between BL-and RL-grown plants increased with age. At all harvest dates, the levels of two cytokinins, zeatin plus zeatin riboside (Z+ZR, Fig. 2.B) and isopentenyladenine plus isopentenyladenosine (iP+iPA, Fig. 2.C) were found to be severalfold higher in hypocotyls of BL-grown plants. The content of IAA in hypocotyls of 11-day-old plants grown under BL greatly exceeded that in RL-grown plants. However, light quality did not influence the content of IAA in hypocotyls of 21-day-old plants.

The above data show that RL dramatically enhanced the content of gibberellins in shoots and promoted the development of above-ground metabolic sink. This, in turn, restricts assimilate flow to developing storage organs. Indeed, the treatment of radish plants grown under RL by paclobutrazol, an inhibitor of gibberellin synthesis, reduced dry matter accumulation in petioles, simultaneously preventing stem formation (Table 1). In paclobutrazol-treated plants, in which the transfer of assimilates from donor leaves to above-ground sink was restricted, light quality had only a small effect on the accumulation of dry matter in storage organ (Table 1). In contrast, treatment of BL-grown plants with exogenous gibberellin increased the weight of petioles twofold. Moreover, exogenous gibberellin initiated development of stem bolting in radish plants grown under BL.

Table 1. Dry weight (g) of different organs of individual 28-day-old radish plants grown under blue (BL) or red (RL) light and treated with either exogenous gibberellin (GA, 80 μM) or with paclobutrazol (PB, 20 μM). Data are the means ± SE of 8-10 plants.

Treatment	Leaves	Petioles plus stem	Hypocotyl plus roots
BL	0.43±0.02	0.40±0.04	0.54±0.08
BL + GA	0.54±0.11	1.03±0.08	0.60±0.03
BL + PB	0.32±0.02	0.44±0.03	0.27±0.02
RL	0.38±0.02	0.77±0.10	0.18±0.14
RL + GA	0.41±0.04	0.89±0.04	0.07±0.02
RL + PB	0.21±0.08	0.41±0.03	0.38±0.06

In summary, the above results clearly demonstrate a well-coordinated mode of hormone distribution in radish plants, in which hormone levels are adjusted accordingly to environmental conditions in a way that promotes the development of appropriate sinks, either metabolic or storage ones. An increase by RL in levels of hormones promoting the development of the above-ground metabolic sink (gibberellins) was accompanied by a decrease in levels of hormones enhancing the strength of the underground storage sinks (cytokinins and IAA) and *vice versa*.

This study was supported by Russian Foundation for Fundamental Research, grant 01-04-48180

REFERENCES

Aksenova, N.P., Konstantinova, T.N., Sergeeva, L.I., Macháčková, I. and Golyanovskaya, S.A. (1994) Morphogenesis of potato plants *in vitro*. I. Effect of light quality and hormones, *Plant Growth Regul.* **13**, 143-146.

Bukhov, N.G., Bondar, V.V., Drozdova, I.S., Kara, A.N., Kotov, A.A., Maevskaya, S.N., Vasil'ev, A.A., Voevudskaya, S.Yu., Voronin, P.Yu. and Mokronosov, A.T. (1996) Development of storage roots in radish (*Raphanus sativus*) plants as affected by light quality, *J. Plant Physiol.* **149**, 405-412.

Drozdova, I.S., Bondar, V.V. and Voskresenskaya, N.P. (1987) Co-action of light-controlled reactions induced by red and blue light on photosynthesis and morphogenesis in radish plants, *Sov. Plant Physiol.* **43**, 786-794.

Metzger, J.D. (1988) Gibberellins and light regulated petiole growth in *Thlaspi arvense* L., *Plant Physiol.* **86**, 237-240.

Palmer, C.E. and Smith, O.E. (1969) Cytokinins and tuber initiation in the potato *Solanum tuberosum* L., *Nature* **221**, 279-280.

Short, T.W. and Briggs, W.R. (1994) The transduction of blue light signals in higher plants, *Annu. Rev. Plant Physiol. Plant Mol. Biol.* **45**, 143-157.

Smith, H. (1995) Physiological and ecological function within the phytochrome family, *Annu. Rev. Plant Physiol. Plant Mol. Biol.* **46**, 289-315.

HORMONAL REGULATION OF SOMATIC EMBRYOGENESIS ON MAIZE

YU. I. DOLGIKH, T. N. PUSTOVOITOVA AND N. E. ZHDANOVA
K. A. Timiryazev Institute of Plant Physiology Russian Academy of Sciences, Moscow, Russia

1. INTRODUCTION

The ability of plants to regenerate from cultured cells is known to depend on the genotype and the age of the maternal plant, the type of the organ from which the explant was isolated, the composition of the nutrient medium, and the conditions of culturing (Phillips *et al.*, 1988). In maize the formation of embryogenic callus is a key stage for the subsequent plant regeneration. The highest effectiveness in callusogenesis was found for immature embryos. In turn the frequency of embryogenic callus formation by embryos varied from zero to 100% depending on the genotype (Morrish *et al.*, 1987). Taking into account the significance of endogenous phytohormone balance in determination of regeneration capacity (Butenko, 1975), it can be supposed that the level and the ratio of endogenous growth regulators essentially influence cell competence for morphogenesis.

The objective of this investigation was to clarify the relations between the endogenous content of phytohormones in embryos and their ability to form an embryogenic callus.

2. MATERIALS AND METHODS

Two inbred maize (*Zea mays* L.) lines, A188 and Gk26 were used. The plants were grown under field conditions or in a greenhouse in pots with soil.

Two approaches were used to elucidate the interrelation between the explant hormonal balance and its morphogenic potential. (1) The contents of endogenous phytohormones were determined in embryos of the maize lines that markedly differed in the frequency of embryogenic callus formation. (2) Explants were subjected to treatments affecting phytohormone concentrations, and then their capacity for somatic embryogenesis was evaluated.

Immature embryos were excised from kernels harvested on the 10-12th days, additionally, in the case of line A188, on the 17-20th days after self-pollination. A portion of isolated embryos was placed on agar-solidified MS medium with 60 g/l sucrose and 1 mg/l 2,4-D. About 80% of embryos of morphogenic line A188 excised on the 10-12th days after pollination produced an embryogenic callus. The

I. Macháčková and G.A. Romanov (eds.), Phytohormones in Plant Biotechnology and Agriculture, 243-247

embryos of nonmorphogenic line Gk26 excised at the same developmental stage produced embryogenic callus with frequency close to zero. Embryogenic potential of line A188 embryos excised on the 17-20th days after pollination was on average 30%.

Other portion of embryos was fixed, and the concentrations of hormones were determined by the ELISA technique using standard kits for immunoassay (Farmkhiminvest, Russia) (Veselov and Kudoyarova, 1990). The level of cytokinins was evaluated as the content of zeatin-riboside.

In some experiments, p-coumaric acid, 2,4-dichlorophenol or ABA sterilized by filtration were added to the medium. After thee or four weeks, the frequency of embryogenic callus formation was recorded.

3. RESULTS AND DISCUSSION

The enzyme immunoassay demonstrated substantial differences in levels of free phytohormones in embryos of different genotypes and developmental stages. The 12-day-old embryos of line A188 competent to embryogenesis were characterized by a relatively low IAA and a high ABA content. In contrast, a high concentration of IAA and a low concentration of ABA were found in explants on line Gk26 with regeneration capacity close to zero. The differences between lines in cytokinin concentration were less profound, but also essential. When the frequency of embryogenic callus formation decreased with aging of embryos of A188 line, the content of hormones was also changed: the IAA concentration decrased and ABA levels dropped almost to the level characteristic of low-morphogenic line Gk26 (Tab. 1). So, the highest competence for morphogenesis had explants that contained a moderate content of auxin, high concentration of ABA and low level of cytokinin.

*Table 1. The content of endogenous phytohormones in immature maize embryos**

Line	Age of embryos, days after pollination	Frequency of embryogenic callus formation, %	IAA, ng/g fr wt	ABA, ng/g fr wt	ZR, ng/g fr wt
A188	10-12	60-80	130±26	9092±3	16±1
A188	17-20	25-35	8±2	454±35	21±2
Gk26	10-12	0-6	3000±171	283±17	66±12

* The data of one from two independent experiments are presented, means from five recording with their standard errors are given for every experiment.

The ratio beween growth regulators optimal for morphogenesis is evidently species-specific. The contents of auxin and ABA in immature embryos of wheat morphogenic cultivar Tayozhnaya were similar to maize characteristics (Kopertekh and Butenko, 1995). The treatment of wheat callus with compounds favoring IAA destruction enhanced somatic embryogenesis [6]. In contrast, alfalfa explants with a high level of IAA and low level of ABA were found to demonstrate a higher capacity for morphogenesis (Ivanova *et al.*, 1994). In embryogenic carrot suspension the IAA level was 15-fold higher as compared with nonembryogenic suspension (Sasaki *et al.*, 1994).

If hormonal balance is so important for embryo capacity to form embryogenic callus, we can try to affect this balance at the period of callus induction and to check whether it would influence embryo morphogenic ability. One of the ways to decrease IAA concentration is to activate its oxidation (Gamburg, 1976). For this purpose, the cofactors of IAA oxidase, p-coumaric acid and 2,4-dichlorophenol, were added to the medium for callus induction.

The embryos of low-embryogenic line Gk26 placed on the medium with p-coumaric acid demonstrated an increase in their embryogenic potential by two to three times (Tab. 2). In contrast, in high-embryogenic line A188 treatment with p-coumaric acid at concentration of 10^{-4} and 5×10^{-4} M caused a decrease in the frequency of embryos forming embryogenic callus by 22 and 33%, respectively. Perhaps, in low-embryogenic line Gk26 characterized by a high IAA content in immature embryos, IAA-supposed destruction beneficially affected embryo ability for morphogenesis. In a high-embryogenic A188 line with relatively low IAA content, the treatment with p-coumaric acid resulted in a decrease in the morphogenetic potential to the level characterized for the 20-day-old embryos.

*Table 2. The effect of exogenous p-coumaric acid on the frequency of embryogenic callus formation **

Line	Concentration, M	Total no. embryos	Total no. embryogenic calli	Frequency of embryogenic callus formation, %
Gk26	0	99	4	4.0±2.0
	5×10^{-5}	47	4	8.5±2.9
	10^{-4}	109	9	8.3±2.6
	5×10^{-4}	103	13	12.6±3.3
A188	0	89	59	66.1±4.5
	10^{-4}	78	34	43.6±3.6
	5×10^{-4}	92	30	32.8±3.3

*The means from four independent experiments with standard error are presented here and at the tables 3 and 4; 15-45 embryos were analysed in every treatment.

Incubation of Gk26 explants on a medium with 10^{-5} M of 2,4-dichlorophenol resulted in fivefold increase in the frequency of the embryos forming embryogenic callus. Higher concentrations of 2,4-dichlorophenol inhibited callusogenesis (Tab.3).

Table 3. The effect of 2,4-dichlorophenol on the frequency of embryogenic callus formation by embryos of the Gk26 line

Concentration, M	Total no. embryos	Total no. embryogenic calli	Frequency of embryogenic callus formation, %
0	127	4	3.2±1.3
10^{-5}	117	16	13.8±1.4
5×10^{-5}	107	1	0.6±0.6
10^{-4}	104	0	0

One of the methods to decrease intracellular IAA content and simultaneously to increase ABA is to add the latter into the medium for callusogenesis. The exogenous ABA treatment of wheat cell culture three times decreased the endogenous IAA content. The ABA application inhibited dedifferentiated cell growth and rhizogenesis and, at the same time, it induced somatic embryogenesis (Carman, 1988; Brown et al., 1989; Shayakhmetov and Shakirova, 1996). In our experiments the ABA was added to the medium for callusogenesis of Gk26 embryos at concentrations from 4×10^{-7} to 2×10^{-5} M. When the ABA concentration in the medium increased, the frequency of embryos forming embryogenic callus also increased; this increase became ten-fold at the ABA concentration in the medium of 2×10^{-5} M. In contrast, the ABA treatment of high-embryogenic hybrid A654xM320/78 reduced embryogenic callus appearence from 65 to 5% (Tab. 4). The genotype dependence of exogenic ABA effect on morphogenesis was previously described for oil seed explants (Raldugina and Sobolkova, 1994).

The experiments with modifications of concentrations of endogenous hormones confirmed the interrelation between the tissue hormonal status and their competence for perception of external signals.

Table 4. The effect of exogenous ABA on the formation of embryogenic callus

Genotype	Concentration, M	Total no. embryos	Total no. embryogenic calli	Frequency of embryogenic callus formation, %
Gk26	0	140	3	2.1±1.1
	4 x 10^{-7}	65	5	7.7±2.4
	2 x 10^{-6}	125	19	15.1±2.5
	4 x 10^{-6}	121	21	17.3±3.6
	2 x 10^{-5}	86	18	20.8±2.1
A654xM320/78	0	89	58	65.1±4.9
	2 x 10^{-5}	78	4	5.1±1.9

REFERENCES

Brown, C., Brooks, F.J., Pearson, D. and Mathias, R.J. (1989) Control of embryogenesis and organogenesis in immature wheat embryo callus using increased medium osmolarity and abscisic acid, *J.Plant Physiol.* **133**, 727-733.

Butenko, R.G. (1975) *Experimental Morphogenesis and Differentiation in Plant Cell Culture*, 35th Timiryazev Lecture, Nauka, Moscow, pp. 45-48.

Carman, J.G. (1988) Improved somatic embryogenesis in wheat by partial simulation of the in-ovulo oxygen, growth regulators and desiccation environments, *Planta* **175**, 417-424.

Gamburg, K.Z. (1976) *Biochemistry of Auxin and Its Action on Plant Cells*, Nauka, Novosibirsk, pp. 80-82.

Ivanova, A., Velcheva,M., Denchev, P., Atanassov, A. and van Onckelen, H.A. (1994) Endogenous hormone levels during somatic embryogenesis in *Medicago falcata*, *Physiol. Plant.* **92**, 85-89.

Kopertekh, L.G. and Butenko, R.G. (1995) Naturally occurring phytohormones in wheat explants as related to wheat morphogenesis *in vitro*, *Russ. J. Plant Physiol.* **42**: 555-559.

Morrish, F., Vasil, V. and Vasil, I.K. (1987) Development morphogenesis and genetic manipulation in tissue and cell culture of the *Gramineae*, *Adv. Genet.* **24**, 431-499.

Phillips, R.L., Somers, D.A. and Hiberd, K.A. (1988) Cell-tissue culture and *in vitro* manipulation, in G.F.

Sprague and J.W. Duddley (eds.), *Corn and Corn Improvement: Agronomy Monograph*. **V.18**, Am. Soc. Agron., Madison, pp. 345-387.

Raldugina, G.N. and Sobolkova, G.I. (1994) The genotype dependent effect of abscisic acid on callus cultures of *Brassica napus* L., *Russ.J.Plant Physiol.* 41,702-706.

Sasaki,K., Schimomura, K., Kamada, H. and Harada, H. (1994) IAA metabolism in embryogenic and non-embryogenic carrot cells, *Plant Cell Physiol.* **35**, 1159-1165.

Shayakhmetov, I.F. and Shakirova, F.M. (1996) The formation of somatic embryos in suspension culture of wheat in presence of ABA, *Russ.J.Plant Physiol.* **43**, 101-103.

Veselov, S.Yu. and Kudoyarova, G.R. (1990) Immunoenzyme analysis for phytohormone determination, in R.N. Churaev (ed.), *Enzyme Immunoassay of Plant Growth Regulators, Application in Plant Physiology and Ecology*, Ufa, pp. 8-21.

Zaghmout, O.M.-F. and Oliver, M.J. (1995) Differential modulation of root formation in wheat embryogenic callus culture by indoleacetic acid-degrading compounds, *J. Exp. Bot.* **46**, 155-159.

PHYTOHORMONES AND CYTOSKELETON

N. L. KLYACHKO AND M. A. EROKHINA
Timiryazev Institute of Plant Physiology, Russian Academy of Sciences, Botanicheskaya ul. 35, Moscow, 127276 Russia;
fax: 7 (095) 977-8018; e-mail: nellaklyachko@mtu-net.ru

Phytohormones are among the principal regulatory molecules in the plant cell, which control essentially all physiological processes in it. Cytoskeleton provides the basics framework for cell architecture, intracellular transport and cell motility, and cell capacity for division. It is the matrix for most biochemical processes and a target for external and internal regulatory signals. Therefore, these two players in the game have *a priori* to be interrelated, however, there is only limited experimental data to support it. We discuss four levels of this interaction: (1) spatial organization of microtubules (MT) and microfilaments (MF); (2) synthesis and modification of cytoskeletal proteins; (3) cytoskeleton involvement in phytohormone action; and (4) cytoskeleton involvement in processes which are under hormonal control.

(1) The effects of phytohormones on the morphology and spatial localization of tubulin microtubules (MT) and actin microfilaments (MF) have been frequently studied. Cortical MT are usually arranged perpendicular to the direction of growth and the newly formed cellulose fibrils in the cell wall are organized parallel to them. Thus, for example, MT in the root elongation zone are predominantly located perpendicular to the root longitudinal axis; when elongation stops, MT become arranged in an oblique and then longitudinal direction (Baluška *et al.*, 1992). Phytohormones strongly affect MT reorientation, and their action has been considered in a number of excellent reviews (Shibaoka, 1994; Baluška *et al.*, 1999; Nick, 1999). Auxins, gibberellins, cytokinins, ethylene, ABA and brassinosteroids exert such an action. Auxins facilitate transverse MT orientation in stems and coleoptiles, where they induce growth acceleration, but cause longitudinal orientation of MT in roots, the growth of which is retarded by auxins. Gibberellins, which stimulate elongation of stems and roots, usually induce transverse MT orientation. Cytokinins, ethylene, and ABA increase the proportion of longitudinal MT in the systems, in which these hormones retard growth. Such effects are not hormone-specific and can be induced by other factors. They are not always tightly correlated with phytohormone effects on growth. The mechanism of MT reorientation is not fully known but the involvement of protein kinases and phosphatases has been demonstrated. The most likely mechanism for these changes is depolymerization of one MT subpopulation and repolymerization of new MT.

I. Macháčková and G.A. Romanov (eds.), Phytohormones in Plant Biotechnology and Agriculture, 249-253

The effects of phytohormones on actin MF has been cmprehensively studied for ABA-induced stomata closure in *Commelina communis* and *Arabidopsis thaliana* leaves, where ABA induced depolymerization of long MF which radiate from the aperture in the open stomata (Eun and Lee, 1997). This effect also was not a hormone-specific one and can be induced by darkness, for example. Ca^{2+} can replace ABA in its effects on MF. Protein kinases, phosphatases, and small GTPase may mediate ABA action. The physiological significance of these rearrangements is obscure, although it has been suggested that actin cytoskeleton can somehow affect the state of ionic channels. These rearrangments are also not specific for the hormone, and can be induced by other factors such as light (e.g. red and blue light).

(2) Effects of phytohormones on tubulin and actin gene expression have been demonstrated in some studies but have not been found in others. Hormone-induced changes in gene expression did not always induce changes in the amounts of corresponding proteins. In discussing this problem, it should be remembered that both tubulins and actins are encoded by small gene families. Phytohormones may very specifically affect only definite protein isoforms; nevertheless, this may substantially change the properties of cytoskeletal structures. Thus, recently in the Meagher laboratory, it was shown that only one of eight *A. thaliana* actin genes, the ACT7 gene, was under auxin control (Kandasamy *et al.*, 2001). Auxin sharply increased the amount of corresponding mRNA and protein but did not affect the remaining 7 actin genes, which, however, provided for a substantially accelerated callus growth.

(3) Until recently, there was no information on the role of the cytoskeleton in hormonal metabolism and/or action. Last year provided a surprise for biologists in studies of auxin polar transport. Earlier, it was believed that such transport was determined by static asymmetric distribution of auxin carriers in the plasmalemma. Robinson *et al.* (1999) and later on Geldner *et al.* (2001) demonstrated that, in fact, these proteins cycled rapidly between the plasmalemma and some unidentified endosomic compartment and actin filaments served as tracks for the movement of vesicles containing this protein. The cases of reversible internalization of other proteins related to the hormone metabolism are known, e.g. that of auxin-binding protein 1, and cytoskeleton is expected to be involved in these processes.

(4) The cytoskeleton can serve as a scaffold for most cellular processes including glycolysis, vesicular transport, mRNA transport, andprotein synthesis. The control of the interaction between macromolecular complexes and the cytoskeleton is one of the mechanisms by which the activity of a process may be regulated. Phytohormones might regulate these interactions, although experimental data in favor of such a supposition are extremely scarce. Approximately 20 years ago, it was established that some or all of the polysomes in the animal cells were attached to the cytoskeleton, and this was beneficial for translation. Later, this was also shown for plants (Davies *et al.*, 1991; Zak *et al.*, 1995). We attempted to answer the question whether the proportion of the cytoskeleton-bound polysomes (CBP) was variable and could serve as a target for physiological (in particular phytohormonal) regulation. To estimate this proportion, we applied an approach developed by E. Davies (Davies *et al.*, 1991), which is based on different sensitivities of membranes

and cytoskeleton to nonionic detergents. The proportion of CBP was a difference between polysomes peak areas in sucrose density gradients of tissue homogenates treated and untreated with Triton X-100 (stretched regions in Fig. 1). This approach was applied to the detached pumpkin cotyledon system growth and protein synthesis in which is very sensitive to cytokinins and ABA. This system was studied in detail by Kulaeva's group (Zak *et al.*, 1995; Kulikova *et al.*, 2001). Cytokinin activated sharply cotyledon growth and induces a rapid monosomes mobilization into polysomes (Klyachko *et al.*, 1979). The analysis of CBP proportion in these cotyledons (Fig. 1) demonstrated that in intact three-day-old cotyledons, all ribosomes were assembled in polysomes and all polysomes were attached to the cytoskeleton. Cotyledon detachment induced polysome degradation and their partial detachment from the cytoskeleton.

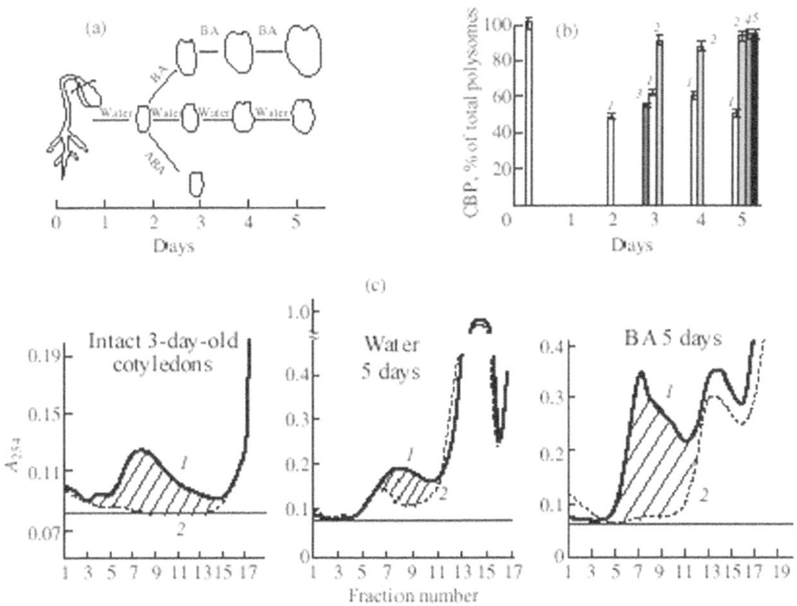

Figure 1. The effect of BA, ABA, and irradiation on the percentage of CBP in detached pumpkin cotyledons.
(a) Experiment scheme; (b) CBP percentage; (c) sucrose density profiles of the ribosomal material.
In (b): (1) Water; (2) BA; (3) ABA; (4) light; (5) BA + light.
In (c): the homogenate (1) untreated and (2) treated with Triton X-100.

Benzyladenine (BA, 5 x 10^{-5} M) treatment resulted in polysome formation and their essentially complete attachment to the cytoskeleton. This effect was not specific for BA and could be induced by other factors, for example illumination (Kulikova *et al.*, 2001) (Fig. 1). It was not coupled with a growth effect because light stimulated cotyledon growth less efficiently but similarly affected polysomes behavior (Fig. 1). The effects of various factors on the CBP were not always correlated with those on total polysome proportion (Fig. 2).

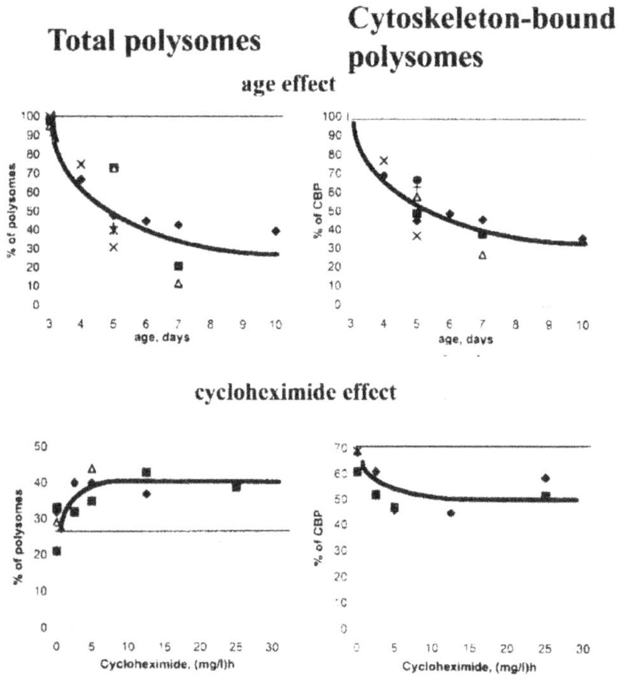

Figure 2. The effects of cotyledon age and cycloheximide on the percentage of total and cytoskeleton-bound polysomes.

Thus, weak concentrations of cycloheximide induced total polysome accumulation but decreased the proportion of the CBP.

In general, the percentage of CBP in the plant cell may vary and serve as a target for physiological regulation. It correlates with translational activity and is evidently an additional level of protein synthesis regulation.

Baluška, F., Parker, J.S., and Barlow, P.W.(1992) Specific pattern of cortical and endoplasmic microtubules associated with cell growth and tissue differentiation in roots of maize (Zea mays L.), *J. Cell Sci.* **103**, 191-200.

Shibaoka, H. (1994) Plant hormone-induced changes in the orientation of cortical microtubules: alteration in the cross-linking between microtubules and the plasma membrane, Annu. Rev. Plant Physiol. Plant Mol. Biol. **45**, 527-544.

Baluška, F., Volkmann, D., and Barlow, P.W. (1999) Hormone - cytoskeleton interactions in plant cell, in K.L. Libbenga, M. Hall, and P.J.J. Hooykaas (eds.), *Biochemistry and Molecular Biology of Plant Hormones*, Elsevier Publ. Co., Leiden, pp. 363-386.

Nick, P. (1999) Signals, motors, morphogenesis - the cytoskeleton in plant development, *Plant Biol.* **1**, 169-179.

Eun, S.O. and Lee, Y. (1997) Actin filaments of guard cells are reorganized in response to light and abscisic acid, *Plant Physiol.* **115**, 1491-1498.

Kandasamy, M.K., Gilland, L.U., McKinney, E.C., and Meagher, R.B. (2001) One plant actin isovariant, ACT7, is induced by auxin and required for normal callus formation, *Plant Cell.* **13**, 1541-1554.

Geldner, N., Friml, J., Stierhof, Y.D., Jurgens, G., and Palme, K. (2001) Auxin transport inhibitors block PIN1 cycling and vesicle trafficking, *Nature* **413**, 425-428.

Davies, E., Fillingham, B-D., Ito, Y., and Ane, S. (1991) Evidence for the existence of cytoskeleton-bound polysomes in plants, *Cell Biol. Int. Rep.* **15**, 973-981.

Zak, E.A., Sokolov, O.I., and Klyachko, N.L. (1995) Plant polysomes bound to the actin cytoskeleton, *Dokl. Acad. Sci.* **344**,549-551.

Klyachko, N.L., Ananiev, E., and Kulaeva, O.N. (1979) Effect of 6-benzylaminopurine and abscisic acid on protein synthesis in isolated pumpkin cotyledons, *Physiol. Veg.* **17**, 607-617.

Kulikova, A.L., Kulikov, A.Yu., Erokhina, M.A., and Klyachko N.L. (2001) The percentage of cytoskeleton-bound polysomes as affected by the physiological conditions of plant cell, *Russ. J. Plant. Physiol.* **48**, 608-614.

Robinson, J.S., Albert, A.C. and Morris, D.A. (1999) Differential effects of brefeldin A and cycloheximide on the activity of auxin efflux carriers in *Cucurbita pepo* L, *J. Plant Physiol.* **155**, 678-684.

REGULATION OF MALE GAMETOPHYTE DEVELOPMENT IN ANGIOSPERMA: THE ROLE OF INTRACELLULAR pH AND TRANSMEMBRANE CHLORIDE TRANSPORT

I.P. YERMAKOV, N.P. MATVEYEVA AND D.S. ANDREYUK
Plant Physiology Department, Biological faculty, Moscow State University, 119899, Moscow, Russia, mail: plantphys@biophys.msu.ru

1. INTRODUCTION

The molecular mechanisms of plant hormone action involve different processes playing a basic role in cell life. Among them are processes controlling intracellular pH value and activity of Cl^--transporters of plasma membrane (Taiz and Zeiger, 1998). Here we show that intracellular pH (pH_i) and Cl^- transport across plasmalemma may be the controlling factors in differentiation and germination of Angiosperm pollen grain. Until now ionic mechanisms regulating pollen grain development are practically unknown. The attention in this field is focused mainly on the role of Ca^{2+} in signaling (Frankling-Tong, 1999). Cl^- dynamics was characterized only during apical growth of a pollen tube (Zonia *et. al.*, 2001).

2. MATERIALS AND METHODS

Plants of *Nicotiana tabacum* L. cv. Petit Havana were used. pH_i was measured by microfluorimetric ratio technique. Cl^--selective electrode was used to reveal changes in intracellular Cl^-. Cl^- contents in anther locular fluid were measured with electron probe microanalysis. O_2 uptake was measured with a Clark-type O_2 electrode.

3. RESULTS AND DISCUSSION

Pollen development in an anther ends in decline of synthetic activities followed by desiccation. The study of in vivo pollen transition from high to low metabolic activity (indicated here by O_2 uptake) reveals a significant decrease of pH_i and simultaneous increase in intracellular Cl^- concentration ($[Cl^-]_i$) to 50 mM (Fig. 1). Analysis of elemental contents of locular fluid surrounding microspores and pollen grains in an anther shows that changes described above are preceded by intensive accumulation of Cl^- and K^+ in the fluid. The data favors the view that the contents of

I. Macháčková and G.A. Romanov (eds.), Phytohormones in Plant Biotechnology and Agriculture, 255-258

locular fluid change at the stage of pollen grain formation: high molecular weight organic solutes are replaced by inorganic ions.

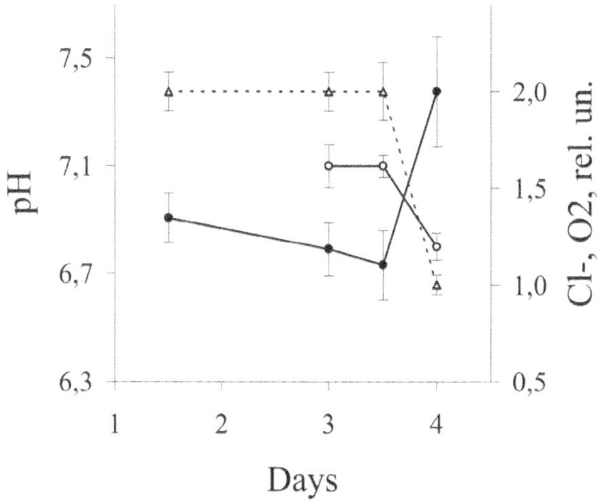

Figure 1. Changes of pH_i (open circles), intracellular Cl^- concentration (closed circles), and O_2 uptake velocity (dotted line) during microspore and pollen grain development in an anther. Abscissa: days after microspore formation.

The role of extracellular Cl^- in the control of pollen state is correlated with data on pollen development *in vitro*. The study reveals the dependence of pH_i on extracellular Cl^- concentration ($[Cl^-]_{out}$): to decrease pH_i to the in vivo value (from pH 7.1 to 6.8) it is necessary to modify standard culture medium by raising $[Cl^-]_{out}$ to 50 mM. This suggests that locular Cl^- accumulation is involved in sporophytic regulation of pollen grain pH_i changes in vivo. About the same time transient accumulation of free abscisic acid (ABA) in tobacco pollen grains occurs (Chibi *et al.*, 1995). ABA is capable to inhibit mature pollen germination, so one can suggest its involvement in the inhibition of precocious pollen germination. But the role of ABA in pollen development as well as its relation to changes in pH_i and Cl^- fluxes remains to be elucidated.

A transition from low to high metabolic activity begins as mature and metabolically quiescent pollen grain lands on a receptive stigmatic surface. It rapidly hydrates, becomes biochemically active, and initiates formation of the pollen tube. The study of the germination in vitro reveals efflux of Cl^- out of pollen grain during the first minutes after hydration (Fig. 2). This rapid Cl^- efflux is followed by an increase in pH_i and activation of pollen respiration (Fig. 2). So the earliest steps of pollen activation are accompanied by significant changes in pH_i and activity of plasmalemma Cl^- - transporters.

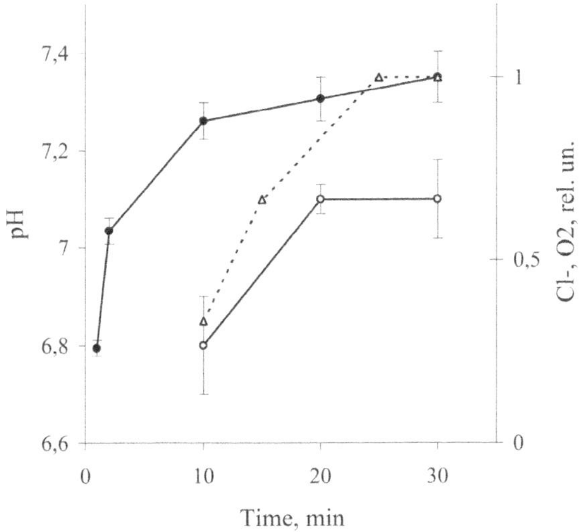

Figure 2. Changes of pH_i, (open circles), Cl^- concentration in culture medium caused by Cl^- efflux out of pollen grains (closed circles), and O_2 uptake velocity (dotted line) during pollen activation preceding the germination. Abscissa: time after hydration beginning.

To examine the role of this pH_i increase in pollen activation the effects of increase/decrease of pH_i on germination frequency were studied. Modifications of pH_i have been induced by different ways: acid-loads using propionic acid (30 mM Na^+ propionate), blocking of Cl^- transporters by ethacrynic acid (1mM), inhibition or stimulation of plasmalemma H^+-ATPase respectively by Na^+ orthovanadate (600 μM) and fusicoccin (15 μM). The results show that the increase of pH_i to 7.2 by fusicoccin, compared to 7.0 in control, doubles the percentage of germinated pollen grains. At pH_i 6.9 (orthovanadate treatment) the germination is the same as in control. However cytoplasmic acidification (pH_i 6.7 compared to 6.8 in nonactivated pollen grains) by ethacrynic or propionic acid completely inhibits the germination. So the data as a whole indicate that pH_i increase is a necessary condition for pollen germination. At the same time the results reveal the dependence of pH_i on activity of Cl^- transporters during pollen grain activation.

The data on Cl^- efflux out of activating pollen grain (Fig. 2) raise the question of what plasma membrane Cl^- transport proteins of plasma membrane are involved in this process. The approach used was to study the effects of known blockers of Cl^- transporters on pollen germination. Antiporter inhibitor oxonol dye $DiBAC_5(4)$ has no effect on pollen germination. Two inhibitors of electroneutral cation-chloride co-transporters furosemide and bumetanide influence slightly only at rather high concentrations (more than 400-500 µM). However,two different blockers of Ca^{2+} -dependent Cl^- channels - niflumic acid and NPPB - completely suppress the germination at 50 µM. The concentration that gives a half inhibition (IC50) is about 10 µM for niflumic acid and between 10 and 20 µM for NPPB. Control experiments show that each of these channel blockers actually suppresses Cl^- efflux out of pollen grain. Thus, Ca^{2+}-dependent Cl^- channels may have a key role in regulating Cl^- flux during pollen grain activation.

Taken together the results provide evidence that Cl^- fluxes have an important functional role during the developmental transition from sporophyte to gametophyte, controlling pollen transition in the state of physiological quiescence. At the whole-cell level Cl^- fluxes control pollen tube formation in that Cl^- channels control pH_i and, as a consequence, pollen activation and germination. So it seems likely that the pollen grain is an example of integration of ionic mechanisms into regulatory networks controlling developmental processes.

REFERENCES

Chibi, F., Angosto, T. and Matilla, A. (1995) Variations of the patterns of abscisic acid and proline during maturation of *Nicotiana tabacum* pollen grains, *J. Plant. Physiol.* **147**, 355-358.

Frankling-Tong, V.E. (1998) Signaling in pollination, *Curr. Opin. Plant Biol.* **2**, 490-495.

Taiz, L. and Zeiger, E. (1998) *Plant Physiology*, Sinauer Associates, Inc., Sunderland.

Zonia, L., Cordeiro, S. and Feijo, J.A. (2001) Ion dynamics and hydrodynamics in the regulation of pollen tube growth, *Sex. Plant Reprod.* **14**, 111-116.

The work was supported by financial help from RFBR, No 02-04-49246

ACTIVITY OF RACEMASE AND APPEARANCE OF N-MALONYL-D-TRYPTOPHAN DURING OSMOTIC STRESS AND DROUGHT CONDITIONS

N.I. REKOSLAVSKAYA[1], O.V. YURIEVA[1], T.V. KOPYTINA[1], R.K. SALYAEV[1], B.A. SHAINYAN[2] AND S. MAPELLI[3]

[1]*Siberian Institute of Plant Physiology and Biochemistry, Siberian Branch of RAS, PO box 1243, Irkutsk, Russia, e-mail: phytolab@sifibr.irk.ru*

[2]*A.E.Favorski Irkutsk Institute of Chemistry, Siberian Branch of RAS, 664033, Irkutsk, Russia, e-mail: bagrat@irioch.irk.ru*

[3]*Istituto Biologia Biotechnologia Agraria, C.N.R., via Bassini 15, Milan, Italy, e-mail: mapo@ibv.mi.cnr.it*

1. INTRODUCTION

D-Tryptophan (DTry) accumulating in wilted plant tissues and seeds may be used simultaneously in IAA biosynthesis and in the formation of N-malonyl-D-tryptophan (MDTry). MDTry synthesis may be considered as a drought inducible step in the regulation of IAA biosynthesis; MDTry synthesis results in a temporary reserve for the inactivated precursor when cells lose water. The stereochemical compartmentation of the indole precursor in the form of DTry or MDTry offers the possibility to use Try only for the IAA biosynthesis but not for the synthesis of proteins.

Natural D-tryptophan (DTry) is synthesized by tryptophan racemase, which converts LTry to DTry during different growth, developmental, or environmental conditions. MDTry has been found in etiolated wheat, barley, pea, tomato seedlings and some other species (Elliott, 1971; Wightman, 1973; Law, 1987; MacQueen-Mason and Hamilton, 1989). Thus, tryptophan racemase is an enzyme of interest whose expression may be crucial for the survival of plants in drought and other natural environmental conditions.

The goal of this study was to investigate the activity of tryptophan racemase in etiolated wheat and tomato seedlings exposed to drought and osmotic stresses of and to evaluate germinating wheat seedlings as target plantlets for the isolation of cDNA of tryptophan racemase .

2. MATERIALS AND METHODS

Spring wheat *Triticum aestivum* L. cultivar Skala and tomato *Lycopersicon*

I. Macháčková and G.A. Romanov (eds.), Phytohormones in Plant Biotechnology and Agriculture, 259-265

esculentum Mill. cultivar Moscovskii osennii and red beet *Beta vulgaris* L. were used in this study. Batches of 200 wheat seedlings were harvested on the 3rd and 5th days of germination. Seedlings were ground in liquid nitrogen in a 0.66 M KH_2PO_4/Na_2HPO_4 buffer, pH 8.3, containing 20 μM pyridoxal phosphate, 1 mM Na EDTA, 4 mM $MgCl_2$, 1 mM phenylmethylsulfonylfluoride, 20% glycerol and 0.1% mercaptoethanol. The homogenate was centrifuged at 4° C and 10000 g for 20 min. The fraction of pellet enriched with etioplasts was employed as an enzyme source. The reaction mixture for the determination of tryptophane racemase activity, which consisted of 5 μmoles of DTry or LTry and enzyme preparation (15-20 mg of protein), was incubated 1 h at 37° C. DTry and LTry were quantified by the use of D-amino acid oxidase or L-amino acid oxidase in separate experiments.

Seedlings of tomato and red beet were also used for measuring the activity of tryptophan racemase.

The synthesis of MDTry and MLTry was performed according to the following reactions:

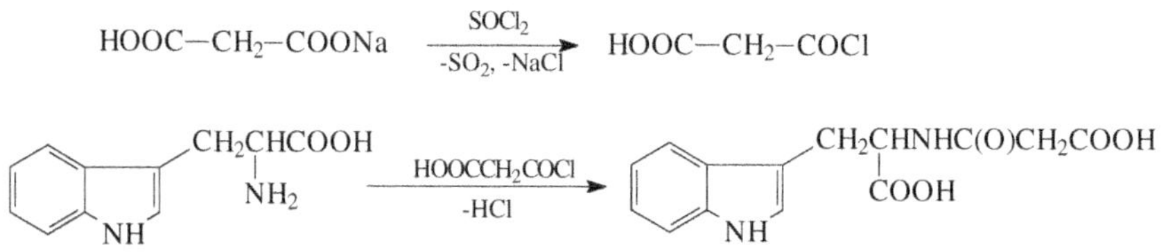

Embryos, which had been excised from dry seeds were grown on a modified Norstog nutrient medium deprived of casein hydrolysate and amino acids (Norstog, 1973). Synthetic MLTry and MDTry were added as auxin precursors to the agar medium.

The configuration of the endogenous MTry was determined by chromatography on "TLC Plates C18-Silica on glass" plates.

The amount of IAA was determined by HPLC with the spectrofluorimetric detector in extracts isolated and purified according to Rekoslavskaya *et al.* (1997). The amount of MTry was estimated after reaction with Ehrlich reagent and the absorption was measured at 564 nm.

The cDNA synthesis was performed according to protocols and kits "RNAgents Total RNA Isolation System", "PolyATract mRNA Isolation Systems","RiboClone cDNA Synthesis System M-MLV RT (H^-) NotI", " Packagene Lambda DNA Packaging System" and "RiboClone EcoRI Linker Ligation System I" from "PROMEGA" (USA).

Each experiment was repeated at least twice. Data in tables are presented an average of two or three analytical runs with calculated standard deviation.

3. RESULTS AND DISCUSSION

3.1. Effects of drought and osmotic conditions on the MDTry accumulation and the activity of tryptophan racemase

When leaves of tomato were exposed to drought or maintained on a solution of mannitol, MDTry accumulated in leaves. NaCl did not stimulate an accumulation of MDTry (Tab. 1), perhaps because NaCl did not evoke the water loss from tissues.

Table 1. The effect of drought and osmotic stresses conditions on the formation of MTry in tomato leaves

Variant	MTry, $nmol \cdot g^{-1} fw$
Fresh leaves	0
Wilted during 3 days	827±30
3 days on 0.5 M mannitol	677±9
3 days on 1(2) % NaCl	18±14

The increase in activity of tryptophan racemase, which was isolated from seedlings of wheat and tomato was demonstrated in both directions after placing seedlings in solutions with 0.5 M mannitol (Tab. 2) or 0.5 M sucrose (Tab. 3) for 1 day. In the L->D direction the activity of tryptophan racemase was higher when the osmotic stress was stronger.

Table 2. The effects of 0.5 M mannitol on the activity of tryptophan racemase from plastid fractions of tomato leaves

Variant	LTry->DTry, nmol	DTry->LTry, nmol
Fresh non-wilted leaves	0	0
Leaves on 0.5 M mannitol 1 day	2152±54	1382±69

Table 3. The effect of 0.5 M sucrose on the activity of tryptophan racemase from plastid fractions of etiolated wheat seedlings

Variant	LTry->DTry, nmol	DTry->LTry, nmol
Seedlings on water during 3 days	349	559
2 days on water + 1 day on 0.5 M sucrose	1506	1721
2 days on water + 2 days on 0.5 M sucrose	5324	3092

Tryptophan racemase was purified on DEAE-cellulose and D-tryptophan agarose columns by using cytosol and etioplast fractions from 4-day-old wheat seedlings. The activity of tryptophan racemase was found as a band of 74 kDa, using native PAAG electrophoresis.

3.2. Effect of MLTry and MDTry on the rhizogenes of wheat embryos in vitro

Chemically synthesized MDTry and MLTry were compared as auxin sources during the *in vitro* growth of isolated wheat embryos. Isolated wheat embryos were grown on agar medium with addition of 50, 100, or 200 μM of MLTry and MDTry. MDTry stimulated growth, root formation, and an increase in auxin activity in contrast to MLTry, which did not stimulate *in vitro* growth of isolated wheat embryos. Synthetic 200 μM MDTry stimulated the formation of roots, which made up 90% of the total mass of embryos grown on agar medium for 20 days. Unlike MDTry, MLTry retarded the growth of all parts of the embryos in all concentrations used. After being placed on the medium with 100 or 200 μM MLTry, about 60% -80 % of the embryos perished or remained unsprouted.

There was an increase in IAA content in wheat embryos growing in agar medium in the presence of 200 μM MDTry. The concentration of IAA in embryos grown on MDTry-containing medium was as much as 120 ng/g of fresh weight, and was considerably higher than that in the control and in embryos growing on the MLTry-containing medium which equalled 10 and 40 ng/g of fresh weight, respectively.

3. 3. The activity of tryptophan racemase during germination of wheat

All MTry found in 2-day old wheat seedlings was identified as MDTry and it was located predominantly in the roots, in a concentration of 30 nmol/g of fresh weight. It should be pointed out that the most intensive growth of etiolated coleoptiles was between the 3rd and 4th days after germination when the coleoptile elongated very fast until it was 5-6 cm in length. The activity of tryptophan racemase was investigated in 3, 5 and 7 day old wheat seedlings during the growth in darkness after germination.

Table 4. The activity of tryptophan racemase during germination in darkness, nmol · mg^{-1} protein · h^{-1}

Days of germination	Etioplasts	Etioplasts	Cytosol	Cytosol
	L→D	D→L	L→D	D→L
3rd day	223 ± 9	34±4	31±3	31±1
5th day	46±0	12±0	10±0	16±0
7th day	18±6	11±2	27±13	12±1

The highest activity of tryptophan racemase was observed on the 3rd day after germination and was, in the L->D direction, up to 6.6 fold more than in the D->L direction. On the 5th day the activity of tryptophan racemase was lowered up to 7.2

fold and on the 7th day it was up to 55.8 fold greater in comparison to that at the 3rd day. The content of IAA in seedlings determined by HPLC was 3.97±0.34 nmol/g of fresh weight on the second day of germination. On the 5th day the amount of IAA diminished to 0.69±0.19 nmol/g of fresh weight. There was a distinct correlation between the IAA content and the activity of the conversion of LTry to DTry during the initial 2-3 days of the growth after germination of wheat seeds. It was found that tryptophan racemase was more active in 7 day old red beet seedlings in the L->D direction. This was correlated with the rise of IAA in beet after drought upto 100 times.

Both IAA and MDTry were located mainly in the upper 7-8 mm apexes of wheat coleoptiles (Tab. 5).

Table 5. IAA and MDTry contents in apices of wheat coleoptiles on the 3rd day after germination, nmol·g^{-1}fw

Variant	Upper part	Basal part
MDTry	242±68	5±0
IAA	138±36	17±15

Therefore, the apical region of etiolated wheat coleoptiles might be a target tissue and could be used as a source for the isolation of cDNA of the gene encoding the synthesis of tryptophan racemase.

3.4. Isolation of the gene tprcm encoding tryptophan racemase

The wheat cDNA library was created using a lambda gt11 vector according to protocols of kits of PROMEGA (USA).

Antibodies to the cytosol tryptophan racemase were raised in serum of rabbit blood. With the use of Western blotting, a cDNA from the wheat gene library, which had been placed in lambda phage was isolated and, by transfection, introduced into the host strain *Escherichia coli* strain Y 1089. The restriction analyses with Bam H1 (Fig. 1A) and NotI EcoRI of DNA (Fig. 1B) isolated from recombinant phage particles revealed an insert of about 2 kb which corresponded to the molecular weight of tryptophan racemase. Experiments with lysogenized cells of E.coli Y 1089 with cloned recombinant phage inserts revealed that the osmotolerance was higher and about twice the amount of cells survived during growth on the medium with 10 mg/ml mannitol in comparison to intact cells. The osmotolerance of recombinant cells of *E.coli* Y 1089 was increased due to the expression in-frame of the cloned insert with cDNA driving under placZ promoter (Sambrook *et al.*, 1989). According to this, the activity of tryptophan racemase was found to be 2769 nmol·mg^{-1}protein in the L→D direction and 3589 nmol·mg^{-1}protein in the D→L direction in comparison to the activity in intact cells, which was less than 100 nmol per mg of protein in both directions. The higher activity of tryptophan racemase of recombinant cells was inducible by the nonsubstrate inductor isopropyl-β-D-galactopyranoside (IPTG), which was added to cell suspensions during the growth of bacteria.

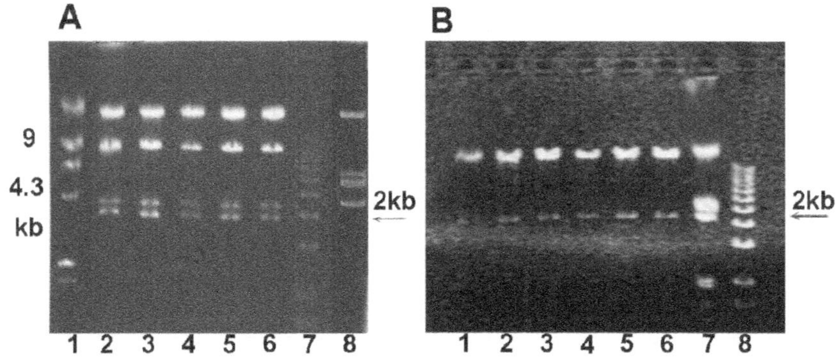

Figure 1. The restriction of DNA of recombinant λgt11 with the insert of the gene tprcm. A - 1- DNA of λ phage restricted with HindIII; 2-6 – DNA of λgt11 restricted with BamHI;

7- ladder DNA 1 kb; 8- DNA of λ phage restricted with BamHI.

B - 1- 6 – DNA of λgt11 restricted with NotI EcoRI; 7 – DNA of λ restricted HindIII EcoRI;

8- ladder DNA 1 kb.

4. CONCLUSION

The conversion of LTry to DTry took place in seedlings of wheat, tomato and red beet under drought and osmotically stressful conditions. DTry might be used in IAA biosynthesis, and in this way it could participate in the growth and the development of plants and to impart resistance to the stress. MDTry is thought to be the storage form of the precursor for the IAA biosynthesis, which is not used for the synthesis of protein. Thus, tryptophan racemase might be a key enzyme responsible for drought resistance.

The gene encoding the synthesis of tryptophan racemase may have biotechnological applications.

REFERENCES

Elliott, M.C. (1971) α–N–Malonyl-D-tryptophan in seedling wheat roots, *New Phytol.* **70**, 1005-1015.

Law, D.M. (1987) Gibberellin-enhanced indole-3-acetic acid biosynthesis: D-tryptophan as the precursor of indole-3-acetic acid, *Physiol.Plant.* **70**, 626-632.

McQueen-Mason, S.J. and Hamilton, R.H. (1989) Plastid localized IAA biosynthesis from D-tryptophan in Alaska pea seedlings, in C.D. Boyer, J.C. Shannon and R.C. Hardison (eds.) *Physiology, Biochemistry, and Genetics of Non-green Plastids*, Amer.Soc.Plant Physiol., pp. 277-280.

Norstog, K. (1973) New synthetic medium for the culture of premature barley embryos, *In Vitro* **8**, 307-308.

Rekoslavskaya, N.I., Yurieva, O.V., Shibanova, L.A. and Salyaev, R.K. (1997) Synthesis and physiological function of D-tryptophan during wheat germination, *Russ. Plant Physiol.* **44**, 196-203.

Sambrook, J., Fritsch, E.F. and Maniatis, T. (1989) *Molecular Cloning. A Laboratory Manual.*, 2nd ed., Vols. 1, 2 and 3, Cold Spring Harbor Laboratory Press.
Wightman, F. (1973) Biosynthesis of auxin in tomato shoots, *Biochem. Soc. Symp.* **38**, 247-275.

GPSR Compliance

The European Union's (EU) General Product Safety Regulation (GPSR) is a set of rules that requires consumer products to be safe and our obligations to ensure this.

If you have any concerns about our products, you can contact us on ProductSafety@springernature.com

In case Publisher is established outside the EU, the EU authorized representative is:

Springer Nature Customer Service Center GmbH
Europaplatz 3
69115 Heidelberg, Germany

Batch number: 10755466

Printed by Printforce, the Netherlands